What was Mechanical about Mechanics

BOSTON STUDIES IN THE PHILOSOPHY OF SCIENCE

Editors

ROBERT S. COHEN, *Boston University*
JÜRGEN RENN, *Max-Planck-Institute for the History of Science*
KOSTAS GAVROGLU, *University of Athens*

Editorial Advisory Board

THOMAS F. GLICK, *Boston University*
ADOLF GRÜNBAUM, *University of Pittsburgh*
SYLVAN S. SCHWEBER, *Brandeis University*
JOHN J. STACHEL, *Boston University*
MARX W. WARTOFSKY†, *(Editor 1960–1997)*

VOLUME 224

J. Christiaan Boudri

What was Mechanical about Mechanics

The Concept of Force between Metaphysics and Mechanics from Newton to Lagrange

Translation: Sen McGlinn

KLUWER ACADEMIC PUBLISHERS
DORDRECHT / BOSTON / LONDON

A C.I.P. Catalogue record for this book is available from the Library of Congress.

ISBN 1-4020-0233-5

Published by Kluwer Academic Publishers,
P.O. Box 17, 3300 AA Dordrecht, The Netherlands.

Sold and distributed in North, Central and South America
by Kluwer Academic Publishers,
101 Philip Drive, Norwell, MA 02061, U.S.A.

In all other countries, sold and distributed
by Kluwer Academic Publishers,
P.O. Box 322, 3300 AH Dordrecht, The Netherlands.

Printed on acid-free paper

All Rights Reserved
© 2002 Kluwer Academic Publishers
No part of the material protected by this copyright notice may be reproduced or
utilized in any form or by any means, electronic or mechanical,
including photocopying, recording or by any information storage and
retrieval system, without written permission from the copyright owner.

Printed in the Netherlands.

To the memory of Casper Hakfoort
6 Januari 1955 – 4 March 1999

Salviati: We can concern ourselves with new speculations, further removed from our goal, after we have solved the difficulties that have already been identified.

Sagredo: But if our deviations from the beaten track were to make us aware of new truths, what is to prevent us following these side-paths, to study the questions that we have encountered by chance? We are not forced to follow some concise and efficient method, we meet for our own pleasure. Moreover, who is to say that we may not stumble upon things that are more enchanting and interesting than the propositions with which we began?

Galileo Galilei, *Discorsi* (1638), First Day

What does it matter, if our concern for the historical is a western bias, providing that we, with this bias, are at least making progress, and not standing still.

Friedrich Wilhelm Nietzsche, *Vom Nutzen und Nachteil der Historie für das Leben; zweite unzeitgemäße Betrachtung* (1874)

Contents

List of Illustrations xiii
Acknowledgements xv

CHAPTER 1: INTRODUCTION

1.1 The Invisible Truth of Classical Physics 1
1.2 Historiographic Orientation 5
    1.2.1 The Horizon of the Scientific Revolution (5); 1.2.2 Mechanics in the Eighteenth Century (9); 1.2.3 Three Controversies over the Concept of Force (15)
1.3 The Place of Philosophy in this Book.. 21
    1.3.1 History of Science or Philosophy of Science? (21); 1.3.2 A Working Definition of Metaphysics (22); 1.3.3 Explicit and Implicit Metaphysics (26); 1.3.4 Substance and Structure: Basic Concepts of Metaphysical Change (28)

PART A. THE UNITY OF THE CONCEPT OF FORCE

CHAPTER 2: FORCE LIKE WATER

2.1 A Brief Archeology of the Concept of Force in the Seventeenth Century 32
2.2 Quantity and Quality 35
    2.2.1 The Distiction according to Aristotle (35); 2.2.2 Scholastic Considerations (37); 2.2.3 Quantification of the Concept of Force (39)
2.3 The Point of Departure: Dead and Living Force... 41
    2.3.1 Galileo (42); 2.3.2 Descartes (47); 2.3.3 Impact, Lifting and Weight (50)
2.4 Impetus: The Goal Attained.. 51
2.5 A Related Distinction: Inertia and Resistance 52
    2.5.1 Impetus, Living Force, and Inertia (52); 2.5.2 Galileo's Concept of Inertia (54); 2.5.3 Descartes's Concept of Inertia (56); 2.5.4 Newton's Concept of Inertia (57)
2.6 The Unity of Newton's Concept of Force... 59
    2.6.1 Instantaneous and Continuous Forces (59); 2.6.2 A Practical Example (62); 2.6.3 Force like Water (65)
2.7 Conclusion 68

CHAPTER 3: LEIBNIZ: FORCE AS THE ESSENCE OF SUBSTANCE

3.1 Introduction 70
   3.1.1 Leibniz's Influence on the Eighteenth Century (70); 3.1.2 Leibniz's Significance for Mechanics (73); 3.1.3 The Structure of this Chapter (74);
3.2 Leibniz's Early Concept of Force 75
   3.2.1 Leibniz's Criticism of Descartes (75); 3.2.2 The Measure of Force (77); 3.2.3 Force as the Contens and Cause of Transference (80)
3.3 Leibniz's Differentiation of the Concept of Force 81
   3.3.1 Aristotelianism and Atomism (81); 3.3.2 From Moving Force to True Unity (83); 3.3.3 A Further Analysis of the True Unities (87); 3.3.4 Force as the Essence of True Unity (89)
3.4 Leibniz's Concept of Mechanical Force 91
   3.4.1 Fundamental and Derivative Forces (91); 3.4.2 Active and Passive Forces (93); 3.4.3 Conservation of Living Force (94); 3.4.4 The Integration of Dead Force (95)
3.5 Conclusion: Similarities between Newton's and Leibniz's Concepts of Mechanical Force 99

PART B. TOWARDS A NEW METAPHYSICS

CHAPTER 4: FROM CAUSE TO PHENOMENON

4.1 Introduction 104
4.2 Between Leibniz and d'Alembert: A Change in the Definition of the Problem 105
   4.2.1 Atomism and Conservation Theories (105); 4.2.2 Experimental Modes of Response (107); 4.2.3 The Importance of d'Alembert in the *Vis Viva* Controversy (109)
4.3 Causality in Mechanics 111
   4.3.1 D'Alembert's Purification of Metaphysics (111); 4.3.2 Final Causes (113); 4.3.3 Mechanical Causes (114); 4.3.4 Necessity and Contingency (117)
4.4 D'Alembert's Foundation of Mechanics 118
   4.4.1 D'Alembert's Method (118); 4.4.2 The Subject of Mechanics: Moving Matter (120); 4.4.3 The Principles of Statics and Dynamics (121); 4.4.4 Mechanics as a Science of Structure (126)
4.5 D'Alembert's Conception of Force 126
   4.5.1 Property, Function and Cause (129); 4.5.2 The Measure of Living Force (131); 4.5.3 D'Alembert's Transformation of the Concept of Force (132)
4.6 Conclusion: Structuralization and Instrumentalization 132

CHAPTER 5: FROM EFFICIENT TO FINAL CAUSES: THE ORIGIN OF THE
PRINCIPLE OF LEAST ACTION

5.1 Introduction 134
  5.1.1 The Greatest Possible Scandal at Court (134); Intermezzo: The
  Mathematical Formulation of the Principle of Least Action (136); 5.1.2 The
  Significance of the Principle of Least Action for Mechanics (137)
5.2 Maupertuis's Newtonian Background 140
  5.2.1 From Soldier to Scholar (140); 5.2.2 Maupertuis's Defense of Newton's
  Theory of Attraction (141); 5.2.3 The Expedition to the Pole (144)
5.3 The Birth of the Principle of Least Action, in Maupertuis and Euler 145
  5.3.1 Minimal Principles in Statics (145); 5.3.2 The Principle of Least Action
  in Optics (149); 5.3.3 Extending the Principle of Least Action to Mechanics
  (153); 5.3.4 Euler's Response to Maupertuis's "Law of Rest" (155)
5.4 The First Steps of the Principle of Least Action 159
  5.4.1 Maupertuis's "Lois du mouvement" (1746) (159); 5.4.2 Euler's Insights
  (163); 5.4.3 Euler's Foundation of the Laws of Collisions (167)
5.5 Conclusion: Teleology and Structure Metaphysics 170

PART C. BETWEEN METAPHYSICS AND MECHANICS

CHAPTER 6: THE CONCEPT OF FORCE IN THE 1779 BERLIN ESSAY COMPETITION

6.1 Introduction 174
6.2 The Berlin Academy's Philosophical Competitions 174
  6.2.1 An Innovation for the Academic World (175); 6.2.2 The Status of the
  Philosophical Competition at about 1779 (177)
6.3 The Significance of the Competition on the Foundation of Force 180
  6.3.1 Parisian Arrogance (180); 6.3.2 Formulation and Analysis of the
  Question (181); 6.3.3 Comparison with Wolff's Concept of Force (185)
6.4 The Competition Essays 189
  6.4.1 General (189); 6.4.2 Selection (189)
6.5 Overview of the Selected Essays 192
  6.5.1 The Prize-Winning Essay (Pap de Fagaras) (192); 6.5.2 Essay I–M726
  (193); 6.5.3 Essay I–M729 (195); 6.5.4 Essay I–M731 (Rehberg) (196); 6.5.5
  Essay I–M733 (198); 6.5.6 Essay I–M734 (Hißmann) (200); 6.5.7 Essay
  I–M736 (201)
6.6 Conclusion: Divergence of Metaphysics and Mechanics 203

Annex to Chapter 6: The Original Formulation of the Berlin Academy's
  Competition Question for 1779 206

CHAPTER 7: LAGRANGE'S CONCEPT OF FORCE

7.1 Introduction 207
7.2 Lagrange's Mathematical Reduction of Mechanics 208
    7.2.1 The 'Peak of Perfection' (208); 7.2.2 The Early Meaning of the Principle
    of Least Action for Lagrange (210)
7.3 Lagrange's 'True Metaphysics' 216
    7.3.1 A Puzzling Remark (216); 7.3.2 Lagrange's Concept of Metaphysics
    (216); 7.3.3 Recognition and Rejection of Metaphysics (218)
7.4 Lagrange's Analytic Foundation of Mechanics 218
    7.4.1 "Recherches sur la libration de la lune" (1764) (218); 7.4.2 The
    Méchanique analitique (221)
7.5 Conclusion: Force and Structure in Lagrange's Thought 224

CHAPTER 8: METAPHYSICS CONCEALED

8.1 Exposition: A Paradox in the History of Science 229
8.2 Development: The Significance of Metaphysical Premises 230
8.3 Climax: The Concept of Mechanical Force between Substance and
    Structure 232
8.4 Resolution: Separation and Transformation around 1780 235
8.5 Epilogue: History and Metaphysics in Modern Natural Science ... 236

Bibliography 241
Index 263

List of Illustrations

Figure 2.1	Galileo's analogy between the effect of an impact and that of the pressure of a weight	43
Figure 2.2	Galileo's analogy of the process of impact with a mechanical tool	44
Figure 2.3	Galileo's analogy between the dynamic problem of an impact and a static situation	46
Figure 2.4	Descartes's comparison of raising a weight with and without a pulley	48
Figure 2.5	Descartes's distinction of action and power	49
Figure 2.6	Illustration of Newton's proof of Kepler's law	64
Figure 2.7	Metaphor of the action of force with water streaming out of a barrel	69
Figure 3.1	Schematic presentation of Leibniz's structure of substances	89
Figure 3.2	Impetus arises from infinitely many infinitely small conatuses	97
Figure 3.3	Parallels between Newton's and Leibniz's concept of force	101
Figure 3.4	Metaphor of the action of force with the pouring of water from one barrel into another	102
Figure 4.1	The principle of composite motion	124
Figure 5.2	Lever subject to a central force working at an infinite distance	148
Figure 5.1	A particle in motion under the influence of external central forces	136
Figure 5.3	System with one degree of freedom	149
Figure 5.4	System with two degrees of freedom	149
Figure 5.5	Euler's analogy with a static system of springs	166
Figure 7.1	Variation δ and differential d of $y(x)$	209
Figure 7.2.	The equilibrium of forces according to the principle of virtual velocities	219

ACKNOWLEDGEMENTS

In the Dutch PhD tradition, it is customary to present a list of propositions along with the actual dissertation. One of the propositions of my own thesis was: "Gaining time by hurrying is a waste of time." I did not mean that hurrying often has a contrary effect, as in the proverb 'more haste, less speed', but that hurrying *in itself* is a waste of time, because one cannot hurry and live at the same time. Hurrying entails a narrowing of the mind, but unlike concentration, it involves losing awareness of what one is actually doing. Although awareness is a *sine qua non* for science, even present day scientists are sometimes the victims of a lack of time, and feel forced to produce their results within strict deadlines. This is a pity for both science and the scientist. I am very grateful for the opportunity that I have had to write this book without such time pressure. Whether this has resulted in a good book, I leave to the judgement of the reader. But it certainly has made this book part of my life.

I would like to thank the Department of the History of Science and Technology at the University of Twente and its members during my days there: Floris Cohen, the late Casper Hakfoort, Marius Engelbrecht, Paul Lauxtermann, Hans Sparnaay, Tomas Vanheste and Péter Várdy, for their constructive comments on parts of the text. The ideas underlying this book came into being amidst this fine and stimulating group of scholars of varying backgrounds.

I am especially grateful to my supervisor during my PhD and friend afterwards, Floris Cohen, without whom the English version would never have appeared. Throughout the process of translation and publication Floris has stimulated me not to lose sight of reality, and to make the book come true. Floris, I thank you greatly for all this and much more.

Without doubt my second supervisor, the late Casper Hakfoort, would have rejoiced over the publication as well. I recall his enthusiasm and sympathy with gratitude, and am happy to be able to dedicate this book to his memory.

Moreover I would like to thank Frans van Lunteren, Louk Fleischhacker, and Peter Kluit for discussions concerning parts of this book. Henri, my brother, I thank you for your assistance in the back-breaking work of creating an index.

The employees of the Archive of the Academy of the Berlin-Brandenburg Academy of Sciences and Humanities gave me a warm welcome and help during my visit in November 1988 (when nobody expected the forthcoming changes, and the institution was still called the Academy of Sciences of the German Democratic Republic.) I also thank the Archive of the Academy for permitting me to quote from the prize essays that are preserved there. Thanks too, to the librarians of the University of Twente for their kindness in dealing with the large numbers of requests for literature from outside the university.

I would like to thank Kluwer Academic Publishers for giving me a hearing when I presented myself with one chapter, some hope, and as yet no great standing in the field. Special thanks are owing to Mrs. Annie Kuipers, Mrs. Stephanie Harmon, and Mrs. Jolanda Voogd.

An excellent translator has been found in the person of Sen McGlinn. He has done a wonderful job. His ability to understand the text, and his willingness to think along with me and make suggestions for improvements often astonished me. I thank him sincerely.

The translation was made possible financially by a generous grant from the Netherlands Organisation for Scientific Research (NWO). I would like to thank especially Mrs. F. Grootoonk of NWO for her patience and her kind help in dealing with all the necessary administrative tasks.

Sander, my last and warmest thanks go to you, who never stops showing me how to seize the day.

Wageningen, the Netherlands

CHAPTER 1

INTRODUCTION

> *The Age of Reason is left the Dark*
> *Ages of the history of mechanics.*
> Clifford A. Truesdell[1]

1.1   THE INVISIBLE TRUTH OF CLASSICAL PHYSICS

There are some questions that physics since the days of Newton simply cannot answer. Perhaps the most important of these can be categorized as 'questions of ethics', and 'questions of ultimate meaning'. The question of humanity's place in the cosmos and in nature is pre-eminently a philosophical and religious one, and physics seems to have little to contribute to answering it. Although physics claims to have made very fundamental discoveries about the cosmos and nature, its concern is with the coherence and order of material phenomena rather than with questions of meaning. Now and then thinkers such as Stephen Hawking or Fritjof Capra emerge, who appear to claim that a total world-view can be derived from physics. Generally, however, such authors do not actually make any great effort to make good on their claim to completeness: their answers to questions of meaning often pale in comparison with their answers to conventional questions in physics.[2] Moreover, to the extent that they do attempt to answer questions of meaning, it is easy to show that they draw on assumptions from outside physics.[3] However, it seems that the fact that physics, ethics and religion are sometimes mixed, should not be attributed to anything inherent in these fields, but to incorrect *understandings* of these fields and their interrelationships. Goethe said that Newtonian science puts "nature on the

---

[1]   Truesdell, "A Program toward Rediscovering the Rational Mechanics of the Age of Reason" (1960), 87.
[2]   For Hawking, see: Hakfoort, "Fysica en wereldbeeld van Descartes tot Hawking" (1992); for Capra and other New Age representatives, see: Vanheste, *Copernicus is ziek* (1996).
[3]   Hösle, "Über die Unmöglichkeit einer naturalistischen Begründung der Ethik" (1989). The energeticism of Wilhelm Ostwald also provides excellent material for study. See, for example, Hakfoort, "Science Deified: Wilhelm Ostwald's Energeticist World-View and the History of Scientism" (1992).

rack," using a metaphor that he borrowed from Bacon.[4] We can give it a new twist and say that physics only makes ethical and religious statements under the compulsion of an inquisitor. Physics, ethics and religion appear to be separated from each other by 'natural' limits.

There are some questions of other types that physics since the days of Newton has not *tried* to answer, not the least of which is the search for the ontology, or essence, of things. At first glance this search might seem closely related to the aim of physics, if we formulate that as achieving a complete mathematical description, in a single system, of all material processes. After all, does not such a description also imply that we know the ontological basis of all that can be experienced? And is this not the answer to the old question into the essence of things? Nevertheless, questions such as: "what is time?"; "what is force?"; "what is a photon?"; "what is the basis of the laws of nature?" can only be answered in physics by showing how these basic concepts function in the entirety of the mathematical-physical system. Time, mass and distance are defined by using a measure that is nothing other than a reproducible unit of the same quantities: time, mass and distance. As a result, the relevant basic concept can be assigned a value, but the concept itself remains elusive. Force, seen until far into the eighteenth century as the *cause* of an effect, is now *defined* by this effect, which is acceleration: $\underline{F} = m \times \underline{a}$. It is natural, therefore, that mathematical relationships are considered as the most important result of physics. In the eyes of the modern physicist, the fact that mass can be transformed into energy and *vice versa*, according to the formula $E = mc^2$, and that experimental measurements of time and energy are fundamentally limited in their precision by Heisenberg's uncertainty principle, means more, for example, than Kant's opinion that time is the *a priori* form of sensibility or the relationship of physical to psychological and biological time. The physicist does not in fact feel any need to speak about fundamental concepts as such, and therefore rejects questions about the essence of nature.

When it comes to questions about the essential nature of phenomena, as distinct from questions of meaning, it is much less clear whether the boundaries defining what physics may deal with are natural or artificial. The Enlightenment and nineteenth century positivism, in particular, have advocated the need to free natural science from metaphysics and theology, the traditional disciplines to which questions of meaning and essence belong. Although this opinion may no longer be maintained in such an extreme form, its general implication—that the creation of modern natural sciences has entailed the gradual elimination of metaphysics and theology from physics—continues to be widely held even among historians who consider the intellectual context of the history of science.[5] For example, in Eduard Dijksterhuis's

---

[4]   Goethe, *Zur Farbenlehre* (1810), 69.

[5]   One problem in evaluating the relationship that metaphysics had, and still has, with modern physics is undoubtedly the fact that this relationship is emotionally charged for some people with the memory of the hostile, or at least dogmatic, position of the Church with regard to science. This association is not accidental, as can be seen from the polemic that E.W. Beth directed at the neothomist Hoenen in Holland just after the Second World War. The issue in this polemic was the

continued on next page

famous history *The Mechanization of the World Picture* (Dutch original 1950), mechanics is primarily a special form of the mathematical description of material reality.[6] In his 1976 essay on the development of early modern physics, Thomas Kuhn distinguished only the mathematical and experimental traditions, and not the traditions of natural philosophy or of metaphysics.[7] However, this approach is scarcely satisfying as an understanding of the past, at least for anyone with a broad vision of metaphysics and theology.

This book will question the low opinion of metaphysics in physics by showing that the metaphysical dimension played a significant, indeed a guiding role in the development of eighteenth century mechanics. Although metaphysics was all too often openly scorned, from the Enlightenment onward, this does not imply that it was absent. Metaphysical arguments continued to play an important role, sometimes not recognized as such, sometimes implicitly assumed. This almost always related to questions of essence; my argument will leave theology, and with it, most questions of ultimate meaning, almost untouched, except where metaphysical themes are involved.

As previously stated, I will focus here on mechanics in the eighteenth century, and specifically on the period between Leibniz and Newton on the one hand and Lagrange on the other. Controversies provide a good starting point for understanding the relationship between metaphysics and mechanics, especially where the metaphysical aspect is concealed. During the relevant period, there were several important controversies concerning the concept of force in mechanics, in which metaphysics and mechanics were closely interrelated. Three of these will be dealt with here:

1. The controversy about the true measure and the conservation of living force (1686–1743)
2. The controversy about the principle of least action (1734–1781)
3. The Berlin Academy's competition question about the foundation of force (1779)

---

continuation of former page
significance of neothomism for understanding modern physics (Hoenen, *Philosophie der anorganische natuur* (1938); Beth, *Natuurphilosophie* (1948)). Metaphysics was easily portrayed as a veiled form of Church dogmatism. Just as the Church wants to found the moral order of the world on Revelation, metaphysics would seek to found the ontological order on dogmatic and unverifiable concepts.

Although this sketch is a caricature, it is important to take account of this emotional charge when conducting historical research. After all, many scientists from the eighteenth century were explicitly anti-Church, although they were still religious. In deism and eighteenth-century freethinking we can observe an ambivalent attitude to belief that is somewhat difficult for our modern secularized society to understand. Their rejection of metaphysical considerations must therefore be taken with a grain of salt.

[6] E.J. Dijksterhuis, *The Mechanization of the World Picture* (1961, translated from the Dutch *De mechanisering van het wereldbeeld* (1950)).

[7] Kuhn, "Mathematical versus Experimental Traditions in the Development of Physical Science" (1976).

There are a number of reasons for limiting the present study to questions surrounding the concept of mechanical force. In the first place, mechanics was the first natural science to be mathematicized and was also the most extensively mathematicized. The essence of Dijksterhuis's thesis about the mechanization of the world picture is that, in the course of the seventeenth century, mechanics became the single most important model for the mathematization of other natural sciences such as optics and electricity.[8] Thus the relationship between metaphysics and mechanics would be expected to provide at least a partial analogy to the relationships between metaphysics and other mathematical natural sciences, although the historical developments were not simultaneous.

Secondly, the concept of force is particularly suited for the articulation of metaphysical suppositions. Until well into the eighteenth century, the concept of force was applied to many different phenomena which would later be differentiated as energy, weight, momentum, etc. As I will show in Chapters 2 and 3, this was not the result of confused thinking, but was due to the fact that the similarities were much clearer than the differences. *There was a unity underlying the various meanings, and this unity was inherited from the substantial concept of force in scholasticism.*

Chapters 4 to 7 show how the concept of mechanical force changed in several ways during the eighteenth century. First of all, the experiential basis shifted from an *internal-bodily* experience of strength, power or force to an *external-sensory* experience of resistance, which meant that 'force' became a qualitative rather than quantitative concept. Secondly, the concept changed in an ontological sense from a substantial to a structural one. These two shifts and their interaction over time are the common thread that runs through this book.

The *first aim* of this book is to defend the general proposition that the historiography of science should take account of metaphysical suppositions. I will show that the *historical comprehensibility* of the developments in eighteenth century mechanics mentioned above is increased if metaphysical suppositions are included in the account, and that such metaphysical suppositions played not only a negative, but also a positive role in the development of mechanics. Note that the terms 'positive' and 'negative' in this context refer not to a norm that is raised above actual history (for example, one might take contemporary mechanics as a norm), but rather to the importance of the suppositions for the further expansion and development of the science of mechanics *of that time*. I do not wish in any way to assign a normative significance to metaphysics, as if explicit metaphysical considerations were a precondition for the successful development of mechanics. But I do want to defend the proposition that metaphysics, during the historical process of discovering the 'secrets of nature', actually functioned as the soil in which the tree of science took root.

The *second aim* is to unravel the broad lines of development in metaphysics itself and in its relationship to mechanics. To return to the metaphor of the tree and the soil, we will deal with changes in the properties and composition of the soil itself

---

[8] Dijksterhuis, *Mechanization*, 500-501 (section V.9).

and how the tree is rooted in the soil, whether with visible, aerial roots or those hidden in the soil. With regard to the *content* of metaphysics, I will emphasize the *transition from substance to structure*, and with regard to its relationship to mechanics, I will emphasize the *transition from an explicit to an implicit form*.[9]

1.2 HISTORIOGRAPHIC ORIENTATION

1.2.1 The Horizon of the Scientific Revolution

In the historiographic tradition of the Scientific Revolution, it has come to be generally accepted that the rise of modern physics was accompanied by the rejection of the Aristotelian-scholastic metaphysics of substance and quality, form and matter, potential and act. There is, by contrast, deep disagreement about whether a different metaphysics has taken its place—and if so, what it is.

The historians Burtt and Koyré have answered this question affirmatively. In 1924 Edwin Arthur Burtt concluded that the metaphysics of modern natural science consisted essentially of ascribing "ultimate reality and causal efficacy to the world of mathematics, which world is identified with the realm of material bodies moving in space and time."[10] For him, mathematics was the language of the reality underlying the phenomena.[11] Alexandre Koyré also saw the historical significance of the language of mathematics, not only as a *description* of reality, but also as an expression of its *essence*.[12] However, Koyré and Burtt's theme was the transformation of Aristotelian-scholastic natural science into Newtonian science, so their statements about post-Newtonian physics should be treated as a 'working hypothesis' at best.

The most radical refutation can be found in the positivism of Auguste Comte, previously mentioned. According to Comte, the father of positivism, physics rose above the levels of theology and metaphysics at the beginning of the seventeenth century, and since then has deserved the title "positive science."[13] The irrepressible ascent of the star of positive philosophy from that time was to be attributed to Bacon, Descartes, and Galileo.[14] During the following centuries, there was nothing left for physicists to do but to scrape away the last remains of metaphysical (and *a forti-*

---

[9] I will return to the key concepts used here in sections 1.3.3 and 1.3.4.
[10] E.A. Burtt, *The Metaphysical Foundations of Modern Physical Science* (1924), 300.
[11] H.F. Cohen, *The Scientific Revolution. A Historiographical Inquiry* (1994), 94.
[12] *Ibidem*, 79.
[13] Comte describes the positive phase of science as follows: "(...) in the positive phase, the human mind (...) [applies itself] solely to discovering, by the well-integrated use of reasoning and observation, the actual laws [of phenomena, JCB], i.e. the invariable relations of succession and correspondence" ("(...) dans l'état positif, l'esprit humain (...) [s'attache] uniquement à découvrir, par l'usage bien combiné du raisonnement et de l'observation, leurs lois effectives, c'est-à-dire leurs relations invariables de succession et de similitude") (Comte, *Cours de philosophie positive. Vol. I. Philosophie première*, 21-22).
[14] *Ibidem*, 27.

ori theological) varnish. After all, the theological, metaphysical and the positivistic methods "*differ* essentially, *oppose* one another radically, and *exclude* each other."[15]

The instrumentalistic views of Ernst Mach are a scarcely milder variant of rejection. Mach, who in fact classified metaphysics under the heading of theology, agreed with Comte's rejection of metaphysics and theology, but in less absolute terms. Theology and metaphysics do, after all, start at the right point, because they focus on the all-embracing whole.[16] However, they have served science only as a rather inadequate crutch on the road towards the ideal, which is, the most economical formulation of the experimental relationships between phenomena.[17] Mechanics has come so close to achieving this ideal that the theological or metaphysical viewpoint provides no more than a *coloring* that has little to do with the *content* of mechanical principles.[18] In this interpretation, metaphysics again appears as a foreign body within modern natural science.

The work of Max Jammer and Eduard Dijksterhuis, among others, provides an interesting variation on instrumentalism. This view emphasizes the functional representation and symbolism of the principles, while theories about the substantial nature of reality, such as mechanical philosophy, are interpreted as heuristic analogies. Dijksterhuis's historical analysis of the Scientific Revolution shows that this can result in a fundamental tension. In the closing pages of his standard work, he provides an explicit statement of the central concept in his characterization of the rise of classical natural science, the concept of the mechanization of the world picture.

> The mechanization of the world picture, which took place during the transition from the science of antiquity to classical natural science, consisted of the introduction of a description of nature with the aid of the mathematical concepts of classical mechanics.[19]

Dijksterhuis's central thesis is at the same time a clarification, and no clarification. The clarification lies in his interpretation that, during the Scientific Revolution, classical mechanics became central to our world-view. Because its mathematization proceeded so quickly and so successfully, classical mechanics became the model for the other natural sciences. Just as Copernicus had placed the sun at the center of the planetary orbits, during the Scientific Revolution classical mechanics became the center around which the other sciences moved and thus became their point of reference in their own development.

An objection to Dijksterhuis's thesis is that the ordering which it provides is merely an *instrumental* one: classical mechanics is the new instrument for under-

---

[15] "*différent* essentiellement, *s'opposent* radicalement, *s'excluent* mutuellement" *(ibidem,* 21n.1).
[16] Mach, *Die Mechanik, historisch-kritisch dargestellt* (1883), 440.
[17] *Ibidem,* 429-444.
[18] *Ibidem,* 440. Although the need to draw a strict separation between theology and physics "was always clear for the greatest thinkers, such as Newton" ("den größten Geister, wie Newton, (...) immer klar war") *(ibidem,* 438), the separation did not become definitive until Lagrange's *Méchanique analitique (ibidem,* 437).
[19] Dijksterhuis, *Mechanization,* 501 (section V.9).

standing the world around us. However, the Scientific Revolution entailed not only a central orientation of the natural sciences and perhaps of the associated world-view, but also a transformation of the center itself. After all, mechanics itself changed just as radically as the orientation of the other sciences. If we ask about the center of the new orientation, Dijksterhuis's thesis provides no answer. His description indicates how our world-view is oriented, but does not reveal the nature of this world.[20]

Classical mechanics is at the center of the new world-view. Here, classical mechanics is "(...) the doctrine of the motions of material bodies in accordance with Newton's system."[21] In his analysis, Dijksterhuis completely dissects this description. To begin with, the etymological association of the word 'mechanics' with tools is entirely misplaced: the science should actually be called 'kinetics'.[22] In that case, where does Newton's mechanics differ from that of Aristotle, who also discussed motion? The difference, Dijksterhuis answers, is mathematical language. Classical mechanics employs mathesis and is ultimately a form of mathematics itself.[23] The characteristic difference between classical and Aristotelian mechanics therefore appears to be the mathematization of the basic concepts.

However, the basic concepts of mechanics have not only become more precise in the course of their development, they have also changed qualitatively. In other words, in addition to mathematization, there was also an evolution of the concepts with respect to content, as can be seen from the development of the concept of inertia and the distinction between external and internal forces. As the concept of inertia developed, the Aristotelian distinction between activity and passivity had to be reformulated (see Chapter 2). The distinction between *external* and *internal* forces—i.e., between pressure and gravity (the 'Newtonian forces') on the one hand, and the quantity of motion and living force on the other—highlighted the question of the relationship between cause and effect (see Chapters 2 and 3).

Dijksterhuis's description of the concept of 'mechanization' implies that the objects of natural science have not changed, only the instruments that natural science employs. His thesis says only that the modern world-view is oriented around classical mechanics. Due to his rejection of metaphysical interpretations, he ignores the fact that, in the process of mechanization the object of mechanics is also changed. His answer, that the basic concepts have been mathematized, takes us back to the beginning of his analysis and leaves us with the question of what mechanization is, in addition to mathematization. If we are to understand the nature of the modern

---

[20] It would be mistaken to evaluate the concept of 'mechanization' on the basis of these last pages alone, and to ignore the rest of the book in which, as it were, the actual content of his thesis was sown. The fact that Dijksterhuis had great difficulty in harvesting what he has sown does not mean that the harvest was poor, but that he used an improperly forged tool for the harvest. I highlight the deficiencies of his description only to show that it is precisely here that one sees the tension that is inherent in his instrumental interpretation.
[21] *Ibidem*, 498 (section V.6).
[22] *Ibidem*.
[23] *Ibidem*, 499 (section V.7).

world-view, we need also to pose a question at the conceptual level: "what was mechanical about mechanics?"[24]

Let us return to the question of whether the development of modern natural science was accompanied by a renewal of metaphysics, or liberation from metaphysics. One possible explanation of the deficiency of the historical interpretations mentioned above lies in the premature delimitation of the Scientific Revolution. Both Dijksterhuis and Burtt consider Newton to be the climax of the new classical physics. Classical physics was formed *in principle* with the realization of the mathematical-experimental method in the *Principia mathematica* (1687). For Burtt, this also means that the metaphysical foundation was then completed:

> In Newton the Cartesian metaphysics, ambiguously interpreted and stripped of its distinctive claim for serious philosophical consideration, finally overthrew Aristotelianism and became the predominant world-view of modern times.[25]

In his search for the essential differences between scholastic and modern physics, he assigns too much importance to Newton himself here.[26] He also implies that the developments in the eighteenth century did not involve any essential changes in the metaphysical suppositions of physics.[27]

---

[24] The fact that this question was not asked, prevented the concept of mechanization going beyond the function of a convenient heading: all sorts of developments could be placed under it, without much advance in the *interpretation* of the Scientific Revolution. Some popular literature even confuses the historical phenomenon of Cartesian mechanical philosophy with Dijksterhuis's concept of mechanization, using the term 'mechanization' as a mere heading. The initial negative reactions to Newton's theory of universal gravitation, especially in France, show that there was a great deal of tension between the mechanistic and mathematical approaches. The historical phenomenon of mechanical philosophy is, after all, an explanatory natural philosophy which was at odds with descriptive mathematics (Westfall, *The Construction of Modern Science; Mechanisms and Mechanics* (1971), for example 42). Compare, for example, the agnostic defense of action-at-a-distance by Dutch scholars such as 's Gravesande and Van Musschenbroek (Van Lunteren, *Framing Hypotheses: Conceptions of Gravity in the 18th and 19th Centuries* (1991), Chapter 3; the present book, Chapter 5). Dijksterhuis ignores this distinction between explanatory and descriptive theories, suggesting that there is no difference (*Mechanization*, 500 (section V.9)). This means in fact that he ignores the explanatory aspect. Only the aspect of motion then remains: *cause and effect are ontologically equivalent to motion*. We can thus explain how Bohm, Capra and others could treat both Descartes and Newton under the same heading of mechanical philosophy.

[25] Burtt, *Metaphysical Foundations*, 237.

[26] The reason for this was that, "[w]e must grasp the essential contrast between the whole modern world-view and that of previous thought and use that clearly conceived contrast as a guiding clue to pick out for criticism and evaluation, in the light of their historical development, every one of our significant modern presuppositions" (*ibidem*, 16).

[27] Strictly speaking, he states this only in a general sense. However, he suggests that the space that does remain open for development had no fundamental meaning. For example: "Of course, these men [meaning the Newtonians, JCB] do not accept Newton as gospel truth—they all criticize some of his conceptions, especially force and space—but none of them subjects the whole system of categories which had come to its clearest expression in the great *Principia* to a critical analysis" (*ibidem*, 22). Or : "These changes [in the prevailing conception of reality, of causality and of the human mind—*ibidem*, 300] have conditioned practically the whole of modern exact thinking" (*ibidem*, 301).

1.2.2 Mechanics in the Eighteenth Century

However, we cannot conclude from what has been said in the previous section, that we should also look to the eighteenth century in our search for the foundations of classical physics. After all, the eighteenth century has gone into history as the century of the Enlightenment, the century of political revolutions—the American in 1776 and the French in 1789—and, according to some, even the century of mathematics, but certainly not as a century of scientific revolutions. In contrast to the preceding period of creation, and the subsequent period of expansion to many other physical and chemical domains, the eighteenth century does not appear to play a fundamental role in the development of modern science. This is the opinion of Stephen Mason, for example, in his *A History of the Sciences*. The first half of the century played no great role in the development of science in general, and virtually the entire century can be passed over with respect to Newtonian science in particular.[28] As for eighteenth century developments in mathematics (which implicitly includes rational mechanics), Struik writes: "Mathematical productivity in the Eighteenth Century concentrated on the calculus and its application to mechanics."[29] More recently, in his overview of eighteenth century Europe, Jeremy Black concluded that while the Enlightenment thinkers did develop the general concept of a 'scientific revolution', "there was little in the way of scientific revolution during the period," except for the fields of electricity and chemistry at the end of the century.[30]

An interesting historiographical overview of eighteenth century natural science, *The Ferment of Knowledge,* was published in 1980.[31] In this work, the relative lack of later interest in eighteenth century natural science is explained partly because the history of the period lacks a clear distinctive character, as compared to that of the seventeenth and nineteenth centuries. According to some of the authors, this problem could be solved through a contextualizing approach, in which eighteenth century natural science is explicitly considered as part of the Enlightenment process.[32] One has to ask whether this solution was inspired more by the general trend of relativism and contextualization in historiographical methods than by the actual nature of eighteenth century natural science. The seventeenth century, which did have its own character in the historiographical tradition, has nevertheless become subjected to a similar kind of contextualizing approach by two of these authors, an approach which has become famous as a model of social constructivist historiography.[33] Al-

---

[28] Mason, *A History of the Sciences* (1956), 279-288 and 289-301, respectively.
[29] Dirk J. Struik, *A Concise History of Mathematics* (1948), beginning of Chapter VII: "The Eighteenth Century."
[30] Black, *Eighteenth Century Europe* (1990), 275.
[31] Rousseau and Porter, *The Ferment of Knowledge; Studies in the Historiography of Eighteenth-Century Science* (1980).
[32] See also Golinski, "Science *in* the Enlightenment" (1986).
[33] Shapin and Schaffer, *Leviathan and the Airpump: Hobbes, Boyle, and the Experimental Life* (1985). For a critical discussion of the programmatic function of social constructivism for historiographical research into the seventeenth century Scientific Revolution, see H. Floris Cohen, *The Scientific Revolution* (1994), Chapter 3.

though I certainly do not want to play down the relationship between science and the Enlightenment, I think there is a serious danger that this approach will see history once again borrowing its interpretations *a priori* from modern concepts derived from the philosophy and sociology of science. This would result in an artificial unity that is likely to stand in the way of true insight. If we want to find unity in eighteenth century natural science, we must base our search on the material itself.[34] However, I believe that we do not yet have sufficient insight into the objectives and the questions that were being asked, to formulate this unity now, or even to evaluate whether and to what extent such a unity could be present. Moreover, the periodization could turn out to be quite different to what some authors, on the basis of our decimal calendar, have supposed.

If eighteenth century science lacks a clear, distinctive character, this is not an absolute handicap: there may be interesting developments in the individual disciplines. All eighteenth century research in mechanics—and in fact this applies also to the other natural sciences, then referred to as 'philosophia naturalis'—has long been seen as no more than an elaboration or application of Newton's *Principia mathematica* from 1687. However, during the Enlightenment itself this was not the case.

Before the publication of Lagrange's *Méchanique analitique*[35] in 1788 and Kant's *Kritik der reinen Vernunft* in 1781, scholars generally thought that they were participating in the rise of science—frequently referred to as a renewal of science—and contrasted themselves with the dark period of medieval scholasticism.[36] The whole of modern time was still seen as a single continuity beginning with the Enlightenment, Voltaire, for example, thought that Newton's theory of gravitation should be regarded as the first great success of the new method. Especially during the second half of the eighteenth century, Newton was almost unanimously admired.

It was only after the end of the eighteenth century that the thought that the seventeenth century was a *completed* phenomenon arose, with thinkers such as Kant, Whewell, and Mach. Lagrange's *Méchanique analitique* had a great influence on the image of the mechanics of the decades following Newton that formed then and afterwards. For example, his historical introductions confirmed Mach in the belief that nothing fundamental occurred in mechanics between Newton and Lagrange.[37] Mach even concluded that:

---

[34] Here I am also distancing myself from solutions such as those that would create a 'new synthesis' based on the concept of 'power' in all its diversity (Christie, "Aurora, Nemesis and Clio" (1993), 404-405).

[35] The now uncommon spelling of the title I have used here is that of the first printing. Beginning with the second edition, the title was written as *Mécanique analytique*.

[36] See, for example, J.E. Montucla, *Histoire des mathématiques*, the first work in the history of science to deal with the eighteenth century. In the second edition of this work (1799–1802), two of the four volumes were devoted to the eighteenth century. The title of Kästner's *Geschichte der Mathematik seit der Wiederherstellung der Künste und Wissenschaften bis an das Ende des 18. Jahrhunderts* (Göttingen 1796–1800) is also an expression of this perception of continuity. Also compare Hankins, *Science and the Enlightenment* (1985), 1-2.

[37] Truesdell, "Program toward rediscovering," 86.

> The Newtonian principles are sufficient to explain every mechanical case that occurs in practice, without any need to add a new principle. If difficulties appear during this process, these are always of a formal, mathematical nature, and not in any way of a fundamental character.[38]

In his opinion, the task facing the eighteenth century was only to produce an economical reorganization of the principles and their derivations, and their ultimate formalization; in other words the formation of Euler and Lagrange's analytical mechanics.[39] This is certainly not a very interesting challenge!

The nineteenth century subordination of eighteenth century mechanics to the *Principia mathematica* continued during the twentieth century, and may well have been strengthened when the concept of the Scientific Revolution took shape, as a conceptual instrument, in the 1950s.[40] This concept focused attention on the beginning and the climax, and sketched the time before and after in lines converging on these perspective points. This picture suggests that the new science developed rather suddenly and was complete after the climax, or at any rate only required some elaboration. This contradicted the long-standing view that there were a variety of preparatory developments before Galileo, before Kepler, and even before Copernicus, and also that aspects of certain traditions, that were outdated according to modern thinking, continued to play an important role after Newton. However, the concept had an asymmetrical effect, leading to little change in our understanding of preceding developments, but with a significant effect on research into what had taken place after the climax. It made it more difficult to treat developments after the climax as being just as fundamental as the preceding developments. The great historians of the Scientific Revolution, such as Eduard Dijksterhuis, Edwin Arthur Burtt, Alexandre Koyré, and I. Bernard Cohen, all considered eighteenth century mechanics—with regard to its foundations—to be only an elaboration of Newton's doctrine.[41] Even Thomas Kuhn, who had negated the old image of steadily progressing science in his *Structure of Scientific Revolutions* (1962), had a strikingly static view when it came to the eighteenth century. He even takes dynamics, in the

---

[38] "Die Newtonschen Prinzipien sind genügend, um ohne Hinzuziehung eines neuen Prinzips jeden praktisch vorkommenden mechanischen Fall (...) zu durchschauen. Wenn sich hierbei Schwierigkeiten ergeben, so sind dieselben immer nur mathematischer (formeller) und keineswegs mehr prinzipieller Natur" (Mach, *Mechanik*, 272). See especially Chapter 2, section 3 "Newton's Leistungen," and Chapter 4 "Die formelle Entwicklung der Mechanik."
[39] *Ibidem*, especially 409-471. See also Bos, "Mathematics and Rational Mechanics" (1980), 333.
[40] The concept of the Scientific Revolution acquired its modern form especially through the work of Koyré, Butterfield and Hall (H.F. Cohen, *The Scientific Revolution*, 81).
[41] Regarding Dijksterhuis, it must be noted that in the book *A History of Science and Technology; Nature Obeyed and Conquered* (1963), co-authored with R.J. Forbes, Dijksterhuis took a more balanced standpoint. In this work he admitted that Newton's foundation of mechanics was incomplete to the extent that the formulation of the general principles was only completed, or tentatively completed, in the *Méchanique analitique* of Lagrange. Moreover, in the nineteenth century "the foundation upon which the whole structure rested was constantly being critically examined, and discussions on these matters never really ceased" (Dijksterhuis and Forbes, *A History of Science and Technology*, 342). However, their analysis is too short to show to what extent this relativizes the fundamentals of Dijksterhuis's previous views.

time after Newton, as an example of a 'normal science'.[42] The effects of such an approach can be seen in the following passage:

> These problems of application [such as determining the equivalent length of a solid pendulum, and describing the simultaneous motion of more than two bodies, JCB] account for what is probably the most brilliant and consuming scientific work of the eighteenth century.[43]

Although Kuhn's concept of normal science does in principle allow room for fundamental developments, i.e., "paradigm articulation," but in the case of the Newtonian paradigm, in contrast to other paradigms, there was only a reformulation in an "equivalent but logically and aesthetically more satisfying form."[44]

Only recently have attempts been made to assign more intrinsic value to eighteenth century mechanics. As early as the 1950s, Truesdell called for more attention to be paid to the development of rational mechanics, and especially to the development of its principles. He called for a rediscovery of eighteenth century mechanics, a project which he had already begun with a series of introductions to parts of the *Opera omnia* of Leonhard Euler.[45] Truesdell fiercely attacked the idea that Newton's *Principia mathematica* completed the development of rational mechanics. He argued instead that the actual foundation was only laid around 1750 by Euler.[46] The *Principia mathematica* left much work undone, such as the mathematical formulation of the three-body problem (a problem to be solved analytically up to now). Newton's difficulty was not the inadequacy of the mathematics he used, but rather the preceding step: the *mathematization* of the object of mechanics. Newton's 'three laws', in the form in which he had formulated them, were far from adequate for such a complicated problem.

> The first to go substantially beyond Newton in the three-body problem was the man who found out how to set up mechanical problems once and for all as definite mathematical problems, and this man was Euler. The year in which the 'Newtonian equations' for celestial mechanics were first published is not 1687 but 1749 (...)[47]

Moreover, Truesdell goes on to say, the second book of the *Principia*, which deals with motion in viscous media, is more of an exploration than a systematic method of solving problems. For Truesdell, the cause of this lack lay in the inability to write

---

[42] Kuhn understands 'normal science' to be: "research firmly based upon one or more past scientific achievements, achievements that some particular scientific community acknowledges for a time as supplying the foundation for its further practice" (Kuhn, *Structure of Scientific Revolutions*, 10).
[43] *Ibidem*, 32.
[44] *Ibidem*, 33. Pulte provides extensive criticism of Kuhn's view in: *Das Prinzip der kleinsten Wirkung und die Kraftkonzeptionen der rationalen Mechanik* (1989), 13-19.
[45] Truesdell, "Rational Fluid Mechanics, 1687–1765" (1954), "I. The First Three Sections of Euler's Treatise on Fluid Mechanics (...)" (1956) and *The Rational Mechanics of Flexible or Elastic Bodies, 1638–1788* (1960).
[46] Truesdell, "Program toward Rediscovering."
[47] *Ibidem*, 90.

differential equations for mechanical systems. In his view, Newton *began* the formulation of mechanical principles rather than *completing* it.

As a result of Truesdell's work, writes Henk Bos in *The Ferment of Knowledge*, "the very programme of the science of mechanics appears as a creation of the eighteenth century."[48] But at the same time, Bos points out an objection to Truesdell's method: it implies that the goal of rational mechanics must be to achieve an axiomatic mathematical science. But many aspects are not covered by a goal defined in this way, including the aspect of the metaphysical foundation of the principles of mechanics.[49] Perhaps Truesdell's interpretation of the fundamental concepts of mechanics—'force', 'mass', etc.—as 'undefined concepts' blinded him to philosophical problems of a conceptual nature. This means that Truesdell's approach is unfortunately unable to help us in finding the meaning of these and other aspects of eighteenth century mechanics, and may even constitute an obstacle to understanding.

In 1982, Peter M. Harman published *Metaphysics and Natural Philosophy*, in which he sought to show that metaphysical views regarding the nature of reality were fundamental to the establishment of new theories. He opposes the view that metaphysics is no longer relevant to classical physics, and that classical physics has emancipated itself "from the concern with the essential nature of substances in favor of the symbolic representation of the mathematical relations between material entities."[50] I am of two minds about this work. On the one hand, I am following his lead, since I also believe that the metaphysical foundation is an essential part of classical physics and, equally, that "the structure of natural philosophy in the eighteenth and the nineteenth centuries cannot be represented as conforming to the principles of Newton's ontology and metaphysics."[51] On the other hand, on the basis of the disputes that will be treated in this book, I would want to interpret the nature of this metaphysical foundation more broadly than Harman does. He uses a rather narrow definition of the concept of metaphysical foundation, despite his broad formulation of the concept of metaphysics.[52] In practice he focuses primarily on *substantial* arguments, arguments that relate to the substantial nature of reality. I, on the other hand, want the concept of metaphysics to refer to other aspects of reality as well, such as quality, quantity, purpose, and relation. We will also need to examine the tendency in eighteenth century mechanics to make increasingly less *explicit appeals* to metaphysical arguments (in this respect, eighteenth century mechanics may differ from

---

[48] Bos, "Mathematics and Rational Mechanics," 334.
[49] *Ibidem*, 340.
[50] Harman, *Metaphysics and Natural Philosophy: The Problem of Substance in Classical Physics* (1982), 1.
[51] *Ibidem*, 6.
[52] *Ibidem*, 3. Harman formally defines metaphysics as "the attempt to justify the conceptual rationale of a scientific theory by appeal to regulative maxims such as the law of causality or to criteria of simplicity, analogy or continuity; as well as (...) attempts to justify the intelligibility of a theory by an explanation of the meaning of concepts of matter or force." In the second part of this definition, the printed text reads 'as well as *to* attempts to justify'. However, the semicolon before 'as well as' indicates that the latter does not refer to 'appeal to' but to the first 'attempt to justify'. In other words: 'to' is a typographical error.

eighteenth century physics in general). If 'metaphysics' is conceptualized only in terms of explicit argumentation, this would limit the scope of the argument. A much wider field can be investigated by including *implicit presumptions* as well. This will be considered further below, in sections 1.3.3 and 1.3.4.

In 1985, an interesting attempt to link eighteenth century natural science to the broad context of the Enlightenment was published: Thomas L. Hankins's *Science and the Enlightenment*. Hankins succeeds in showing that the history of natural science in the eighteenth century is more than the sum of the histories of the individual sciences.[53] One of the most important developments of eighteenth century science, according to Hankins, was the formation of new categories of scientific knowledge. In this sense, the Enlightenment is a transitional period.[54] Disciplines that are sharply differentiated today had a great deal of overlap at that time. For example, the modern distinction between natural philosophy and natural science did not yet exist. Hankins cites the example of d'Alembert's "Système figuré des connoissances humaines" from 1751, in which 'philosophy' is the general term for knowledge that is within the scope of reason, and 'science' is the specific term for subdivisions of knowledge, such as ontology, natural science, theology, and the humanities (psychology, logic, ethics).[55]

But is not only in the differentiation of scientific disciplines that the Enlightenment constituted a transitional period: something comparable can also be seen in research itself. Philosophical questions concerning the soul, the will, the existence of God, and the nature of matter, for example, were not strictly differentiated according to discipline, but were addressed by everyone who sought knowledge. Examples of such an overlap can be seen in Descartes's mechanical philosophy, Leibniz's best world theory and the *vis viva* controversy. It was only in the course of the eighteenth century that the various disciplines separated from one another and became specializations.

The connections between the various scientific disciplines which Hankins notes lead one to suspect that such connections are also present in the knowledge itself. As this book will show, the concept of mechanical force is a *Fundgrube* for such connections, both on the level of the disciplines themselves and on the level of specific scientific knowledge. This is precisely because the fissure that was developing between philosophy and natural science went right through the concept of force.[56]

Up to now, this outline of the historiographic tradition has only discussed the characteristics of eighteenth century natural science in general and of mechanics in

---

[53] This development in the historiographic tradition puts the criticism of Golinski, in "Science *in* the Enlightenment" (1986), into perspective. Golinski's criticism is that Hankins has not succeeded in treating eighteenth century science by work *from* the idea of the Enlightenment, i.e. as part of the total process of cultural and social progress. It is true that Hankins does not treat eighteenth century science as a *component* of the Enlightenment, but is it not a merit that he has, as the title of his book announces, placed them *side by side*?
[54] Hankins, *Science and the Enlightenment* (1985), 11.
[55] D'Alembert, "Discours préliminaire des editeurs" (1751), xlvii-li.
[56] See also Hankins, *Science and the Enlightenment*, 13-16.

particular. What can we deduce from this outline? Truesdell makes it clear that the Scientific Revolution was certainly not completed with Newton's *Principia mathematica* and that mechanics, seen from an axiomatic mathematical viewpoint, had only just come into existence. Harman explored the importance of explicit metaphysical arguments, especially those concerning substances, in the formation of physical theories. Finally, Hankins indicated that natural philosophical, natural scientific, and theological viewpoints were closely related, because some fundamental questions were addressed by all these disciplines.

In my study, I confirm Truesdell's thesis (but taking an entirely different route, and one which he would say was missing the point entirely) that, in an important sense, the foundations for classical mechanics were not completed until the eighteenth century. In my argument, however, this foundation is metaphysical in nature. Like Harman, I want to show that metaphysics played an important role in eighteenth century mechanics, but because I use a broader concept of metaphysics, I refer here primarily to changes in the content of the metaphysical foundation and in the relationship of that foundation to mechanics. The distinction between explicit and implicit metaphysics is an articulation of Hankins's sketch of the Enlightenment, in a way that allows us to recognize the link between metaphysics and mechanics without reference to the transformation of scientific fields that Hankins depicts.

1.2.3   Three Controversies over the Concept of Force

1.2.3.1 *Vis Viva*

In the historiography of mechanics, gravitation occupies a comfortable position among the multitude of forces. In both classical and modern mechanics, the seventeenth-century criticism of occult qualities has been a continually recurring source of dissatisfaction with the peculiar phenomenon of action-at-a-distance.[57] Perhaps the inexplicability of force-at-a-distance is the reason why the tendency to apply the physics of a later age as a norm had a less serious effect on the historiography of the concept of gravitation than of the controversies concerning the true measure of living force ("vis viva") and the principle of least action ("principium minimae actionis"). Living force and the principle of least action were seen primarily as predecessors of the modern formulas for kinetic energy and Hamilton's principle, respectively. The controversies about these topics were regarded as rearguard actions against the metaphysical tangle of Leibniz and of Maupertuis, respectively, with the controversy about the principle of least action also being seen as a fine specimen of

---

[57] In 1957, Max Jammer published an erudite and impressive history of the concept of force: *Concepts of Force: a Study in the Foundation of Dynamics*. Despite its title, this is almost entirely devoted to the predecessors of the concept of gravity. Its strength is in its breadth of treatment, extending from mythological ideas to the views of modern physics. An historiographical objection is, however, that he allows his epistemological standpoint to weigh too heavily in his historical evaluations. For the period between roughly 1700 and 1900, therefore, a better source is Van Lunteren's elegant overview of hypotheses to explain gravity (Van Lunteren, *Framing Hypotheses*).

a political power struggle, and rather amusing as well. Recognition was given only to the proposition that living force is *conserved*, and to the purely *mathematical impetus* that appeared in the development of the principle of least action. The fact that Leibniz was given the credit in both cases must be attributed to the reverent idea that important thinkers from the past would naturally have had a premonition of the correct understandings of physics. However, such an attitude does not benefit historiography much better than the radical anti-metaphysical attitude of someone like Ernst Mach or even, in 1977, István Szabó.[58]

Until the 1960s the historiography of the controversy about living force was trapped in an instrumentalist net in which d'Alembert, with his *Traité de dynamique* from 1743, appeared as the great hero. D'Alembert was supposed to have closed the debate by stating that the entire controversy was a 'battle of words'. In the 1960s however, historiography was freed from this net by Thomas Hankins, Carolyn Iltis and Laurens Laudan.[59] These authors showed that d'Alembert's criticism of the *vis viva* concept was not the *first* in history, nor was this criticism historically *decisive* for thought about *vis viva*. In short, d'Alembert was quite rightly dethroned. His dethroning removed the pressure to provide a physically correct interpretation of d'Alembert, and made it possible to take a fresh look at the *content* of his solution.

Wilson Scott did something similar in 1970. He treated the controversy, or at least the period following the Leibniz-Clarke controversy of 1717, as part of the conflict between atomism and conservation theories.[60] In his view, d'Alembert provided a partial solution to this conflict.[61]

In this book I place d'Alembert's solution in a different perspective, namely that of his 'de-metaphysicalization' of force. My argument begins on the one hand with the original problem, and on the other hand with the entirety of d'Alembert's foundation of mechanics.

In one respect, Papineau preceded me in 1977, by searching for a common background for this controversy. According to Papineau, both standpoints—that either mv or mv^2 was the true measure of the force of a moving body—should be seen as "alternative modifications of a common system of mechanical thought."[62] An objection to his analysis, however, is that it limits itself primarily to argumentation based on Quine's linguistic philosophy. In concrete terms, Papineau wants to relinquish an unequivocally determined *meaning* of scientific terms, and to view these instead as

---

[58]   Szabó, *Geschichte der mechanischen Prinzipien und ihrer wichtigsten Anwendungen* (1977).
[59]   Hankins, "Eighteenth-Century Attempts to Resolve the *Vis Viva* Controversy" (1965); Iltis, "D'Alembert and the *Vis Viva* Controversy" (1970) and "Leibniz and the *Vis Viva* Controversy" (1971) (both stem from her dissertation, *The Vis Viva Controversy: Leibniz to d'Alembert*. University of Wisconsin 1967, but I have not consulted this); Laudan, "The *Vis Viva* Controversy, a Post-Mortem" (1968). This of course does not preclude historians even today from clinging to the old ideas, such as Szabó, *Geschichte* (revised edition 1987), and Speiser [Introduction to Daniel Bernoulli, *Werke* III, Mechanik] (1987).
[60]   Scott, *The Conflict between Atomism and Conservation Theory (1644–1860)* (1970).
[61]   *Ibidem*, 47-64.
[62]   Papineau, "The *Vis Viva* Controversy" (1977), 118.

arising from the "structure of generalizations and observational procedures."[63] Most of his article is devoted to explaining this philosophical viewpoint, so that his analysis of the *vis viva* controversy itself does not get very far. Papineau ended by concluding that the conflicting standpoints had a common ground in the view that force is conserved and is transferred from one body to another during a collision.[64] A great deal more can be done with this conclusion, however. Chapters 2 and 3 will show that the concept of force, though later differentiated, was originally based on a unity that comprised not only the force of inertia and Newton's external force, but also both forms of the force of a body in motion. Thus the original concept of force was fundamentally different to the concept currently used in analytical mechanics.

Based on this analysis, I will show in Chapter 4 that d'Alembert's solution consists of removing a residue of the scholastic tradition by transforming the concept of force from a causal into a phenomenal concept. However, removing this residue does not necessarily imply that every form of metaphysics is excluded from mechanics. It is in the first place a *transformation* of both the *contents* of metaphysics and its *relation to mechanics*. Although d'Alembert himself believed that he was in fact removing metaphysics from mechanics, I will show that this was an over-reaction to an outmoded and unusable form of substance metaphysics.[65] While the rejection of this received form of substance metaphysics can be seen as linked to the historical development of classical mechanics, the denial of having any metaphysical foundation would not do justice to the same classical mechanics. In fact, d'Alembert does not really deny every form of metaphysical foundation: he actually seeks to establish one of a different sort. However, although the germ of the positive aspect, the development of a new form of metaphysics, is present in d'Alembert's work, he does not make it explicit.

1.2.3.2 The Principle of Least Action

A great deal has been published about the genesis of the principle of least action. Most of this literature relates to the sensational controversy between Maupertuis and König and the resulting court scandal, in which Frederick the Great of Prussia and that self-willed man of letters, Voltaire, played such important roles. Much of the remaining literature suffers from a one-sided emphasis on the mathematical aspect or is strongly anti-metaphysical in tone, so that the history of the developments preceding the full mathematical formulation of the principle by Lagrange in his *Méchanique analitique* from 1788 has remained rather obscure.[66] The tone for such

---

[63] *Ibidem*, 115.
[64] *Ibidem*, 141.
[65] For an explanation of this term, see below, page 28.
[66] For a summary of the literature about the principle of least action, see: Brunet, *Étude historique sur le principe de la moindre action* (1938); Fleckenstein, [Preface to Euler's *Commentationes mechanicae; principia mechanica*] (1957), xlvii-l; Pulte, *Das Prinzip der kleinsten Wirkung* (1989), 26n.90.

historiography was set in the previous century by the re-discoverers of the principle, such as Adolf Mayer and Hermann von Helmholtz.[67]

Two works that approach the action principle from a broader historical perspective have been published in recent years. The first is Mary Terrall's *Maupertuis and Eighteenth-Century Scientific Culture* (1987). The action principle is only part of the theme of Terrall's book, and she treats it primarily from Maupertuis's point of view. She proceeds primarily by clarifying the metaphysical aspects of the action principle in terms of their coherence with his biological theory of epigenesis and his political ideas. Her approach is in the tradition of contextual historiography, in which "[t]he practice of science must necessarily form part of a complex culture in which scientists live and work." [68] The merits of this approach are evident in her work. She is one of the few who has not fallen into the trap of interspersing Maupertuis's thoughts with anti-metaphysical remarks or denigrating comments on his mathematical abilities. The connections she has made, such as her interpretation of his theories as a transition from Newtonian physics to a universal teleological physics, deepen our understanding of the development of Maupertuis's theories. Terrall's approach links Maupertuis's work in widely divergent fields—mechanics, chemistry, biology, ethics and politics—that had previously always been treated separately, even in Pierre Brunet's authoritative biography from 1929.[69] Although my study deals only with Maupertuis's ideas on mechanics, and although I certainly do not follow Terrall in her modern-contextualistic historiography, her insights into the inner coherence of Maupertuis's theological, biological, and mechanical ideas have had a considerable influence on my interpretation of Maupertuis's thought and work.

The second work is Helmut Pulte's *Das Prinzip der kleinsten Wirkung und die Kraftkonzeptionen der rationalen Mechanik* (1989). Pulte chooses an entirely different approach from Terrall, but one which is equally valuable:

> This study does not focus on the mathematical development in itself (...) or on the much-discussed physico-theological issue, (...) but rather on the 'intermediate' field of epistemology, from which we [may] understand the frequently criticized 'speculative' character of Maupertuis's philosophy.[70]

---

[67] Mayer, *Geschichte des Princips der kleinsten Action* (1877); Helmholtz, "Ueber die physikalische Bedeutung des Princips der kleinsten Wirkung" (1886), "Rede über die Entdeckungsgeschichte des Princips der kleinsten Action" (1887), "Zur Geschichte des Princips der kleinsten Action" (1887), "Das Princip der kleinsten Wirkung in der Elekrodynamik" (1892), "Nachtrag zu dem Aufsatze: Ueber das Princip der kleinsten Wirkung in der Elektrodynamik" (1894).

[68] Opening sentence of Terrall, "The Culture of Science in Frederick the Great's Berlin" (1990), 333. This aim is also present in her dissertation, although it was not yet a dominant theme.

[69] Brunet, *Maupertuis. Vol. I: Étude biographique. Vol.II: L'œuvre et sa place dans la pensée scientifique et philosophique du XVIIIe siècle* (both volumes from 1929).

[70] "[n]icht die isolierte mathematische Entwicklung (...) und nicht die oft diskutierte physikotheologische Problematik (...) stehen im Mittelpunkt dieser Arbeit, sondern die 'dazwischenliegende' wissenschaftstheoretische Ebene, von der her etwa der vielkritisierte 'spekulative' Charakter der Philosophie Maupertuis' (...) verstanden werden [könnte]" (Pulte, *Das Prinzip der kleinsten Wirkung*, 27).

By epistemology, Pulte means "the ideas about what constitutes science (its rationality, its methods, its style, etc.)."[71] He wants to use Lakatos's concept of a 'research program' as a basis for his historiography, without wanting to utilize all the details of Lakatos's methodology, since this would "impose normative criteria of rationality" on the process of the development of science and reduce the model to a caricature of itself.[72] With this modification, a 'research program' has three elements: metaphysical assumptions about the nature of the world; empirical, fundamentally applicable laws; and epistemological premises. The development of eighteenth century rational mechanics in general, and the principle of least action in particular, are then seen by Pulte as a confrontation ("Auseinandersetzung") between the three great programs of Descartes, Leibniz, and Newton, resulting in a tentative victory for Newtonianism in the form of the physics of Laplace and Poisson.

Pulte's work is intended to unite the history of science and the philosophy of science by making the latter more historical. However, although he rejects the idea of 'rational reconstruction', its spirit is still present in his approach. I believe this is due primarily by the emphatic line of development sketched above, making history rather one-dimensional.

In Chapter 5 I will describe the approach employed by Maupertuis and Euler as an attempt to find an alternative foundation for mechanics. Their enthusiasm arose from the hope offered by a teleological principle, when it had become increasingly clear that a foundation in terms of efficient causes was not practicable, due in part to the controversy about living force. However, in the case of the principle of least action we can also distinguish another level. Maupertuis and Euler discuss only the aspect of final causation, but I will demonstrate that their approach is also based on a recognition of the structural character of the object of mechanics. Thus it appears that the initial history of the principle of least action can be interpreted as a metaphysical speculation about the *structure* of reality. However, the principle does not continue in its teleological form beyond this initial phase, but is transformed by Lagrange into a mathematical principle. For Maupertuis and Euler, the structure was still grounded in a 'highest substance', which for Maupertuis is God, and for Euler is Nature. God, or Nature, shapes the laws of motion according to their respective purposes. Lagrange rejects the possibility of acquiring insight from such speculations, but also does not make any clear choice between readings at an instrumental or agnostic level. For him, the principle of least action is not a metaphysical principle but a consequence of the first principles of mechanics. Nonetheless, his work contains traces of a structural metaphysics, as I will show in Chapter 7.

---

[71] "Vorstellungen darüber, was Wissenschaft (ihre Rationalität, ihre Methoden, ihren Stil usw.) ausmacht" (*ibidem*, 23).
[72] *Ibidem*, 22.

1.2.3.3 The Berlin Essay Contest for 1779

In the eighteenth century, contests were an important way to stimulate and direct scientific activity, as can be seen for example from the important essays written by Johann I Bernoulli for the Paris contests on the laws governing the collision of hard and elastic bodies (1724 and 1726 respectively), or from the famous contest of the Dijon Academy for 1754, on the basis of the inequality between people, won by Jean-Jacques Rousseau.[73] Contests formed a link between the organization and content of eighteenth century science, but they have hardly been a popular subject for study, especially if one excludes institutional and biographical historiography.[74] In 1900, Adolf Harnack published an outstanding and detailed overview of the Berlin contests in his *Geschichte der Königlich Preussischen Akademie der Wissenschaften zu Berlin*. After that, research of the Berlin contests was conducted primarily by the *Akademie der Wissenschaften der DDR*, the formal successor to the academy founded by Frederick the Great. However, this research shows a clear ideological emphasis.[75]

In addition to the previously cited overview by Harnack, there are two other works that are important in relation to the philosophical contests. In 1980, Laurence Bongie edited an anonymous prize essay from 1747 about Leibnizian monadology. In his introduction, Bongie claimed that this essay was written by Etienne Bonnot de Condillac, but did not provide any evidence for this conclusion. However, his introduction to the history of the Berlin Academy and the position of the contests in this history is very valuable.[76]

---

[73] Johann I. Bernoulli, *Discours sur les lois de la communication du mouvement* (1724–1726) (see also Chapter 4 of the present book) and Rousseau, *Discours sur l'origine et les fondemens de l'inégalité parmi les hommes* (1754).

[74] McClellan III, *Science Reorganized: Scientific Societies in the Eighteenth Century* (1985), 11n.22 (298). McClellan states that there is still no overall study of the essay contests. Yet some initial work has been done for particular academies. McClellan provides a number of references for French-speaking areas in Europe. In addition, an inventory of the contests of the *Hollandsche Maatschappij der Wetenschappen* can be found in *Inventaris van de Prijsvragen* (J.G. de Bruijn, ed., 1977). As examples of individual historical biographical studies, Buschmann lists K. Fischer, "Immanuel Kant und seine Lehre," in: *Geschichte der neuern Philosophie* IV and V (Heidelberg 1909–1910, especially Volume IV, 213 and the following pages); and R. Haym, *Herder*, (2 Volumes) Berlin 1954 (^{1}1885), especially Volume I, 428-439 and 689-713 (Buschmann, "Philosophischen Preisfragen," 167n.6).

[75] A study of the mathematical contests of the Berlin Academy was published in 1964 by Kurt-R. Biermann (Biermann, "Aus der Geschichte Berliner mathematischer Preisaufgaben"). The agroeconomic contests were inventoried in 1975 in Hans-Heinrich Müller's *Akademie und Wirtschaft im 18. Jahrhundert*. A selection of several essays from the remarkable contest for 1780 ("Est-il utile de tromper le peuple?") was published in 1966 by Werner Krauss, with a commentary (see also Krauss, "Eine politische Preisfrage im Jahre 1780" (1963)). In addition, Andreas Kleinert has discussed the award of the prize for 1746 to d'Alembert. This award had a very positive influence on the social status of the Berlin contests (Kleinert, "D'Alembert et le prix de l'Académie de Berlin en 1746" (1989)).

[76] [Condillac], *Les monades*. Prize essay, published anonymously in: Justi a.o., *Dissertation* (1748) (ed. 1980).

By far the most important study in relation to the philosophical contests is Cornelia Buschmann's "Die philosophischen Preisfragen und Preisschriften der Berliner Akademie der Wissenschaften im 18. Jahrhundert" (1989). In this article, the result of a doctoral study, Buschmann provides a critical analysis of the contents of all the Berlin Academy's philosophical contests.[77] She provides a valuable interpretation of the competition questions and a selection of the essays submitted. There are a few serious objections to her work, however. Firstly, the socialist ideology is sometimes rather overriding. This also causes some of her interpretations and positions to be obscure. In addition, it is frequently impossible to verify her interpretations because she provides the reader with little or no information on how she selected the essays that she discussed.

One important line that Buschmann has found is the gradual increase of utilitarian themes proposed for the competitions.[78] The first competition questions, for 1745 and the following years, were mainly speculative, whereas the competition questions that were proposed in the sixties and seventies of the eighteenth century are "both interesting and useful."[79] It is therefore remarkable that the contest dealt with in the present study, concerning the foundation of force, does not appear to fit this development at all. However, the controversial character of that competition question at the time gives me a handhold to locate the significance of the question and its answers as part of the historical development of the concept of mechanical force, as set out in Chapter 6. Although the essays in themselves have little value for the history of ideas, the competition question itself, which was speculative and thought by philosophers to be important, reveals the tension that existed even then between mechanics and metaphysics. The reactions to the competition question, both from those who ridiculed the question and those who responded to it, show the extent to which the development of the modern concept of mechanical force was accompanied by a division between the practitioners of mechanics and the general public.

1.3   THE PLACE OF PHILOSOPHY IN THIS BOOK

1.3.1   History of Science or Philosophy of Science?

This book is historical in nature, but philosophy plays a role, in the sense that it supports the historical clarification. Although the main concepts of my interpretation are metaphysical in origin, I will not attempt to clarify the nature of these concepts in a philosophical sense. For the task that I have set myself here, it is sufficient to explain

---

[77] Buschman, "Philosophischen Preisfragen."
[78] "(...) the contests sponsored by the Berlin Academy in the following years [i.e. after 1751, JCB] emerged from a coherence of their contents and, in a manner of speaking, arranged themselves concentrically around the previously discussed central questions in the development of Enlightenment ideology" ("(...) daß die Preisfragen, die die Berliner Akademie in den Folgejahren stellte, aus einem inhaltlichen Zusammenhang erwuchsen und sich sozusagen konzentrisch um die oben dargelegten Kernfragen aufklärerischer Ideologiebildung gruppierten") (*ibidem*, 193-194).
[79] Paraphrased remark of Frederick the Great in 1777 (*ibidem*, 220).

them to the extent that they can convey my interpretation of the development of mechanics. The support that philosophy provides for the historical research in this study relates to the choice and formulation of basic concepts that correspond to historical reality.

The attempt made in this book, to interpret changes in the metaphysical aspect of mechanics, appears to suppose, *a priori*, the presence of a metaphysics in even a modern form of physics. Is this not a philosophical position, hidden behind the historical interpretation? Such an *a priori* will certainly be resisted by an instrumentally or positivistically oriented reader. Is it not equally or more defensible, says the instrumentalist, that in principle *any assumption* can be made in the formation of scientific theory, that no *a priori* restrictions can be made, and that it is only the success of the theory that can set any limitations? In that case, one begins with experience of the subjects and a methodology for ranking these experiences, without having an *a priori* determined understanding of the object of study. Is not every metaphysics then an improper limitation of theory? A positivist would agree, while adding that true knowledge—i.e. knowledge that we can systematically derive from our experience—can only change in nature if experience itself leads to this, and can never change because of assumptions on our part.

My answer to these objections has two parts. First, I should clarify what exactly I understand metaphysics to be. Second, I should explain why a historical interpretation of the content of this metaphysics and its relationship to mechanics is independent of a philosophical evaluation of mechanics. The importance of these points, and particularly of the second point, is that they will allow the historical interpretation I provide here to be followed, as much as possible, by anti-metaphysicists and a-metaphysicists. It is important to delay a parting of the ways as long as possible (to the extent that this parting is caused by differences of position in the philosophy of science) in order to maintain as far as possible the independence of the discipline of the history of science.

1.3.2 A Working Definition of Metaphysics

In the history of western philosophy, 'metaphysics' does not have an unequivocal meaning. Etymologically speaking, it is the title given posthumously to Aristotle's work about the highest science, the first philosophy. Aristotle defined this at the outset as a speculative discipline that seeks to discover the principles and primary foundations of reality.[80] In later chapters, he specified the subject in a broader sense as being as being, and in a narrower sense as absolute being.[81] Over the course of time, Aristotle's definition has been joined by many others, but it can still serve as the starting point for every discussion of metaphysics. However, for historical studies in which metaphysical changes are central, it would be undesirable to use a

---

[80] Aristotle, *Metaphysics*, 3-4 (book A (1), 982a).
[81] *Ibidem*, 59-85 and 122-129 (book Γ (4) and book E (6)).

definition that is linked to a specific form of metaphysics defined by its contents. We need a definition that expresses a more general form. In this section, therefore, I will develop and formulate a *working definition* of the concept of metaphysics, to be used in this study.

The definition I use can be introduced by indicating several divergences from Aristotle's definition. In the first place, Aristotle viewed metaphysics as the most general discipline that deals with being as being. But the question addressed here deals with a sub-area of the whole of being: being in the form of matter-in-motion. I will therefore adopt only Aristotle's general definition and will expand his specifications by including sub-areas of being as its subject. Since modern sciences usually do not concern themselves with the systematic study of their principles and primary foundations, this expansion of the concept of metaphysics, which I am certainly not the first to apply, is justified.

Secondly, for Aristotle, being was, in a primary sense, 'substance'. After all, anything that is called a being refers to a substance, whether it is substance itself, a property of a substance or relates to a substance in some other way.[82] As long as the object of modern physics—matter—was understood as a substance, Aristotle's definition of being could be retained in modern physics. This is what happened in the seventeenth century, and it continued into the eighteenth century, as I will show later. But the result of this perseverance was that metaphysics was driven to the margins of modern science, where it gradually withered away, practiced by only a few eccentrics. This is because the core of the modern concept of matter lies not in its substantial nature, but in its embedding in structures.

There are two parts to this embedding. Firstly, matter is *compounded*, and it itself *compounds*. For example, molecules of water form water, and they in turn are compounded of atoms of hydrogen and oxygen. Secondly, the modern concept of matter is characterized by the *laws* of motion and other changes in matter. Any definition of the concept of being that is also to be usable in modern physics must therefore not be fixed to the concept of substance, but must also allow space for the core concept of modern matter: structure.[83]

Finally, Aristotle's definition is based on an ontological realism, without concerning itself with epistemological reflection. During the eighteenth century the need for epistemological reflection became increasingly urgent, and this development prevented the naive identification of thought and being. To allow room for more skeptical views concerning the relationship of thought and being, such as an instrumentalist or social constructivist view of science, I should state explicitly that I am not concerned with the nature of reality in itself, nor with the relationship of scientific knowledge to reality, but with the *form of knowledge by which we know reality*.

The meaning of metaphysics in this context, therefore, does not entail the specific contents of Aristotle's metaphysics, but only its general form. Immanuel Kant's

---

[82] *Ibidem*, 59-61 (book Γ (4), 1003b).
[83] I will return to the concepts of substance and structure in section 1.3.4.

Metaphysische Anfangsgründe der Naturwissenschaft (1786) provides a good starting point for a working definition of metaphysics that is adequate to my purposes. Here Kant defines a transcendental metaphysics comprised of that part of natural science which encompasses "pure intellectual knowledge based on mere concepts."[84] Stated more precisely:

> All true metaphysics originates from the nature of the capacity to think itself, and (...) comprises pure acts of thought, including concepts and *a priori* principles, that first link the complexity of empirical notions in such a systematic way that they can become empirical knowledge, i.e., experience.[85]

For example, in the section on dynamics, he develops the idea of matter—that which can be moved in space—as the substance of space, and the idea of the fundamental forces ("Grundkräfte") as that which constitutes matter.[86] The transcendental character of Kant's concept of metaphysics makes it independent of an ontological interpretation, because metaphysics says nothing about the thing-in-itself ("Ding an sich"). It only indicates the direction and limits of our understanding of reality.

This transcendental turn in the concept of metaphysics is a great strength of Kant's definition. After all, whether concepts such as space, time, mass, force, and energy represent a reality in the world outside ourselves, or not, one can still ask *how* reality is approached in a specific science. Nevertheless, Kant's definition needs to be supplemented to make it suitable for my study.

Historical mutability requires that Kant's rigid assessment of *a priori* concepts and principles should be replaced by a more dynamic 'apriori-zation', following the neo-Kantian Albert Görland.[87] The three-dimensionality of space and the constitution of material substances by fundamental forces are only historical forms of such an apriori-zation: they are not the necessary *a priori* of the ideas of space and mobility, respectively.[88] Historical research is not concerned with the *necessary* conditions of knowledge for natural science, but only with the fundamental presumptions found *in historical reality*, or those which we can assume played a role behind the scenes.

---

[84] "[r]eine Vernunfterkenntnis aus bloßen Begriffen" (Kant, *Metaphyische Anfangsgründe*, 13 (A VII)).
[85] "Alle wahre Metaphysik ist aus dem Wesen des Denkungsvermögens selbst genommen, und (...) enthält die reinen Handlungen des Denkens, mithin Begriffe und Grundsätze a priori, welche das Mannigfaltige *empirischer Vorstellungen* allererst in die gesetzmäßige Verbindung bringt, dadurch es *empirisches* [sic!] Erkenntnis, d.i. Erfahrung, werden kann" (*ibidem*, Vorrede A XIII).
[86] *Ibidem*, 55-82 (A 43-80).
[87] In 1930 Görland designated this as the process of principalisation ("Prozeß der Prinzipierung") (Görland, *Prologik; Dialektik des kritischen Idealismus*, 14-17). The Dutch philosopher Pieter Tijmes explains this as follows: "In Görland's hands, the rigid apriori of Kant changes into 'apriori-zation'. In this way, Görland wants to allow for the fact that, in the course of its history, science itself replaces an initial hypothesis with another hypothesis, generally more encompassing, in which the older hypothesis is seen, for example, as a specific case of the new one" (Tijmes, "Albert Görland, een systematisch denker van formaat" (1985), 208).
[88] Compare Kant, *Metaphysische Anfangsgründen*, 25-99 (sections "Phoronomie" and "Dynamik").

According to this definition, 'metaphysics' is all existing *a priori* understandings and assumptions regarding both the nature of reality—whether this concerns ontology, ethics, or aesthetics—and possible ways of knowing this reality. In short, this encompasses the presumptions and first principles of our interpretation of reality. As for Kant, the term 'a priori' does not necessarily mean that the relevant understandings precede *in time* all insights that are partly based on experience. They can even be *discovered* as insights that are *inherent* to all knowledge, both possible and actual, of the relevant discipline or disciplines in that time. Therefore it would be preferable to call them *hypothetical* understandings.

Since the definition is to be applied in historical research, 'possible knowledge' should be understood not as *logically* possible knowledge, but as *historically* possible. The historically possible can be defined here as that which is *not in conflict with the accepted insights of a specific time and within a specific discipline*. Its most important characteristic is that it is still open to further research. For example, a teleological explanation is unacceptable in modern mechanics, while in biology it is a possibility, albeit a controversial one. While this further delimitation of the meaning of 'possible' makes historical research more limited than philosophical research, it also adds a dimension to it, since time becomes a factor.

Using this specification of the concept of metaphysics, the relationship between metaphysics and mechanics need not be deductive at all. In other words, it is not necessary that metaphysical principles should come first and that the form or content of mechanical propositions should be derived from them. To take an example from the history of thermodynamics: both the perseverance in searching for experimental proof of the equivalence of heat and work (to the extent that this is based on a non-experimental belief) and the gradual awareness of the fundamental consequences of this equivalence for our world-view, are metaphysical aspects of physics.

Once the concept of metaphysics has been formalized and made historical, another crucial difference between this use of the term metaphysics and the philosophical use becomes evident. Görland's modification allows room for changes in knowledge. When 'possible' is restricted to the 'historically possible', the philosophical reference point is detached from the present and transferred to the context of the period being considered. The resulting description of the concept of metaphysics now makes it possible to arrive at a historical interpretation of the relationship between metaphysics and mechanics without having to link this directly to a viewpoint in the philosophy of science. In other words, it is possible to provide a systematic examination linking explicit and implicit metaphysical elements, *on a factual* basis, with the subject concerned, without implying that this link is *necessary and general*.[89] The principles and fundamental consequences that emerge in this study can therefore be viewed either as *conjectures and assumptions*, or as *true principles and concepts*.

---

[89] Compare Kuhn's argument about the distinction between the history of science and the philosophy of science in his "The Relationship between the History and the Philosophy of Science" (1968).

The working definition of metaphysics given here is so broad that a logician might feel that my promise to show that the metaphysical dimension played a leading role in the development of eighteenth century mechanics is dangerously close to a tautology.[90] In terms of historiography, however, this is far from the case. The issue, after all, is to explain *how* the metaphysical dimension played this role and how our *historical understanding* is enlarged by this explanation.

A parting of the ways should only become inevitable when it comes to the philosophical *implications* of my research; where I would say that classical mechanics is essentially linked with the transformation of metaphysics described here, Mach's adherents might refer to a still incomplete liberation from metaphysics.

1.3.3   Explicit and Implicit Metaphysics

From what has been said, it is evident that a distinction must be made between metaphysics that is presented as such and the metaphysics that lies hidden in the practice of science. I indicate this distinction in the relationship between metaphysics and science with the terms *explicit* and *implicit* metaphysics respectively. 'Explicit metaphysics' is the metaphysics used by the science concerned when it presents itself to the outside world. In Aristotelian physics, for example, the purposiveness of natural existence is assumed to hold, in the Darwinian theory of evolution it is denied. This explicit metaphysics may be unreflective, and is then the metaphysics of a science that is not, or is not yet, aware of itself. It can also be reflective, as the metaphysics of a science that is aware of itself. 'Implicit metaphysics' is the metaphysics that lies hidden in the assumptions and methods of the relevant scientific discipline.

There will usually be a tension between the explicit and the implicit metaphysics associated with a specific scientific discipline. This can be clarified by analogy with transsexuality. Explicit metaphysics is analogous to one's physical gender identity and implicit metaphysics as psychological gender identity. One presents oneself, as it were, to the outside world with one's physical gender identity, and one is consequently expected to adopt the accompanying gender role and place in the world. Psychological gender identity, on the other hand, is determined by the inner desire for a specific gender role. Up to a certain age, the psychological gender identity remains unconscious (and consequently the physical identity is also, in a certain sense, unconscious). During puberty, or sometimes before, one becomes conscious of both. If the conscious psychological identity turns out to be different to the physical gender identity, this would be called transsexuality. The consciousness of this deviating 'inner gender' is sometimes so strong that surgical help is sought to modify the physical characteristics to match the psychological ones. In the same way, becoming aware of implicit metaphysics may also give rise to transformation of the explicit metaphysics, not from the outside (as in medical treatment), but from within.

---

[90]   See above, page 3.

The distinction between explicit and implicit metaphysics enables us to analyze the development of the relationship between metaphysics and mechanics at a deeper level than would be possible looking only at the metaphysics that mechanics uses in its self-presentation. After all, the implicit metaphysics will only be roughly the same as the explicit if the development of metaphysics is in step with the development of the relevant scientific discipline. This was the case, for example, with the physics of Aristotle and of Leibniz.[91] In such a situation, an examination of the explicit metaphysical arguments alone will generally be sufficient to gain an understanding of the presuppositions.

This situation changed during the eighteenth century. It is true that it was normal for a physicist to explicitly discuss the metaphysical foundations of his science; he was called a 'natural philosopher' after all. But at the same time, the discipline of philosophy was critical of mathematical natural scientists, for example, as if they did not have sufficient philosophical training for such a task. Wolff and d'Alembert, for example, contested Euler's competence to give his opinion on monadology and the metaphysical foundation of gravity. We can see, in the course of the development of mechanics, that this situation is internalized with Lagrange. He makes no statements about the foundations of mechanics, but bases his arguments only on principles, which are then elaborated mathematically. He prefers his analytical approach to remain without explicit metaphysical considerations. His approach can be seen as a model for the generations of the nineteenth century. Nonetheless, there is some metaphysics in Lagrange, at an implicit level. Chapter 7 will describe how this is so.

This further specification of the distinction between 'explicit' and 'implicit' metaphysics enables me to describe the second objective of my study more clearly. This objective was formulated above as "to unravel the broad lines of development in metaphysics itself and in its relationship to mechanics."[92] Regarding the relationship of metaphysics to mechanics, I stated at that point that during the century following Newton's *Principia mathematica* we can observe a transition from an *explicit* to an *implicit* form of metaphysics. I can now add that the explicit metaphysics that accompanied mechanics at the beginning of the eighteenth century was not compatible with its implicit metaphysics, and that this became clear through the controversies that will be described here. The result was a transformation of the explicit metaphysics. However, in terms of the analogy with transsexuality, this was like someone with an incomplete awareness of his gender identity, who believes himself to be asexual. This led to an operation from which a creature emerged that was neither male nor female.

---

[91] It is not my intention to put the physics of Aristotle on a par with modern physics: what matters is only the *analogy* of the *relation* between physics and metaphysics with Aristotle and with modern physics.
[92] See above, page 4.

1.3.4   Substance and Structure: Basic Concepts of Metaphysical Change

We can describe the formal objective of classical mechanics, by analogy to the objective of modern physics,[93] as the complete mathematical description in a single system of all motion processes of material particles. The first aim of this book, however, is to show that this description is not an adequate characterization of mechanics, as long as nothing is said about the metaphysical assumptions concerning the nature of the reality being studied.[94] The question here is not the further delimitation of the possible *subjects of study* such as atoms, heat, particles, energy quantities, etc., but first of all the *orientation* of the research. After all, if we ask about the meaning of the laws and particles that have been discovered, we assume there was a *route* that we followed to learn about these laws and particles, and therefore suppose that our knowledge had an *intention*.

In this book, I interpret the development of the concept of force within the science of mechanics, from Newton and Leibniz on the one hand to Lagrange on the other, as a *transition in intentionality*. During the seventeenth and eighteenth centuries, mechanical force can be defined in general terms as the cause of the action of particles on one another. This is a general definition and covers not only Newton's action-at-a-distance and Descartes's force of impact, but also the "substantial powers" of Locke.[95]

This definition shows that there are two routes by which forces can be given a foundation in reality. The first route is the distinct particles, the second is that of the structure of the whole. In the first, the particles are seen as the bearer and source of the action of forces. The action is therefore in accordance with, and can sometimes even be derived from, the nature of the particles. Metaphysically speaking, this means that the relationship between substance and property that was developed in scholasticism and was a commonplace well into the eighteenth century[96] is the starting point for the mathematization of force. I therefore refer to this route as a substantial route, and the corresponding general form of metaphysics I call *substance metaphysics*. Leibniz's metaphysics is an outstanding example of such a substance metaphysics (see Chapter 3). But, quite unexpectedly, we can also find expressions of this in Newton's view of force (see Chapter 2). Force links a variety of material particles, and while it has its own properties (it is for example symmetrical), it is ultimately founded in discrete substances and their properties. This foundation of the concept of force entails one important limitation: that which is transferred by

---

[93] See above, page 2.
[94] See above, page 4.
[95] Locke interpreted the so-called secondary qualities such as color, heat, sound, etc., as the action of material substances on the human mind. This substantial interpretation gained a large following at the end of the eighteenth century. See Heimann and McGuire, "Newtonian Forces and Lockean Powers: Concepts of Matter in Eighteenth-Century Thought" (1971).
[96] See Chapter 6 concerning the Berlin contest for 1779. See also Moses Mendelssohn, who in his essay *Abhandlung über die Evidenz in metaphysischen Wissenschaften* (1764) defends the thesis that properties follow from the complete concept of things.

a force from one substance to another must be expressible in the properties of both substances separately. Newton's and Leibniz' views concerning the action of forces can therefore be illustrated by the material analogy of the transfer of water between two vessels.

The second route along which force can be founded is that of structure. By structure, with respect to mechanics, I mean the way the particles are arranged in space and time. The particles themselves are now no longer substances in a proper sense, but elements of a structure comprised of specific quantities that are linked in mathematical relationships. In this case, a force is not attributed to a single particle, but to multiple particles simultaneously, or more precisely, to their structure. This opens up the possibility of a different law of force, such as the modern form of Newton's second law of mechanics (see Chapter 2). It also lays the foundation for another modern insight, the interpretation of force as a *function* of temporal and spatial coordinates. Insights such as the principles of symmetry and of invariance, which are general principles concerning the form of physical laws, are coherent with this foundation. The principle of least action can also be viewed in this way, as the concretization of a general principle: the distinct forces have been put aside and the actions are mathematicized along a structural route. Structuralization in this sense is more than mathematization alone, because it determines in more detail *how* the mathematization takes place: proceeding from an insight into, or with an eye to, the structure of the whole.

This book is concerned with the transition from seventeenth century to nineteenth century mechanics. This transition entailed saying goodbye to scholastic metaphysics and formulating new basic principles for mechanics. These basic principles then proved to constitute another metaphysics, characterized not by the concepts of 'substance' and 'quality', but by the concept of external structure. Mechanics in its modern, Lagrangian form presupposes a structure of space and time, whose elements are masses with a velocity and a position. But it contains no reference at all to what lies hidden behind the elements: the substances. Within this new world of analytical mechanics, a new domain has arisen, containing a new unity of material reality, to which mathematics is the key.[97] In this way, the open end in Dijksterhuis's characterization of modern physics can be closed by asking "what was mechanical about mechanics?" This will also bring us one step closer to a clarification of those magical words: 'the mechanization of the world picture'.

---

[97] Of course, this does not mean that substance no longer plays a role here, but that it has been subordinated to external structure. For further analysis about the change in the function of substance and the relationship of substances to structure, see the epilogue in Chapter 8.

A

THE UNITY OF THE CONCEPT OF FORCE

CHAPTER 2

FORCE LIKE WATER

> *It remained for the eighteenth century to define the concept of force with adequate rigour.*
> Richard S. Westfall[1]

2.1 A BRIEF ARCHEOLOGY OF THE CONCEPT OF FORCE IN THE SEVENTEENTH CENTURY

Force occupies a central position within classic mechanics. Seventeenth-century mechanics had restricted the use of forces to statics and the laws of impact (other motions being regarded solely as kinematic), but after Newton, mechanics linked forces and motion almost by definition. For example, d'Alembert defined mechanics in 1765 as the science "that considers movement and motive forces, their nature, their laws and their effects within machines."[2] Newton's and Leibniz's work in the field of mechanics can be viewed as the motive force that moved the concept of force to center stage. However, this shift in the concept of force from the realm of everyday experience and the peripheral regions of occult phenomena to the center of mechanics did not take place without difficulty. Distinctions within the concept of force had to be introduced or refined, such as the distinctions between 'inertia' and 'cause of acceleration', between 'momentum' and 'kinetic energy' and between 'contact force' and 'action-at-a-distance'. The process of making these distinctions was often accompanied by an extensive battle in which conflicting viewpoints were defended tooth and nail.

---

[1] Westfal, *Force in Newton's Physics: The Science of Dynamics in the Seventeenth Century* (1971), 476.

[2] "qui considere le mouvement & les forces motrices, leur nature, leur loix & leurs effets dans les machines" (D'Alembert, "Méchanique" (1765), 222). Lagrange provides us with another example, giving the following definition of the two branches of mechanics—statics and dynamics: "Statics is the science of the equilibrium of forces" ("La Statique est la science de l'équilibre des forces") (Lagrange, *Méchanique analitique* (1788), 1); "Dynamics is the science of accelerating and retarding forces, and the various movements which they can produce" ("La Dynamique est la science des forces accélératrices ou retardatrices, et des mouvements variés qu'elles doivent produire") *(ibidem,* 207).

The best-known example of a dispute of this kind can be found in the history of the force of gravity. Ever since Newton's cautious postulation of the universal force of gravity as action-at-a-distance, in his *Principia mathematica* of 1687, there had been an unremitting struggle between those who wanted to explain gravity using ether theories and those who regarded a force of this kind as fundamental and irreducible.[3] This struggle moved to the background only after the appearance of field theories at the end of the nineteenth century.

During the development of mechanics, the distinctions made within the concept of force did indeed become generally accepted, but the concept of force also acquired an entirely different meaning. Except for their subject matter, Lagrangian and Hamiltonian formulations of mechanics have little in common with the ponderous and laborious way in which Newton related aspects such as the action of compounded forces to the action of the individual forces.[4] Although the advantages of the modern formulation are patently obvious, it also has its drawbacks; it has become more difficult to understand the historical development, if only because the original meanings seem strange or even bizarre from the viewpoint of the modern formulation.

Faced with the term 'force', the modern physicist thinks only of the modern formulation of Newton's second law, in which force is the product of mass and acceleration: $\underline{F} = m \times \underline{a}$. For him, the fact that the inertial force has the formal name of 'force' is no more than a residual trace of a very early philosophy of nature, in which forces were quite simply considered responsible for everything that moved. After all, it is as clear as daylight that inertia is a property that needs no cause! The same applies to the concept of 'living force' which today is, at best, regarded as an old name for kinetic energy[5], but which no longer has anything in common with Newton's concept of force.

Naturally, this kind of attitude is inadequate if one is to understand what happened to the concept of force during the development of mechanics; a physicist would also acknowledge this. Nevertheless, it appears to be very difficult to gain a real sensitivity to the meaning of questions connected with previous (and in a way outdated) conceptions of force. The historiography concerning the controversy about the true measure of living force, for example, all too often gives the impression that the dispute was about essentially non-physical matters, involving mere wars of words and power struggles, rather than the subject itself. Lack of sensitivity to the meaning of the old concept of force appears to be a major cause of this reduction.

In the history of ideas, it is not unusual to find that very different ideas originate historically from a common concept. Frequently, the common term is then reserved

---

[3] For an excellent survey of this conflict, see Van Lunteren, *Framing Hypotheses; Conceptions of Gravity in the 18th and 19th Centuries* (1991).
[4] Newton, *Philosophiae naturalis principia mathematica* (1687), Liber I, Leges Motus, Corollaries I and II, 14r.21-17r.10.
[5] For which a conversion factor of ½ has to be applied to the formula for *vis viva*.

for one of these ideas, while the others acquire new names. The relationship between the historical unity and historical diversity may be a simple species/genus, as in the distinction between various zebra finches where previously only a single species was recognized. But it need not be restricted to this. Cases in which there has been a *change in viewpoint* are much more interesting. For example, the meaning of the Greek ψυχή (psyche) in the *Iliad* was a unification of breath and soul: it was the force that breathes in life, the 'soul-breath'.[6] The distinction between these two meanings only came into existence later, when breath was connected with the body, and the soul with consciousness. This distinction does not reside in a difference between 'species', but in another view of life in general, and of humankind in particular. Other examples include the development of adolescence as an intermediate stage between 'child' and 'adult'[7] and the distinction between a word and the thing it indicates (resulting in an extremely troubled relationship between modern man and the biblical myth of Creation). Every one of these distinctions is taken for granted in our own times, yet each retains its own assumptions. The historical approach offers the possibility of revealing the transition from the original unity to the modern distinction that we take for granted; by this means we can turn something that goes without saying into something that can *truly* speak for itself.

In order to realize this potential, I shall devote a good deal of this chapter to an investigation of the original meaning of the concept of mechanical force. In other words, I will search for the historical basis of the identical, or at least analogous, approach that was taken to kinds of force that are, according to the modern conception, entirely different from each other. These kinds of force include energy, power, and the forces of inertia, impact, and gravity. My starting point is the end of the seventeenth century, when Leibniz made his famous distinction between dead and living force. From this vantage point, I shall return to earlier times, to Galileo and Descartes, and even to the Paris nominalists so that, just as in an archeological expedition, we may uncover shards and fragments from the deeper layers that can clarify the original unity of Leibniz's distinction. Finally, I will clarify the relationship between this unity and substantial-metaphysical considerations.

After this I will take the reader back to the eighteenth century by another route, via different stages of the concept of inertia. During the return journey it will become clear that the original unity of the concept of mechanical force also percolated through to the eighteenth century by another, completely different, route: in Newton's concepts of inertia and of force.

I now invite you to accompany me on this expedition. As we proceed I will frequently draw attention to a variety of matters that are important to our objective. I have found outstanding guides for the expedition, although not for its objective, in

---

[6] Derived from 'ψύχω' ('psycho'): 'to blow'. Compare also with the Sanskrit 'prana', meaning not only 'breath' but also 'life' and 'soul'.

[7] See Philippe Ariès, *Centuries of Childhood* (1960), especially part 1; and Marc Kleijwegt, *Ancient Youth. The Ambiguity of Youth and the Absence of Adolescense in Greco-Roman Society* (1991).

the historians E.J. Dijksterhuis, R.S. Westfall and I.B. Cohen.[8] Their work places the history of the concept of force firmly within its intellectual context.

However, to ensure sufficiently discerning observation on our journey of discovery, I will first ask you to accompany me on an exploratory mission to examine the background of the relationship between 'quantity' and 'quality' in the seventeenth and eighteenth centuries. During this mission we shall also discuss the concept of 'measure', a discussion which some of you will certainly have been expecting to encounter.

2.2 QUANTITY AND QUALITY

2.2.1 The Distinction According to Aristotle

The heritage of Aristotle and the scholastics included the distinction between 'quantity' and 'quality'. In his *Metaphysics,* Aristotle defined these two as categories of being. For him, the quantitative (ποσόν) was that which can be divided up in such a way that each part can have independent existence according to its nature. For example, length is a quantity because every part of it is also a length. Within the category of quantity he further distinguished between 'size' and 'number' as the measurable and the countable, respectively.[9]

Unfortunately, where the category of quality (ποιόν) is concerned, he did not succeed in providing a clear definition, and had to limit his attempt to descriptions and examples. The term 'quality' indicates firstly the essential difference in substances, and secondly the definitions of the changeable as changeable and the differences in these changes.[10] "Quality is one of the factors that determine form: a difference in quality is a difference in essence, in condition or in motion (in the general Aristotelian sense)."[11] Therefore a man and a horse are qualitatively different because the first is bipedal and the second is a quadruped.[12] Examples of qualities indicating a condition are: good and bad; hot and cold; white and black. Heaviness and lightness, on the other hand, are examples of qualities pertaining to motion.[13]

An important difference between 'quantity' and 'quality' is the way in which one can speak here of 'more' and 'less'. The length of a piece of wood or the price of a

---

[8] It goes without saying that this list is not complete. Jammer, Pierson, Whiteside and Kutschmann must certainly be included. However, the present chapter does not pretend to be comprehensive: it should be taken by the reader together with Chapter 3 as a preparation for the central chapters of this book. Moreover, later authors have not come significantly farther than the three cited above in relation to the interpretation of the relationship between force and substance.
[9] Aristotle, *Metaphysics,* 105-106 (book Δ (5), 1020a7-32).
[10] *Ibidem,* 106-107 (book Δ (5), 1020a33-1020b25).
[11] "De qualiteit is een der factoren, die den vorm bepalen: verschil in qualiteit is verschil in wezen, in toestand of in beweging (in algemeenen aristotelischen zin)" (Eduard Jan Dijksterhuis, *Val en worp; een bijdrage tot de geschiedenis der mechanica van Aristoteles tot Newton* (1924), 89).
[12] *Aristoteles, Metaphysics,* 106 (book Δ (5), 1020a).
[13] *Ibidem,* 107 (book Δ (5), 1020b).

loaf can be reduced, the number of employees can be increased. Quantities are thus comparable in terms of size or number. But what about differences in qualities: the whiteness of laundry, the heat of two ovens, the weight of two hammers? We can perceive the difference, but only as 'stronger' or 'weaker', not as 'larger' or 'smaller', although in some cases we can relate a qualitative difference to a quantitative one. We can get a better sense of why quantification was problematical if we consider the limits it has today. Within certain limits, quantification is generally accepted; outside these limits a few people may attempt it, but others fight against it all the more fiercely. Happiness and beauty, for example, are beyond the scope of quantification.[14] The value of the environment in prosperity is a quality that verges on the limit itself: it is still not clear whether modern economists will succeed in bringing it within the scope of quantification.[15] The quizzical looks that usually greet attempts to quantify such concepts now, were probably just as common in the fourteenth century for qualities such as velocity and heat.

Clearly, where the possibility and range of quantification are concerned, we cannot simply lump all qualities together. We could define a complete quantification (based on Aristotle's definition of quantity) as one in which the *parts* of the quantity bear the same relationship to the quality in question as the *whole* quantity. In those cases one can posit an absolute linear scale in which values can be added and subtracted without losing the relation to the original quality. Duration would then be a 'completely quantified' quality: a time scale can be divided into parts, each of which also represents a duration. The sum of two parts of the time scale is equal to the total of the corresponding durations. The success of this quantification explains why, in modern physics, the duration of time is identified with its quantification.[16]

One example of a less far-reaching quantification is Celcius' temperature scale. Heat and cold—or as we say now, 'temperature'—is converted into a linear scale by measuring the expansion of the volume of a material such as mercury. The freezing point and boiling point of water at a pressure of one atmosphere serve as the cali-

---

[14] This does not alter the fact that history reveals many attempts to do so. An elegant example concerning the cocnept of happiness is provided by the chemist and philosopher Wilhelm Ostwald (see Hakfoort, "Science Deified: Wilhelm Ostwald's Energeticist World-View and the History of Scientism" (1992)). In the eighteenth century Moses Mendelssohn tried to adapt mathematics to ethics and esthetics (see Mendelssohn, *Abhandlung über die* Evidenz in metaphysischen Wissenschaften (1764)).

[15] See for example the method, developed under Hueting's guidance, for quantifying the value of the environment in national prosperity by means of an adjusted Gross National Income, the 'Sustainable National Income' (R. Hueting, P. Bosch and B. de Boer, *Methodology for the Calculation of Sustainable National Income*, Central Bureau for Statistics of the Netherlands (1991)).

[16] Despite this success, something important appears to be lost in the quantification of time: what a time scale represents is time as it appears in physics, and not as it is experienced in daily life. The French philosopher Henri Bergson and the Dutch philosopher Jan Hendrik van den Berg in particular have focused on this problem (for example, see Henri Bergson, *Essai sur les données immediates de la conscience* (1989) and Jan Hendrik van den Berg, *Metabletica van de materie I. Meetkundige beschouwingen* (1968)).

bration points. Although the resulting scale is fairly linear, this is true only in a relative sense. For example, if the temperatures of two glasses of water are 15 and 30 degrees centigrade, then one is hotter than the other. But how much hotter? Strictly speaking, the answer '15 degrees centigrade' is wrong: the *difference* in temperature would then be just as great as the *level* of the temperature in one of the glasses, which would make the quality of temperature in one glass *twice as large* as that in the other! The absurdity of this reasoning derives from the fact that the zero point of the scale does not coincide with the 'absence' of a temperature. The Celsius scale is not an absolute linear scale, but an ordinal one, in which addition and subtraction are impossible.[17] Therefore, the level of the *quantity* 'temperature' cannot simply be identified with the intensity of the *quality* 'temperature'.

2.2.2 Scholastic Considerations

As if history wanted to make things easy for us, during the scholastic period the interest in the problem of the quantification of qualities was sparked in connection with a concept that is precisely of the sort that is regarded as unquantifiable today. In the twelfth century Peter Lombard commented, in a passage in his *Libri Sententiarum,* that a person's 'caritas' can increase or decrease, and can gain or lose intensity at different times.[18] Later, this increase and decrease was called the 'intensio' and 'remissio' respectively of a 'forma'. The views that developed as a result of Peter Lombard's comment can be broadly divided into two different schools. The first, defended by Thomas Aquinas, the Paris school of nominalists and the Oxford calculatores, proposed that the 'intensio' and the 'remissio' of a 'forma' are analogous to an increase or decrease of a quantity. The second viewpoint upholds the basic Aristotelian distinction between qualitative and quantitative changes.[19]

Where kinematics is concerned, the reflections of Nicole Oresme in the fourteenth century on the concept of motion played a major role. Today we are inclined to think that the definition of momentaneous velocity is self-evident: the ratio of the distance covered to the time required, as time approaches the limit of zero. But in the fourteenth century, when the concept of the infinitely small was still unknown in mathematics, it was not possible to perform such an operation. Even the simple definition of velocity as the ratio of distance covered to elapsed time was problematic; strictly speaking, it was impossible to divide quantities of different kinds, such as space and time, by one another. The concrete result of this, as Dijksterhuis notes, was that velocity was treated in much the same way as whiteness, heat, or human love.[20]

---

[17] See also Hoenen, *Philosophie der anorganische natuur* (1938), 203 and 252.
[18] Dijksterhuis, *Val en worp*, 90.
[19] For a more extensive discussion, see *ibidem*, 90-94.
[20] *Ibidem*, 132n.111. See also the comments on dimensional analysis at the end of the present section.

In his *Val en worp,* Dijksterhuis shows how much the development of kinematics depended on Oresme's philosophy of qualitative change. Oresme provided a graphic demonstration of the different grades, or intensities, of a quality by means of line segments or 'extensiones', the ratios of which represent the ratios of intensities of the quality.[21] His application of his method to kinematics showed that velocity (which, after all, was also a quality), can also be represented as an 'extensio'. As long as constant velocities are involved, this produces no difficulties. A constant velocity can be defined in a simple way as the change of position within a given time, i.e. as the ratio of two finite quantities. It is a different matter in the case of changing velocities, since these have to be determined at each instant of time separately, i.e., as *momentaneous velocity.* There is no way forward here, unless one introduces infinitely small units of time and distance. Oresme had no access to these concepts or the manner of thinking they involve, but he was still able to define velocity in free fall:

> (...) the degree of velocity in free fall is greater the further the moving object falls, or rather, would fall, [in a certain amount of time, JCB] if it were to continue in the same way.[22]

In his history of specific gravity during Antiquity and the Middle Ages, Bauerreis provides us with an elegant illustration of how complex and opaque quantitative matters were in those times, and how much room they left for numerous interpretations and speculations.[23] His description of the successive measuring techniques shows repeatedly that complicated descriptions were being used during the Middle Ages, where we would simply refer to 'specific gravity', i.e. weight per volume. In a text dating from before 1017 AD, al-Bîrûnî reports the results of measurements made with the aid of a pycnometer he had developed. For example, he reports the weight of the amount of water displaced by submerging 100 mithqâls of the metal or precious stone under investigation. He also reports the weight of the metal or precious stone that has the same volume as 100 mithqâls of gold.[24] We can see how remote the meaning of these figures is from the interpretation 'weight per volume', or 'volume per weight' by the fact that he relates the two measurements by means of the highest common denominator of both figures. Nowadays, this figure has no significance in physics, but al-Bîrûnî must have hoped that it would put him on the track of a deeper truth.

---

[21] *Ibidem* 95-96.
[22] "[G]radus velocitatis descensus est maior, quo subjectum mobile magis descendet vel descenderet si continuaretur simpliciter" (Oresme, *Tractatus de figuratione potentiarum,* H. Wieleitner ed. 1914, 224; cited by Dijksterhuis, *Val en worp,* 101). Dijksterhuis adds the comment that this definition of momentaneous speed was the only one possible before differential calculus was introduced.
[23] Heinrich Bauerreis, *Zur Geschichte des spezifischen Gewichtes in Altertum und Mittelalter* (1914).
[24] *Ibidem,* 16-17.

2.2.3 Quantification of the Concept of Force

In view of the difficulties presented by Oresme's attempt to quantify the quality 'velocity', it is hardly surprising that the concept of force presented an even bigger problem three centuries later. For force itself cannot be perceived by our senses, only the changes in the material world that we attribute to the actions of this force. The principle for the quantification of such qualities was already known to Thomas Aquinas: one measures the quantitative effects of the quality concerned.[25] But this was still only a *principle* of measurement; it in no way answered the questions of whether there *are* forces, *how many* forces there are, and *what actions* are connected to them.

The major elements of what would later be regarded as the action of mechanical force—velocity, time, and quantity of matter—had been quantified satisfactorily in the period between the fourteenth and seventeenth centuries. However, this did not mean that the path for its mathematization was already prepared and evident when force was reintroduced into mechanics at the end of the seventeenth century.

In the controversy about the true measure of living force, the main concern was for the right connection between empirical, mathematical, and metaphysical elements. In this controversy, what is called the 'measure', 'estimation' or 'determination' of the force of a body in motion is in fact the way empirical elements are combined in a mathematical expression. Thus it is not a 'measure' in the sense of a modern standard measure, but rather a prescription showing how to compare physical quantities with one another.

It is comparable to the modern concept of dimension, but only in the sense that the dimension of a physical quantity is a reflection of the composition of the quantity from basic elements such as mass, length, and force. However this is a treacherous comparison, since it can easily be taken too literally, resulting in the expressions of that period being subordinated to dimensional analysis. However, the modern concept of dimension was still unknown at that time. Aspects such as distance, surface, volume, weight, and time were related rather by *proportionalities*.

One problem that may have contributed here was that mathematical representation was based on Euclid's *Elements*. Referring to the *Tractatus proportionum* (authored by the English mathematician Thomas Bradwardine in 1328) Dijksterhuis states that, strictly speaking, one could only have ratios between quantities *of the same kind*, that is, quantities that are multiples of each other.[26] Dissimilar quantities were like apples and oranges, and could not be divided by one another. Consequently, Bradwardine could not speak of velocity as a ratio of distance to time, but only as a *proportion* between the distance covered and the time that had elapsed.[27]

---

[25] Hoenen, *Philosophie*, 248.
[26] Dijksterhuis, *The Mechanization of the World Picture* (1961), 189-193 (section II.120-125).
[27] Thinking in terms of proportionalities makes it extremely difficult, if not impossible, to construct composite quantities. Stillman Drake suspected that Galileo was the first to do so, in forming the concept of 'breaking strength' with the help of composite ratios (Drake (1974) in: Galileo, *Two New*

continued on next page

However, to use dimensional analysis, one has to be able to discuss *ratios* between *dissimilar* quantities.

One can see in d'Alembert's article "Méchanique" (1765) in the *Encyclopédie*—especially from his explanation of the equations of motion in mechanics—that the idea that only similar quantities can be divided by each other continued into the eighteenth century. These equations refer not only to distances, as in a purely geometric treatment, but also to differences in time. Nevertheless, d'Alembert states that it is, strictly speaking, impossible to compare two things of different natures, such as time and space. But there can still be a mathematical equation for a trajectory, since time and space do not themselves appear in it, but rather the *ratio* of parts of the time, and the *ratio* of parts of the distance. These ratios can properly be compared. The equation of a trajectory therefore expresses "the ratio between the ratio that the parts of time bear to their unity, to the ratio that the parts of the space traversed bear to theirs."[28]

The absence of the concept of dimension means that the approach to the problem of quantification at that time differs enormously from that of later times. With the aid of dimensional analysis, the quantification of qualities becomes much simpler, but conversely the lack of dimensional analysis goes hand in hand with a more qualitative way of thinking. We will encounter this in the ambiguity of Newton's use of the concept of innate force, i.e. as both mass and quantity of motion,[29] and in Leibniz's linking of dead and living force.[30]

Our exploratory mission has thus provided a first impression of the kind of problems that can be expected on the actual journey. We can now identify three elements that were crucial in the development of the concept of mechanical force. Firstly, a qualitative approach was as important as a quantitative one. Secondly, the way that force would ultimately be quantified had not yet been decided. Thirdly, proportionality forms the basis for any quantification.

Now our journey can begin!

---

continuation of former page
Sciences, 121n.9). However, for comparison see Marshall Clagett, who says that the concept of speed as a *quantity*, rather than only as a *proportion*, had already entered mechanics in the thirteenth century, in Gerard of Brussels's *Liber de Motu* (Clagett, The *Science of Mechanics in the Middle Ages* (1959, xxv).

[28] "[L]e rapport du rapport que les parties de tems ont à leur unité, à celui que les parties de l'espace parcouru ont à la leur" (d'Alembert, "Méchanique," 225). It is unfortunate that d'Alembert's clear-cut formulation of this problem has not been used in schools. Many students still have difficulty in understanding how time and distance can be divided by each other. On the other hand, D'Alembert's answer is not entirely adequate, since it allows quantities to have a *value* but not a *dimension*. Consequently, it makes it difficult to justify the fact that differing quantities cannot be added to, or subtracted from, each other.
[29] See below, section 2.5.4.
[30] See below, section 3.4.

2.3 THE POINT OF DEPARTURE: DEAD AND LIVING FORCE

As indicated in section 2.1, the point of departure for our expedition is the distinction between Leibniz's two kinds of force: 'dead force' and 'living force'. In "Specimen Dynamicum" (1695), he writes that dead force is that "in which there exists not yet a motion, but only a tendency to move."[31] Examples of dead force are found in centrifugal force, the force of gravity, and elasticity. Living force is the "ordinary force connected with actual motion."[32] It is the force that is at work in an impact, for example. It is that which "(...) arises from an infinite number of continuous impressions of a dead force."[33] Only later was this distinction to be interpreted as one in which dead force indicates an external, Newtonian force, such as gravity or pressure, while living force (to which Leibniz gave the formula mv^2) corresponds with kinetic energy, $\frac{1}{2}mv^2$, except for the factor ½. This interpretation, however, holds little interest for our archeological investigation, since it shows only the *difference* between Leibniz's two concepts: since the dimensions of external force and energy are different, the two cannot be compared quantitatively.

That both concepts also *agree* in some respect, can be deduced from the terms Leibniz chose, from which it appears that he thought the distinction was that of two species within a genus. In this formulation, the genus is force and the species are indicated by the adjectives 'dead' and 'living'. A force is dead if it entails only a striving towards motion, and living if it results from the action of dead forces. This genus-species distinction shows that Leibniz had some idea of a terminological unity at least, and perhaps even of a conceptual unity. Nevertheless, Leibniz only tells us which concepts he is distinguishing *between,* not *on what* it is that he is subdividing. This 'what', the unity of dead and living force, demands a deeper excavation of the historical soil from which the concept of force grew.

While cautious attempts were being made, in the period before Leibniz, to arrive at a similar distinction within the general concept of force, these attempts were, paradoxically, a result of research originally aimed at expressing different types of forces mathematically in terms of each other. This research initially posited a unity in the concept of force, but this was brought into question as a result of the actual investigation.

We see a similar development in the work of Galileo and Descartes. In their case, two kinds of force were involved, one relating to the tendency to motion (i.e., pressure), the other to the motion itself (i.e., the force of impact). The first is fairly easy to measure; the question is whether, and how, a quantity can be allotted to the second. Let us continue our investigation through these archaeological layers, to see whether the unity of force is more clearly expressed there.

---

[31] "(...) in ea nondum existit motus, sed tantum solicitatio [sic!] ad motum (...)" (Leibniz, "Specimen dynamicum pro admirandis naturae legibus circa corporum vires et mutuas actiones deligendis et suae causas revocandis" Part I (1695), 238).
[32] "(...) alia vero vis ordinaria est, cum motu actuali conjuncta, quam voco *vivam*" *(ibidem).*
[33] "(...) ex infinitis vis mortuae impressionibus continuatis nata" *(ibidem).*

2.3.1 Galileo

Galileo deals with the problem of impact in the Sixth Day, published posthumously and in an unfinished statein 1718, but originally intended to form part of the *Discorsi* (1638). The Third Day and Fourth Day in this work are devoted to his second 'new' science: motion in space. After dealing separately with inertial movement and free fall in the Third Day, in the Fourth Day he discusses the combination of both in projectile motion. As part of the argument, he formulates proofs and several propositions concerning the magnitude of the *impetus* of a moving body. In this argument, 'impetus' is equivalent to the 'force of impact' or 'energy of impact'.[34] Although in the calculations it is proportional to velocity, Salviati (Galileo's spokesman) warns us that it is not identical to it; the impetus depends not only on the velocity of the body that strikes, but also on the velocity of the body that is struck, the angle of impact, the hardness of the objects, and so forth. By this, Salviati means that the impetus is related to the impact as a whole, and that the relationship of the impetus to the moving body can therefore be determined unambiguously only if the other conditions of impact are kept constant according to some fixed agreement. In the *Discorsi* this is done (incompletely, in fact) by supposing that the angle of impact is perpendicular, and that the impact is against a stationary object.[35]

This definition of impetus evokes a query from Sagredo, Salviati's intelligent interlocutor:

> [My] doubt and puzzlement resides in my inability to understand the origin and principle of the immense energy and force that is seen to exist in impact, when, with a simple blow of a hammer that weighs no more that eight or ten pounds, we see resistances overcome that would not yield to the weight of a body exerting its impetus on it without impact, by merely weighing down on and pressing it, though this heaviness may ammount to many hundreds of pounds. I should still like to find a way of measuring this force of impact, which I do not believe to be infinite, but rather think that it has its limit of equalization with, and finally of control by, other forces—pressure, heaviness, levers, screws, and other mechanical instruments, the multiplication of force by which I quite understand.[36]

---

[34] "(…) forza ed energia della percossa" (Galileo, *Discorsi,* Fourth Day, 291)
[35] *Ibidem,* 292.
[36] *Ibidem,* 292: "E 'l dubbio e lo stupor mio consiste nel non restar capace onde possa derivare, e da qual principio possa dependere, l'energia e la forza immensa che si vede consistere nella percossa, mentre col semplice colpo d'un martello, che non abbia peso maggiore d'8 o 10 libre, veggiamo superarsi resistenze tali, le quali non cederanno al peso d'un grave che, senza percossa, vi faccia impeto, solamente calcando e premendo, benchè la gravità di quello passi molte centinaia di libre. Io vorrei pur trovar modo di misurar la forza di questa percossa; la quale non penso però che sia infinita, anzi stimo che ella abbia il suo termine da potersi pareggiare e finalmente regolare con altre forze di gravità prementi, o di leve o di viti o di altri strumenti mecanici, de i quali io a sodisfazione resto capace della multiplicazione della forza loro." An alternative translation for "e finalmente regolare con altre forze" would be: "ultimately to be balanced with other forces."

Galileo's contemporaries were to hear little more on this matter. Salviati did promise Sagredo that he would tell him everything he heard from 'our Academician' (Galileo), but this promise was not kept during the Fourth Day. In the end, his contemporaries were told only of a proposition that Sagredo had acquired from Galileo through hearsay: "that the force of impact is unbounded, not to say infinite."[37]

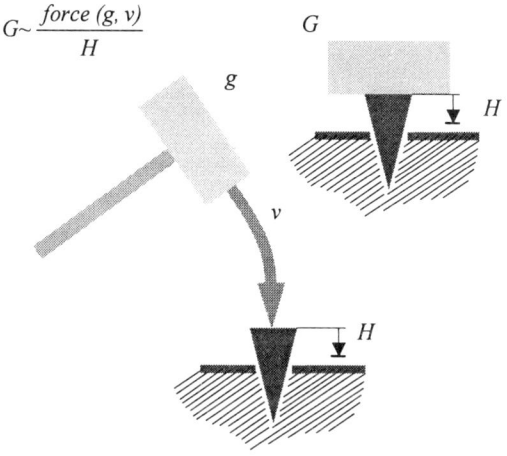

Figure 2.1 Galileo's analogy between the effect of an impact and that of the pressure of a weight

This proposition was developed only in the Sixth Day of the *Discorsi*. This development consists of successive attempts to hold the resistance that stops the effect constant. These attempts are based on two suppositions: (1) that the force of a body during impact is distributed across—or comprised of—the weight of the body and its velocity; (2) that the static force of a weight at rest can serve as the measure of the force of impact of a body in motion.

The simplest of the situations described involves the comparison between the effect of a hammer blow on a post, and that of the pressure of a weight merely resting on that post (see Figure 2.1). Given the weight of the hammer and the velocity with

---

[37] *Ibidem*, 313: "cioè che forza della percossa è interminata, per non dir infinita." The erroneous comparison with this proposition that Leibniz was to make in his "Specimen dynamicum," as if it was an enigmatic expression of the relation between dead and living force, can only be based on the above phrase, and not on the elucidation given in the Sixth Day (See Leibniz, "Specimen Dynamicum," 238). After all, the Sixth Day was not published until 1718 (see below). Westfall gives us an example of the type of reasoning used during the intervening period concerning the force of impact: in his *Cursus seu mundus mathematicus* (1674), the Jesuit Milliet De Chales calculated that the blow of a hammer weighing one pound, with a velocity of 144 feet per second, would correspond with a weight of 5100 pounds. In his reasoning he used both the quantity of motion *(mv)* and the mathematical equivalent of the modern concept of work (weight times height) as a measure of force (Westfall, *Force in Newton's Physics*, 200-203).

which it falls, one seeks to find the amount of weight-at-rest that would be required to achieve the same effect as the hammer blow. Regardless of how small the force of the impact is (or, in the case above, how small the weight of the hammer and its velocity are), it appears that every finite resistance must give way to it: every post can be driven into any kind of ground. Only an infinitely large resistance would be able to withstand the blow. In contrast, a 'dead' weight, in order to have any effect, must be greater as the resistance increases. This led Salviati to the provisional conclusion that "the force of impact [is] infinite—or rather, let us say indeterminate, or indeterminable, being now greater and now less, according as it is applied to a greater or lesser resistance."[38]

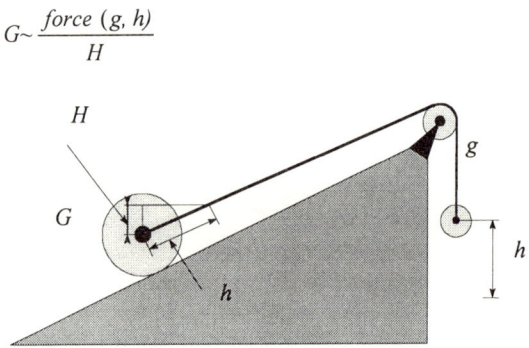

Figure 2.2 Galileo's analogy of the process of impact with a mechanical tool

In his efforts to make some progress towards quantifying the force of impact, Salviati finally arrives at an ingenious analogy between an impact and an imaginary mechanical contrivance. The argument provides a splendid illustration of the way Galileo links two phenomena that are totally different in our modern view. The contrivance involves a frictionless pulley with unequal weights g and G, where the heaviest weight G rests on a frictionless incline, and the lightest weight g hangs vertically downwards (see Figure 2.2). The incline is just steep enough to maintain the two weights in equilibrium. Salviati now deduces that a drop in height h of weight g causes a vertical rise H of weight G according to the ratio $H : h = g : G$. The rise H would be less, *mutatis mutandis,* as the weight G increases.

The analogy is as follows: the fall h of weight g corresponds to the force of impact of a body with a weight g and a velocity v achieved by falling from height h. The vertical rise H of weight G corresponds to the effect of the force of impact, i.e.

---

[38] Galileo, *Discorsi*, 328 "(...) la forza della percossa essere infinita, o vogliamo dire indeterminata o indeterminabile, e farsi ora minore ed ora maggiore, secondo che ella viene applicata ad una maggiore o minore resistenza."

the displacement of the point of action against a certain constant resistance over a certain distance. Finally, weight *G* corresponds to the dead weight necessary to achieve a similar effect to that of the force of impact.

It is important to note that weight *G* plays a double role: besides acting as the element corresponding to *dead weight,* it also acts as the element corresponding to *resistance.* Upon closer inspection, this double role constitutes the center around which the analogy revolves, and in which Galileo, by comparison with the views of modern mechanics, turned the world upside down. However, this criticism is of little interest here: what is interesting is simply that Galileo aimed at reducing the *dynamic* problem of impact to a *static* problem by means of this analogy. A schematic representation of the analogy is given in Figure 2.3.

According to Galileo, the relationships between the corresponding elements are now the same in both situations. Just as in the static situation, where weight *G* (with *g* and *h* remaining equal) is inversely proportional to *H,* in the dynamic situation the necessary dead weight is inversely proportional to the resulting depth of the blow, with *g* and *v* remaining equal. In other words:

> PROPOSITION
> If the effect made by an impact of the same weight falling from the same height shall be to drive a resistent of constant resistance through some space; and [if] to produce a similar effect there is needed a determined qunatity of dead weight [merely] pressing, without impact,
> I say that if the original percussent, [acting] upon some greater resistent, with the given impact shall drive it (for example) through one-half the space that the other was driven, then in order to accomplish this second driving, the pressure of the said dead weight will not siuffice, but there will be required another one, twice as heavy.
> And similarly in all other ratios, when a shorter [constantly resisted] drive is made by the same percussent, then inversely by that much there will be required, to do the same, a greater pressing quantity of dead weight.[39]

Since an ever increasing resistance will always be overcome by the same force of impact, although this takes place across an ever smaller distance, whereas the equivalent dead weight, along with the resistance, would eventually become infi-

---

[39] Galileo, *Discorsi,* Sixth Day, 340: "PROPOZIONE. Se l'effetto che fa una percossa del medesimo peso, e cadente dalla medesima altezza, caccierà un resistente di resistenza sempre eguale per qualche spazio, e che per fare un simile effetto ci bisogni una determinata quantità di peso morto, che senza percossa prema, dico che quando il medesimo percuziente sopra un altro resistente maggiore, con tal percossa, lo caccerà, v.g., per la metà dello spazio che fu cacciato l'altro, per far questa seconda cacciata non basta la pressura del detto peso morto, ma ve ne vuole altro il doppio più grave; e così in tutte le altre proporzioni, quanto una cacciata fatta dal medesimo percuziente è più breve, tanto, per l'opposito, con proporzione contraria vi si ricerca, per far l'istesso, gravità maggiore di peso morto premente."

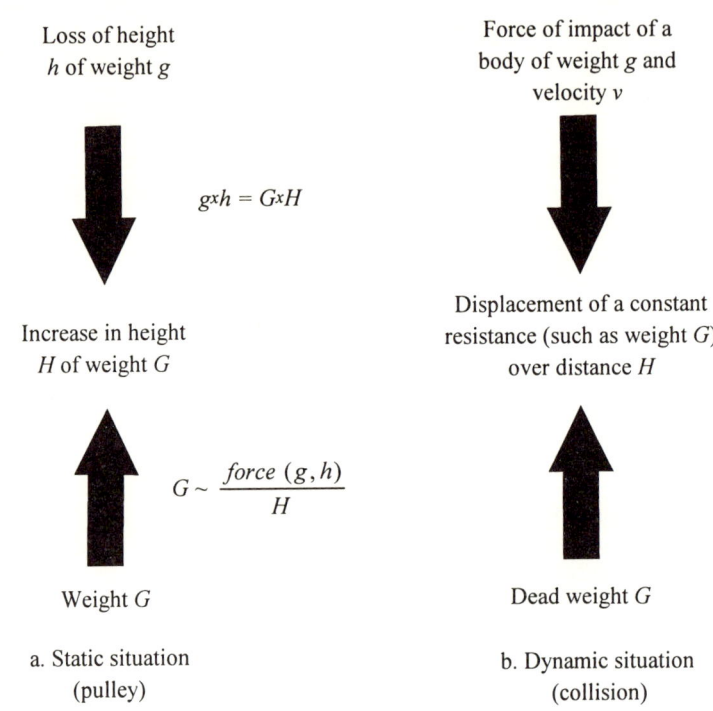

Figure 2.3 Galileo's analogy between the dynamic problem of an impact and a static situation

nitely great, Salviati (speaking for Galileo) concludes that the force of impact is, in a certain sense, infinite in relation to a dead force or weight.[40]

Although the posthumous edition of the Sixth Day ends with a number of fragments in which the relationship between 'force of impact' and 'weight' is developed further, I shall not deal with this here.[41] This is not because these fragments cannot

---

[40] *Ibidem*, 341: "(...) la forza della percossa essere infinita." Westfall conceives this infinity as a 'potential' infinity (Westfall, *Force in Newton's Physics*, 37). This qualification is not made in the *Discorsi*, and it is also not clear whether it would concur with Galileo's understanding of potential and actual infinity. Therefore I only want to qualify the term 'infinity' with 'in a certain sense'.

[41] The first fragment of the text posits that the two kinds of force, pressure and the force of impact, correlate with two kinds of resistance that are assumed to be present in every body (Galileo, *Discorsi*, Fragments after the Sixth Day, 343). Pressure and force of impact are thus each opposed by their respective motive resistances. Pressure can only withstand a certain amount of resistance, but can also propel the body exercising the resistance over an unlimited distance. Impact is stronger than any resistance, but can only propel it over a limited distance. Westfall thinks that the resistance that correlates with pressure contains the indistinct germ of the concept of weight, while the

continued on next page

be attributed to Galileo with any certainty, but rather because they add nothing to the argument above. Galileo was on the verge of making a distinction between 'pressing force' and 'impact force', but was unable to make it clear. He does reveal one aspect of the unity of the two forces—weight as a common measure—yet the distinction itself lay just over the horizon of his mechanical insight.

2.3.2 Descartes

In the same year (1638) that saw the publication of Galileo's *Discorsi,* Descartes made a related distinction. In his correspondence with Mersenne he raised (among other issues) the problem of how to measure the quantity of a weight.[42] In his argument he introduces the "general foundation of all statics': the principle that the force needed to lift a weight through a certain vertical distance is proportional to the product of the weight and height.[43] At Mersenne's request, Descartes later clarified in a letter his use of the concept of force: "Above all, one must bear in mind that I have talked about the force that serves to lift a weight to a certain height (...) and not about that which serves to maintain it at each point (...)"[44] In a subsequent letter, Descartes calls the former force 'action' and the latter one 'power' ("puissance").[45]

It is obvious that his power corresponds to what Leibniz called 'dead force', i.e. a continuously operating pressing or pulling force. The action, the force by which a weight can be raised—whether it derives from a person, a spring, or from another weight—corresponds to the modern concept of work (W), the integral of an external Newtonian force F over the distance covered s: $W = \int F ds$. Descartes's distinction thus corresponds with the modern distinction between 'force' and 'work'. The dif-

---

continuation of former page
resistance that corresponds to the force of impact contains an even more indistinct germ of the concept of mass (Westfall, *Force in Newton's Physics,* 38).
The second fragment presents the idea that the 'momento' (the force of the motion of a body) is comprised of an infinite number of 'momenti', whether natural and internal (as in the case of the weight of a body at rest) or enforced and external (as in the case of a motive force). During the motion of free fall, the 'momenti' (which derive from the weight) accumulate from instant to instant, and the velocity increases accordingly. This view entails an element of the understanding that $mv = \int F dt$, i.e., that the force of impact is the integral of a continuous external force over time. But Galileo himself had certainly not yet achieved this insight. The double usage of 'momento' as both 'weight' and 'motive force' makes this sufficiently clear.

[42] Letter from Descartes to Mersenne, [13 July 1638] (*Œuvres de Descartes* II, 222-245).
[43] "fondement general de toute la Statique" (*ibidem,* 228). There is a certain irony in the fact that Leibniz was to use precisely this principle against Descartes's proposition concerning the conservation of the quantity of motion, thus rebutting Descartes's with his own principle. See below, section 3.2.1.
[44] "Il faut sur tout considerer que i'ay parlé de la force qui sert pour leuer vn poids a quelque hauteur (...) et non de celle qui sert en chacque point pour le soutenir (...)" (letter from Descartes to Mersenne, 12 September 1638. In: *ibidem,* 352-362, this passage at 352-353).
[45] Letter from Descartes to Mersenne, 15 November 1638 *(ibidem,* 419-451, this passage at 432-433). Descartes's definition of action must not be confused with that given by Maupertuis. D'Arcy made this blunder, thereby bogging himself down in futile wordplay (see Chapter 5 below).

ference between these two can easily be understood with the aid of integral calculus. Descartes did not have integral calculus at his disposal and therefore had more difficulty in explaining this difference. To explain the precise nature of the distinction between the force in his static principle and the force that bears or carries something, he uses the concept of dimension: action has two dimensions while power has only one.[46]

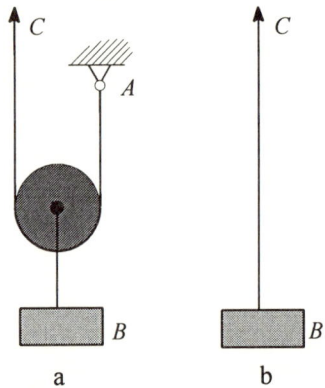

Figure 2.4 Descartes's comparison of raising a weight with and without a pulley

Although we can detect the germ of the modern physical concept of a dimension in Descartes's idea of dimension, we should not read too much of the later concept into it. After all, the issue at stake here is what Descartes managed to achieve with it, in relation to his differentiation of the different forms of force. Indeed, Descartes did not define the concept of dimension. From his use of the concept it appears that it had a physical, rather than a geometrical, meaning: it is the magnitude of a component part of a physical quantity. The concept only acquires a geometrical meaning when a dimension is *represented* by a line segment: action and power, the former a force having two dimensions and the latter one, relate to each other in the same way as a plane and a line.[47]

Descartes uses the idea of a pulley system (see Figure 2.4) to illustrate this geometrical analogy. Point A is fixed to the ceiling; point C is held up by a force (Figure 2.4a). The force required at the free end C to balance the weight of B, is only half the force that is required without a pulley (Figure 2.4b). The other half is of course pro-

---

[46] Letter from Descartes to Mersenne, 12 September 1638 *(ibidem, 352-362, this passage at 352-353)*.
[47] *Ibidem*, 353 and 356-357. This comparison was to continue to play a role in discussions about force for a long time, but at a later stage it acquired the character of an *argument* rather than a *clarification*. For example, when d'Alembert was unable to clarify how a motion could arise from a force, i.e. a mere tendency towards motion, he appealed to the equally incomprehensible mathematical reality of the transition from point to line, line to plane, and so on.

vided by the ceiling at point *A*. Descartes now says that some people conclude from this that half the force would be needed to raise weight *B* over a certain distance. However, he can easily show that this conclusion results in an absurdity: by extending the pulley system in a similar way, the force needed at the free end can again be halved, and so forth. Descartes's solution, on the other hand, supposes that the power as a whole is indeed halved, but that the change of the action implies a halving of only one dimension. The other dimension, the height of lift, is doubled because it refers to the displacement of end *C* rather than displacement of weight *B*. The area formed by the line segments representing both dimensions, the product of both, therefore remains the same (see Figure 2.5): area *KLIF* = area *HMGF*.

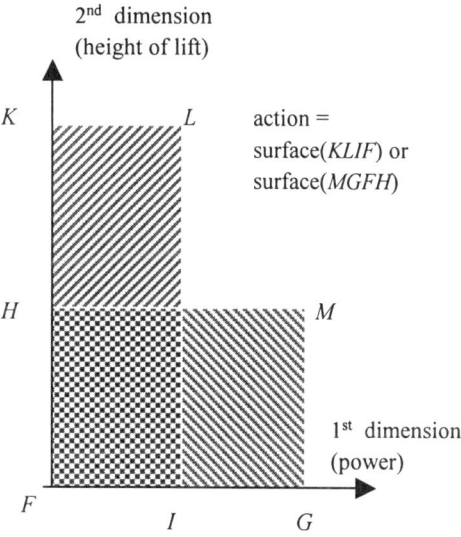

Figure 2.5 Descartes's distinction of action and power

In Descartes's view, one dimension of the action is the supporting force at point *C*, and the other is the distance over which the point of application of this force is displaced. The fact that the supporting force at point *A* does not imply any decrease in the action required at point *C* corresponds to the geometrical fact that a line added to a plane in no way increases or decreases it.[48]

Thus Descartes formulates the difference between 'action' and 'power' with considerably more clarity than Galileo had attained in distinguishing between 'weight' and 'force of impact'. Still, it is clear from Descartes's lengthy explanation, and his complaint about his readers' lack of comprehension, that the distinction was cer-

---

[48]   *Ibidem*, 357.

tainly not yet clear to his contemporaries.[49] We may also conclude, from his continuing use of the concept 'force' rather than 'action' or 'power', that Descartes himself saw a conceptual unity underlying the two forms of force.[50] For Descartes, as for Galileo, the unity between various forms of force was greater than the difference between them.

In fact the direction Descartes was going in, when he made his distinction, is not as simple as it seems in the passages from his letters to Mersenne that were discussed above. In these he defines 'action' as a force proportional to the distance covered, but he would probably have abandoned this position if he had realized that the action, if transferred to dynamic situations, would be proportional to mv^2 rather than to mv.[51] His metaphysical premise of the conservation of the quantity of motion mv would have made this especially problematic. There was a good reason why, fifty years later, everyone believed that Descartes had identified force with quantity of motion.[52] At the time, however, the differentiation could be taken in two directions, corresponding to *momentum* and *work* in classical mechanics ($\int F dt = \Delta mv$ and $\int F ds = \Delta(\frac{1}{2}mv^2)$ respectively). The fact that the dimensions of these quantities differ—$kg \cdot m \cdot s^{-1}$ and $kg \cdot m^2 \cdot s^{-2}$ respectively—is not important here, as we saw in section 2.2.3.

2.3.3 Impact, Lifting and Weight

Although with Descartes and Galileo it is not yet entirely clear where their distinctions in the concept of force were leading, the thing to which they were *relating* those distinctions is certainly clear: pressing weight is the reference point for all forces. They distinguish between forces related to motion itself, and forces (such as weight) that are merely a cause of motion. However, living force was not clearly emancipated from dead force until it was realized that "the capacity of a body to act" (as Westfall defines Descartes's and Galileo's concept of force[53]) depended not only on the characteristics of the actual body, i.e. the quantity of matter and the velocity, but also on the resistance to be overcome. It was no accident, although at the same time not yet necessary, that during this process these terms were differentiated into 'energy', 'momentum' and 'force'.

We can also interpret Descartes's and Galileo's distinction in the other direction: from the later distinction to the original unity. We then see that the analogy between the motive or lifting force and static force was based on the idea that the force of a body in motion was the same kind of force as weight. In this way, weight (for which a method of quantification had long been known in statics) was the primary force.

---

[49] *Ibidem*, 353.
[50] Westfall, *Force in Newton's Physics*, 73.
[51] As Galileo had already shown, in free fall, for example, it is mv^2 rather than mv that equals the height of fall.
[52] For a discussion of this misunderstanding, see Chapter 3, note 17 of the present book.
[53] Westfall, *Force in Newton's Physics*, 64.

Other forces were compared to it and understood by analogy to it. The effect of this analogy is still evident in Leibniz's distinction between dead and living force. This was expressed, among other ways, in Leibniz's conception of the living force of a particle as a property that could be determined unequivocally and that was independent of the properties of other particles.[54]

2.4   IMPETUS: THE GOAL ATTAINED

In the previous section we saw how Descartes and Galileo prepared the way for the modern distinction between energy, momentum and force by making a qualitative, primarily intuitive, distinction within the concept of force. For them, however, the unity was more obvious than the difference. But in order to make the historical unity truly clear, we must expand our archaeological search and look at the tradition in which this unity of weight and force of impact still had a positive meaning. This tradition can be found in the impetus theory from scholasticism, and specifically in the teachings of the Parisian nominalists or terminists.[55] In this school of thought, the positive meaning of unity was tangibly present in the theory of projectile motion.

By way of introduction to this theory, we may begin with the Aristotelian view of motion, according to which everything that moved was necessarily moved by something else, unless the cause of the motion was present in the body itself.[56] This meant, among other things, that the artificial propulsion of a body, such as a handcart, required the continuous application of a force. Such a body does not have the cause of locomotion within itself, but it has a resistance to motion as such. Only natural motions, which are rising or falling motions, have their cause within the body itself. In the Aristotelian view, bodies seek their natural place: towards the center of the cosmos (water and earth) or away from the center (air and fire). In the case of the handcart, it is obvious that as soon as the external force ceases, the cart will stop moving. In this case the theory corresponds to everyday experience. With regard to projectile motion, however, this is anything but the case. Although the force ceases as soon as the projectile leaves the hand or the throwing apparatus, the motion certainly does not cease. In order to explain the continuation of the motion, another cause must be identified.

Aristotle thought that the cause was to be found in the surrounding medium. Most scholastics followed him in this opinion. However, some transferred the moving cause from the surrounding medium to the moving substance itself. The opinion of the fourteenth century Parisian nominalists, preceded by the work of Ioannes Philoponos from the sixth century, was that motion is caused by "a certain

---

[54]   See Chapter 3 of the present book.
[55]   I discuss the theory very superficially here. For more information, see: Dijksterhuis, *Val en worp* (1924) and *Mechanization* (1961, Dutch original 1950); Clagett, *Science of Mechanics* (1959); and Grant, *Physical Science in the Middle Ages* (1971).
[56]   Aristoteles, *Physics*, 127 (book H, 241b 24-26).

impetus or a certain motive force, impressed in the moving body by the thrower."[57] 'Impetus' is therefore the effect of the throwing force, and it was only natural that this concept was initially dealt with in an analogous fashion. Impetus then becomes the internalized motive force or an intangible motive power.[58]

Impetus was often lumped together with weight, 'gravitas'. For example, when Albert of Saxony introduced impetus, he called it "gravitas accidentalis."[59] Conversely, Dominicus Soto considered gravity to be an "impetus naturalis" towards the object's natural position.[60] In this tradition, therefore, the force of a moving object was seen as comparable to its weight. Although the basis of their presence in the body differs, they are essentially the same: both are the cause of the motion of the body. Weight, or external force, and impetus are seen as a direct continuity: the former is transferred, *in its entirety and instantaneously*, to the material body to overcome its resistance to motion for a longer period.

Seen in this way, the analogy of motive force and weight in the impetus theory is a logical consequence of the ancient concepts of force and matter. Force is the cause of motion as such and must overcome the material resistance to this motion. Impetus is required to overcome this resistance after the initial force has disappeared. The material resistance to motion can only be overcome by internalizing the external force. A material body then has both an inherent resistance to motion and a force that overcomes this resistance, the latter being added to it from outside. The force that operates from outside is converted into a force that works from within. The material body is then like a container filled with force.

2.5 A Related Distinction: Inertia and Resistance

2.5.1 *Impetus, Living Force, and Inertia*

We now appear to have reached our destination on this journey of archeological exploration. The impetus theory is the historical basis for Leibniz's unity of dead and living force. Armed with this insight, we can return along the same road to take a fresh look at the controversy about the true measure of living force. However, this return route does not show a linear development; with the result that the relationship between weight and impetus cannot be translated directly into the relationship between dead and living force. This is because of indeterminacy in the concept of impetus. Since impetus overcomes material resistance after the external cause has disappeared, it forms the link between material resistance and external force. There-

---

[57] "(...) motor in movendo mobile imprimit ei quendam impetum vel quandam vim motivam (...)" (John Buridan, *Quaestiones totius libri physicorum*, 12th quaestio, quoted by Dijksterhuis, *Val en worp*, 72).
[58] The thought of an intangible moving power probably has its origin in Philiponos' criticism (from the sixth century AD) of Aristotle's *Physics* (Dijksterhuis, *Val en worp*, 42-44).
[59] Dijksterhuis, *Val en worp*, 85; *Mechanization*, 185 (section II.114).
[60] Dijksterhuis, *Val en worp*, 119.

fore we cannot permit the concept of impetus to evolve, in our thinking, into the concept of living force or—somewhat further along the line of historical development—into the concept of energy within classical mechanics, since these embrace only one aspect of the concept of impetus: the relationship of impetus to external force.

One would expect the unity of the concept of force to have left traces in the development of the relationship between impetus and material resistance. I would therefore like to use our journey back to the eighteenth century to trace the development of a concept that derives from the idea of material resistance: that of inertia. This route was suggested by a problematic issue concerning the theory of impetus within scholasticism. Although this theory is in accordance with the Aristotelian view of *force,* it gave rise to a new problem within the Aristotelian concept of *matter*. Since impetus had to be regarded as an *immaterial motive power*, which in one way or another had become entangled in the mobile object and could therefore never be found in the surrounding matter such as air or some intangible ether of one kind or another, the concept came into conflict with the traditional conception of the resistance of matter to motion. For example, in his critique of Philoponos' theory on projectile motion, Thomas Aquinas says that it conflicts with the character of an enforced motion to attribute its continued existence to an intrinsic principle.[61] The concept of impetus presupposes that material bodies are the seat for a force that enables them to persist in motion, whereas previously only a resistant force was seen. This poses a dilemma: how can contrary forces be found in the same material body at the same time?

The development of the idea of a force of resistance is evident in the various phases of the development of the concept of inertia. As I will demonstrate below, the concepts of inertia in Galileo, Descartes and Newton provide successive solutions to the dilemma described above. They all differ from the modern concept of inertia in that they attempt to establish a relationship between the force of persistence and the resistance to motion *as such.* In contrast, in the modern concept of inertia there is no longer a resistance to motion as such—only a resistance to *change* of motion, expressed in the concept of inertial mass. This follows from the modern definition of inertia: the continuation of a uniform rectilinear motion, once acquired, in the absence of external influences. This change made it possible for d'Alembert in 1757 to view the 'force d'inertie' no longer as a *force,* but simply as a *property.*[62] The development of the concept of inertia between the time of Galileo and Descartes and that of Newton can now be seen as a synthesis of these two forces—resistance to motion and impetus—into a single inherent force as conceived by Newton.[63]

---

[61] *Ibidem,* 67.
[62] D'Alembert, "Force" (1757), 110.
[63] There is a considerable difference indeed between the modern concept of inertia and the scholastical view of impetus, as Anneliese Maier argued in *Die Vorläufer Galileis im 14. Jahrhundert* (1949). Firstly, where the impetus theory is concerned, there is still the need to explain the persistence of the motion, while the theory of inertia simply regards this persistence as self-

continued on next page

In the same way that the living force was the basis of Leibniz's mechanics, inertia was the basis for Newton's. Both concepts can be seen as a solution to the ancient problem of the persistence of motion. While the concept of inertia problematised the relationship between resistance to motion and the persistence of motion, the concept of living force did the same for the relationship between external and innate forces. It would be interesting to use this difference to explain the difference in the concepts that ultimately result from them, i.e. 'inertial mass'—that on which a force *acts*—and 'energy', that in which the action *results*. I will not do this here, but I will demonstrate instead that it is connected with a major problem in Newton's formulation and mathematization of his second law. Although ostensibly very different, Newton's and Leibniz's concepts of force remain within the same unity that is based historically on the impetus theory.

2.5.2 *Galileo's Concept of Inertia*

In some important respects, Galileo's concept of inertia is far removed from the modern concept. Firstly, Galileo's inertia principle is *circular,* where the modern one is *linear:* although in Galileo's view a body that has once achieved a certain velocity will always maintain that velocity (providing that no external forces are involved), this does not mean that the body will take a straight, infinitely long path. To Galileo, the inertial motion was the 'natural' motion, i.e., a path in a circle around the center of the earth.[64] Secondly, Galileo's principle of inertia was restricted to motions on the horizontal plane (for him, this meant a plane parallel to the earth's surface), whereas the modern principle of inertia does not distinguish between directions.[65]

However, the most important difference is the absence of the idea of *resistance* in Galileo's conception of inertia. This distinguishes his concept of inertia from im-

---

continuation of former page
evident. Secondly, the motion caused by an impetus can diminish, as the impetus itself diminishes, while the motion that complies with the axiom of inertia will always remain the same, unless external forces are at work.

History, however, is more ambiguous than would appear from what I have written here. Even in the Middle Ages, solutions were posited in which the persistence of motion required no cause. For example, William of Ockham thought the question about the cause of persistent motion was meaningless because he viewed the projectile as its own motor, and consequently there was no reason to distinguish between motor and mobile object (Dijksterhuis, *Val en worp,* 68). How far is this from the assertion that impetus is produced by the mobile object itself, and that there was no need to look further for the cause of persistent motion? Nevertheless, this and similar refinements barely affect the general line of my account of the seventeenth century; I have therefore left them out of the discussion.

[64] Dijksterhuis, *Val en worp,* 264-271; Westfall, *The Construction of Modern Science: Mechanisms and Mechanics* (1971), 18.
[65] For example, Salviati defines the horizontal in a passage added later to the *Discorsi* as "l'orizontale, che qui s'intende per una superficie egualmente lontana dal medesimo centro" ("the horizontal, which means here a surface [everywhere] equidistant from the said [common] center") (Galileo, *Discorsi,* [215]).

petus, on the one hand, and the modern concept of inertia on the other. For Galileo, 'inertia' meant that "(...) on the horizontal (...) the movable is found to be indifferent to motion and to rest, and has of itself no inclination to move in any direction, nor yet any resistance to being moved (...)"[66] Indeed, if there is a distinction between 'natural' and 'imposed' motion, as Aristotle had said (and Galileo still assumed), is there any option except either a natural *resistance* to imposed motion—which means that a constant external cause is required to maintain a constant speed—or *indifference* to imposed motion? Natural motion and natural resistance to motion are two sides of the same coin. This is made clear by what Galileo says, in the fictional character of Sagredo, about the one situation in which, in his view, there is resistance to motion:

> At present the only internal resistance to being moved which I see in a movable body is the natural inclination and tendency it has to an opposite motion. Thus in heavy bodies, which have a tendency toward downward motion, the resistance is to upward motion.[67]

However, since a material body on a horizontal plane has no preference for any particular direction, i.e. it has no natural motion in a horizontal direction, it also cannot exercise any resistance to any such motion. Consequently, Galileo's theory that matter is indifferent to motion can fairly be regarded as a radical reaction to the Aristotelian and scholastic conception of resistance to motion *as such*. However, although this solved one problem, and was therefore progress in one sense, it was itself a new obstacle. According to the indifference theory, an arbitrarily small force would give a body an instantaneous and arbitrarily great velocity.[68] This is why Westfall says that the concept of inertia had to be transformed into the concept of resistance to *changes* in motion, as a precondition for the concept of inertial mass, and thus also for the formulation of the Newtonian force as the product of mass and acceleration. Galileo's belief that matter exercises resistance only in certain cases, and is not generally characterized by resistance, constitutes a radical divide between his concept of inertia and that of classical mechanics.

---

[66] *Ibidem:* "[on the horizontal] dove il mobile si trova indifferente al moto e alla quiete, e non ha per sè stesso inclinazione di muoversi verso alcuna parte, nè meno alcuna resistenza all'esser mosso (...)"
[67] Galileo, *Dialogo sopra i due massimi sistemi del mondo* (1632), 240: "Io per adesso non veggo esser nel mobile resistenza interna all'esser mosso se non la sua naturale inclinazione e propensione al moto contrario, come ne' corpi gravi, che hanno propensione al moto in giù, la resistenza è al moto in su (...)" (Translation: Drake ed., *Dialogue*, 213).
[68] Father Pardies drew this conclusion in his *Discours du mouvement local* (1673): a particle struck by another particle will begin to move at that same instant with the velocity of the particle striking it (Westfall, *Force in Newton's Physics*, 195).

2.5.3 Descartes's Concept of Inertia

This divide in the concept of inertia was partly bridged by Descartes who, like Galileo, regarded matter as a substance with no resistance to motion. In this he went further than Galileo, by saying that matter in itself cannot exert *any* force whatever. The resistance we experience when we try to set a body in motion is only apparent, and stems entirely from the body's weight and hardness.[69] This absence of resistance to motion enables us to understand Descartes's conviction that bodies begin to move with a finite velocity, both during a collision and in free fall.

Yet bodies do exert a force:

> Here we must precisely identify what constitutes the force that every body has to act upon another, or to resist the action of another, and it is only this: that every thing inclines, as much as in it lies, to perpetuate the same state in which it happens to be (...)[70]

Motion and rest are examples of states which are perpetuated, but so is the direction of a motion and, for two particles, the state of being joined or being separated. Descartes's justification for the existence of a force of this kind is his first law:

> (...) that every thing, as much as in it lies, always remains in the same state. Thus that which moves once, always continues to move (...)[71]

His proof is somewhat hidden away in the last sentence of the section. We can reconstruct it as a syllogism:

> *premise:* nothing passes to a contrary state on its own accord ("ex propriâ naturâ"), since this would be self-destruction ("destructio suî ipsius");
>
> *middle term:* rest and motion are contrary states;
>
> *conclusion:* no material body will, on its own accord, go from a state of motion to a state of rest, or vice versa.

Descartes probably realized that even though a body cannot *change* its own state, this does not mean that the *maintenance* of a state, once attained, does not require

---

[69] *Ibidem*, 69.
[70] Descartes, *Principia philosophiae* (1644), 66 (II.43): "Hîc verò diligenter advertendum est, in quo consistat vis cujusque corporis ad agendum in aliud, vel ad actioni alterius resistendum: nempe in hoc uno, quòd unaquæque res tendat, quantum in se est, ad permanendum in eodem statu in quo est (...)" Here, and below, I opt for the original Latin text because the 1647 French translation is not authorized, and diverges considerably from the original in places (compare Adam, "Avertissement" in: *Œuvres de Descartes* IX-2, XIX-XX).
[71] *Ibidem*, 62 (II.37): "quòd unaquæque res, quantum in se est, semper in eodem statu perseveret, sicque quod semel movetur, semper moveri pergat." In this passage, the French transition adds the condition, generally made nowadays, "provided that nothing changes it" (Descartes, *Les principes de la philosophie* (1647), 84 (II.37).

something similar to a force.[72] This explains Descartes's somewhat half-hearted attitude to the concept of force. Formally, Descartes does not ascribe any forces to matter; in fact, however, he reasons about the laws of motion as if matter does indeed exert forces.

The fact that he believes these forces can assume different forms is apparent from his fourth rule of motion, which involves a smaller body striking a larger stationary body. In Descartes's view, the larger body cannot be set in motion by the smaller; it resists the smaller body, regardless of their quantities of motion.[73] In a body at rest, the force of resistance clearly depends solely on the quantity of matter. On the other hand, the force of resistance in a moving body also depends on the velocity with which it strikes another body and the surface area against which it strikes.[74] So, in Descartes's thought, the way the force of resistance is determined by matter and its motion is ambiguous: it is reminiscent of the modern concepts of both inertial mass and momentum.[75]

Nevertheless, one cannot conclude from this that Descartes regards forces as truly existing. In the example above, the felt resistance to motion is derived from the weight and hardness of the body; elsewhere, he explains it as an illusion resulting from the presence of the quantity of matter in the quantity of motion.[76] We can, however, say that for Descartes inertia is no longer solely an indifference to motion. There is a force of resistance, even though he did not work out how this force results from the tendency to persevere in its state of motion.

2.5.4 Newton's Concept of Inertia

Newton succeeded in reestablishing the idea of resistance by introducing an internal force that is innate ("vis insita") to matter.

> Definition III.
> The innate force of matter is a power to resist, by which every body, as much as in it lies, perseveres in its state, whether this be rest or a uniform rectilinear motion.[77]

---

[72] It is therefore a mistake to state that in Descartes's thought, similar to that of Galileo, the actual relationship between matter and motion is characterized as *indifference*. Remarkably, Westfall does this anyway: "In common with the century as a whole, [Descartes] held that matter is wholly inert, lacking in any internal force and dominated in its motion by external actions upon it. In Galileo's classic phrase, matter is indifferent to motion or rest" *(—, Force in Newton's Physics,* 69).
[73] Descartes, *Principia philosophiae,* 68 (II.49).
[74] *Ibidem,* 67 (II.43)
[75] See also Gueroult, "Métaphysique et physique de la force chez Descartes et chez Malebranche" (1954) and Gabbey, "Force and Inertia in Seventeenth-Century Dynamics" (1971).
[76] Westfall, *Force in Newton's Physics,* 70.
[77] Newton, *Principia mathematica,* 2r.2-4: "Definitio III. Materiæ vis insita est potentia resistendi, qua corpus unumquodque, quantum in se est, perseverat in statu suo vel quiescendi vel movendi uniformiter in directum."

Axiom I.
Every body perseveres in its state of rest or uniform rectilinear motion, except in so far as it is compelled by impressed forces to change its state.[78]

This force expresses itself on the one hand as resistance to any change of motion, and on the other hand as impetus, in a collision with another body. In the latter case, it is an active force able to change the state of another body.[79] Since rest and motion are relative concepts, it is "our manner of conceiving" that determines whether the innate force is perceived in one or the other form.[80] In this way resistance and persistence become one. Once more we can see how little importance dimensional differences had in the development of theory. The innate force is, after all, proportional to the quantity of matter (i.e. mass),[81] whereas the impetus is also proportional to the velocity. Although it is therefore not incorrect to say that Newton is defining the *vis inertiae* here, it is certainly misleading, since the *vis inertiae* constitutes only one viewpoint, one form of expression, of the innate force.

Moreover, there is something else unusual about the innate force. On the one hand, Newton says that a body exerts this force only when an external force acts upon it. It is thus a *force of reaction*. This would mean that, if no other force is at work, no force is needed for the continuation of the state of motion or rest. In other places, however, he again speaks of the necessity of the innate force for the continuation of the motion itself. For example, the following rule appears in his proof of Kepler's law of areas: "Let the body describe the right line AB by its innate force."[82] This indicates that Newton was not entirely sure about the function of the innate force in the persistence of motion.

---

[78] *Ibidem*, 13r.5-7: "Lex I. Corpus omne perseverare in statu suo quiescendi vel movendi uniformiter in directum, nisi quatenus illud a viribus impressis cogitur statum suum mutare." The text in the third printing deviates slightly from that in the first and second printing; in the third printing, 'illud' is added, and the reflexive 'suum' replaced the 'illum' used earlier. Cohen (1967, n.19; 1970, n.21) reports that these changes cannot be found in Newton's manuscripts and that their probable originator is Henry Pemberton, the editor of the 3rd edition.

[79] *Ibidem* 2r.11-12, Liber I, Definitio III: "estque exercitium illud sub diverso respectu & resistentia & impetus." In the first and second printings, 'ejus' was written instead of 'illud'.

[80] *Ibidem* 2r.6, Liber I, Definitio III: "in modo concipiendi."

[81] Newton, in his explanation of Definition III *(ibidem,* 2r.5), writes: "[The innate force] is always proportional to the body [to which it belongs, JCB]." In the scholium in "Leges motus" *(ibidem,* 26r.3-14) we find: "And since those bodies are equally forceful in collision and reflection, of which the velocities are in inverse ratio to their innate forces" ("Ut corpora in concursu & reflexione idem pollent, quorum velocitates sunt reciproce ut vires insitæ"). Because 'idem pollent' is used here to designate the equivalence of *mv* (the quantity of motion), we have to conclude that the innate force is proportional to the quantity of matter.

[82] *Ibidem* 38r.32-33, Liber I, Sectio II, Propositio I: "(...) describat corpus vi insita rectam AB." For the rest of Newton's proof, see section 2.6.2 below. Another example is: "Nam corpus P per vim inertiæ, nulla alia vi urgente, uniformiter progredi potest in recta VP" ("For the body P will proceed uniformly along the straight line VP by means of its inertial force, if no other force is acting") *(ibidem,* 135r.22-24, Propositio XLIV, Corollary 6). It is incorrect to claim, as Westfall does *(Force in Newton's Physics,* 477), that Newton had already ceased to think in such terms about uniform motion, and only referred to a force in this way here because of the use of the parallelogram to represent force and path. See further section 2.6 below.

Newton's exact meaning is important, to show to what extent he overcame the inherent tension within the impetus theory. Dijksterhuis interprets Newton's innate force as a continually operating force, and he concluded from this that it does not differ from the Paris nominalists' impetus and Galileo's impressed force.[83]

However, this characterization does not do justice to the facet of Newton's innate force dealt with here. Newton took the problem of the persistence of motion significantly further. On the one hand he made it clear why a body can be in any state of motion, and is in that sense indifferent to motion (as Galileo and Descartes had posited), while on the other hand he showed how matter can offer resistance to a change of speed. He united indifference to the quantity of motion, and resistance to change in motion, in his concept of innate force.[84]

At the same time, in the concept of innate force Newton retained the memory of an earlier definition, in his unpublished treatise *De motu* of 1684. In this work he still viewed the innate force of a body as, in Westfall's words, "the sum total of the forces that had acted on it."[85] Although he abandoned this definition in his *Principia*, replacing it with a much vaguer one, it is not correct to say that he was thereby applying the modern principle of inertia. It is much more likely that the old definition retained a hidden presence, particularly noticeable where actions are mathematized. The following section will demonstrate that this is the case in Newton's mathematization of action-at-a-distance.

2.6 THE UNITY OF NEWTON'S CONCEPT OF FORCE

2.6.1 *Instantaneous and Continuous Forces*

The modern formulation of Newton's second law is unambiguous: force is equal to the product of mass and acceleration. For a long time there was no doubt as to the intent of Newton's own formulation, not even on the part of Dijksterhuis when he criticized Newton in 1950 for his careless axiomatization. In *The Mechanization of the World Picture*, he identified a lacuna in Newton's axiomatization of mechanics. Newton says in his second axiom:

> The change in motion is proportional to the impressed motive force, and occurs in the direction of the right line in which that force is applied.[86]

Usually, this axiom is read to mean that force is proportional to the product of mass and acceleration: $\underline{F} = m \times \underline{a}$. In order to understand Dijksterhuis's criticism, we first need to find the definitions implicit in the axiom. In the first place, 'motion' is an abbreviation for 'quantity of motion'. The fact that Newton mentioned 'quantity' in

---

[83] Dijksterhuis, *Mechanization*, 466 (section IV.295).
[84] See also Westfall, *Force in Newton's Physics*, 450.
[85] Westfall, *Force in Newton's Physics*, 447.
[86] Newton, *Principia mathematica*, 13r.16-17: "Lex II. Mutationem motus proportionalem esse vi motrici impressæ, & fieri secundum lineam rectam qua vis illa imprimatur."

the definitions, but for the sake of convenience omitted it when making practical use of the concepts, suggests on the one hand that a sense of the original qualitative nature of motion and force was still present in Newton's thinking, and on the other hand that the distance between this qualitative character and its quantitative determination had become extremely small. Newton defines 'quantity of motion' as follows:

> The quantity of motion is the measure of the same, obtained from the product of the velocity and the quantity of matter.[87]

The definition of 'impressed motive force' is more difficult to discover, since Newton does not define it directly. I propose that it is permissible to treat the concept as a compound of 'impressed force' and the modifier 'motive'. First, the definition of 'impressed force':

> An impressed force is the action exerted on a body, in order to change its state, whether this be rest or uniform rectilinear motion.[88]

The meaning of the modifier 'motive' can now be derived from its application to the centripetal force, one of the possible sources of the impressed force, and, in the *Principia mathematica*, its most important manifestation.[89]

> The motive quantity of a centripetal force is the measure of this force, proportional to the motion it generates in a given time.[90]

If we omit the term 'quantity', we can now describe the 'impressed motive force' as 'the measure of the action exerted on a body, proportional to the motion it generates in a given time'.

Well then, says Dijksterhuis, the equation $\underline{F} = m \times \underline{a}$ is sufficient, but not necessary, to the validity of Newton's definition and axiom:

> [t]his case resembles that of the Emperor's clothes in the fairy tale: all people saw them because they were convinced of their existence, until a child said that the Emperor had nothing on. Similarly Axiom II of the introductory chapter of Newton's *Principia* always used to be interpreted in the sense that a constant force produces a constant acceleration, and that their magnitudes are proportional, but if one looks at it impartially, nothing of the kind can be discovered.[91]

---

[87] *Ibidem*, 1r.19-20: "Definitio II. Quantitas motus est mensura ejusdem orta ex velocitate et quantitate materiæ conjunctim."

[88] *Ibidem*, 2r.20-21: "Definitio IV. Vis impressa est actio in corpus exercita, ad mutandum ejus statum vel quiescendi vel movendi uniformiter in directum."

[89] *Ibidem*, 2r.24-25, Liber I, Definitio IV: "Est autem vis impressa diversarum originum, ut ex ictu, ex pressione, ex vi centripeta" ("The impressed force has different origins, however, such as impact, pressure and centripetal force").

[90] *Ibidem*, 5r.2-3: "Definitio VIII. Vis centripetæ quantitas motrix est ipsius mensura proportionalis motui, quem dato tempore generat."

[91] Dijksterhuis, *Mechanization*, 471 (IV:303). In the Dutch text the last sentence is stated much more strongly; in translation it runs: "Let us, however, work our way through the foundation given [by

continued on next page

Equally adequate, of course, would be a constant force that does not result in a constant acceleration, but (for example) in an acceleration proportional to time. In that case the acceleration becomes uniformly greater with time. The formula for the acceleration a then becomes $a = cxt$, in which c is a constant, and t represents the time. Suppose that the constant c is proportionate to the force F according to the equation $F = mxc$, where m represents mass. For the relationship between force F and the resulting momentum mv, it then follows that $Fxt^2 = 2mv$. If we compare the resulting momentum for varying constant forces, but for the same mass and the same time interval, we see that the forces are proportional to the resulting momentum.[92]

Does this mean that we must conclude, along with Dijksterhuis, that the formulation of axioms was not Newton's strong point, since he failed to explicitly state in an axiom something that was self-evident to him?[93] I do not think so. However correct Dijksterhuis may generally have been in his judgment, something else is involved here. That 'something else' is Newton's concept of force.

As I. Bernard Cohen argued at length in 1967 and 1980, and Richard S. Westfall in 1971, behind Newton's formulation there actually lurked another measure of force, and consequently another concept of force.[94] Both writers pointed out that forces in Newton can occur in the form of both force of impact and continuous force.[95] A force of impact is that force which is produced in a collision of bodies. It is assumed to exert its action instantaneously: with no lapse of time. The measure of its action was the change in the quantity of motion: $\Delta(mv)$. Continuous forces—i.e, attraction, repulsion, pressure and traction—do however act within time. The measure of their action is the *rate* of change: $d(mv)/dt$, that is: mass times acceleration, $mx\underline{a}$.[96]

---

continuation of former page

Newton] in the naively unprejudiced manner of a child; that is, by bracketing everything we know and therefore expect to find in it. It then appears that precisely the principal basis of classical mechanics is not contained therein at all" ("Wanneer men echter de gegeven fundering met kinderlijke onbevangenheid, dus met uitschakeling van wat men al weet en daarom verwacht te vinden, doorwerkt, blijkt ze den voornaamsten grondslag voor de klassieke mechanica juist helemaal niet te bevatten") (—, *Mechanisering*, 518).

[92] *Idem, Mechanization*, 471 (IV:302).
[93] *Ibidem*, 473 (IV:304)
[94] I. Bernard Cohen, "Newton's Second Law and the Concept of Force in the *Principia*" (1967 and 1970), and *The Newtonian Revolution* (1980); Westfall, *Force in Newton's Physics* (1971), 471-491.
[95] An interesting detail is that external forces were still being distinguished in this way in 1897. See Klein and Sommerfeld, *Über die Theorie des Kreisels* (1897).
[96] I.B. Cohen, *The Newtonian Revolution*, 171. Cf. Stuart Pierson, "*Corpore Cadente* ...: Historians Discuss Newton's Second Law" (1993). Pierson gives an historiographical overview of interpretations of this dichotomy. However, his review is quite inadequate because he limits himself to interpreting the mathematical problem of integration (see below, in the main text) from a contemporary rather than a contemporaneous view. Where Pierson reproaches his fellow historians with being "something Whiggish," the pot seems to be calling the kettle black. He gives his own interpretation of the dichotomy, in a later article on the same subject, "Two Mathematics, Two Gods: Newton and the Second Law" (1994). In my own argument, I leave out Pierson's explanation, cryptically formulated in the title as "two mathematics, two Gods." At best, it may be a reference to the difference between the ontological status of mass, space, and time, on the one

continued on next page

Aided by this distinction, they then interpreted the ambiguity in Newton's second law as a dilemma about how to unite forces of impact and continuous forces within a single approach. The use of the term 'mutatio' in the formulation of the second law, means that it applies explicitly to forces of impact, yet Newton actually applied it to continuous forces in the remainder of the text.[97]

The analysis, given above, of the term 'impressed motive force' in the second axiom provides an additional argument for this interpretation. The axiom is articulated with regard to impressed forces quantified in a particular way, i.e., as a quantity of motive force that is impressed. Newton, in his *Principia,* only provides a definition of this quantification in the case of the centripetal force. However, in the explanatory note to the definition he states that it is also applicable to impulsive forces.[98] Moreover, although Newton mentions only that he has abbreviated the expression 'quantity of motive force' to 'motive force' in the case of the quantity of motive *centripetal* force, this is no reason for restricting the expression 'impressed motive force' to these forces. It is more plausible that Newton, in using this term, was simultaneously referring to all three sources of the impressed force: centripetal forces, forces of impact, and pressing forces.[99]

2.6.2 A Practical Example

To show how Newton applied his second axiom to continuous forces, I will now discuss his proof of Kepler's law of areas. In this proof, the first real proof in the *Principia,* he applies the second law for the first time to continuous forces. The proposition is as follows:

---

continuation of former page
hand, and centripetal force, on the other. In any case, if we are concerned with the dichotomy in the use of the concept of force, as force of impact and as continuous force, it is irrelevant.

[97] Newton's attempts to reformulate the second law, undertaken in 1692-1693, make this even clearer. He changed the term he had initially used—'mutatio motus'—to 'motum genitus'. While the term 'mutatio' might still suggest the idea of *rate* of change, the term 'genitus' makes it crystal clear that Newton meant an altered *quantity* (see I.B. Cohen, "Newton's Second law" (version 1970), 161-168). Incidentally, as a comparison between sections 303, 304, and 307 of the 1950 Dutch and the 1961 English versions of *The Mechanization of the World Picture* reveals, Dijksterhuis had (in 1950) already explored the extent to which such an interpretation might be useful. By 1961 he had come to defend the view that axiom II is not a *badly* phrased axiom for *continuous* forces, but rather a *well*-phrased one for forces of *impact*. In the 1960s, other authors were also concerned with the ambiguity of Newton's concept of force. The most important are: Max Jammer, *Concepts of Force* (1957), 124-130; Brian D. Ellis, "Newton's Concept of Motive Force" (1962); R.G.A. Dolby, "A Note on Dijksterhuis's Criticism of Newton's Axiomatization of Mechanics" (1966). Recently Stuart Pierson reviewed the variety of views and arguments then put forward in "*Corpore Cadente ...*; Historians Discuss Newton's Second Law" (1993) (see note 96 above).

[98] Newton, *Principia mathematica,* 5: "I likewise call Attractions and Impulses, in the same sense, Accelerative and Motive" ("Porro attractiones & impulsus eodem sensu acceleratrices & motrices nomino"). The same follows from the fact that the centripetal force is presented as just one of three possible sources of the impressed force (see note 90).

[99] See also Cohen, "Newton's Second Law," 137 (version 1967) or 153-154 (version 1970).

The areas described by circling bodies with their radii which are drawn to the immovable center of the forces, lie on immovable planes, and are proportional to the times.[100]

Newton's working method is, first of all, to show that the second part of the proposition is valid for *inertial* motions. He then demonstrates the validity of the whole proposition for *instantaneously acting* centripetal forces. Finally, he allows the number of forces of impact to increase, and the distances between them to decrease, both to infinity, which proves the proposition for *continuously acting* centripetal forces.

1. "Let the time be divided into equal parts, and in the first part of this time, let a body describe, by its innate force, the straight line *AB*."[101] See Figure 2.6.

2. *Inertial motion:* If other force is active, the body moves in the second part of the time across the distance *Bc* (which has the same length as *AB*); the areas of *ASB and BSc* are equal so that the second part of the proposition is proved for inertial motion.[102] It is remarkable that Newton does not complete the proof of the first part here, which would have been a simple task.

3. *Instantaneously* acting centripetal force: suppose that at point *B* a centripetal force imparts an instantaneous but great impulse ("impulsu unico sed magno") to the body, which deflects it from the line *Bc*, so that it moves along *BC*.[103] Newton proves in a simple fashion that the areas of *BCS* and *BcS* are equal, and that the areas *BCS* and *ASB* are therefore also the same. Moreover, they both lie on the same plane. The same reasoning can be used to reach the same conclusion for the areas of *EFS, DES, CDS, BCS* and *ABS*.

4. *Continuously* acting centripetal forces: now let the number of triangles increase and their width decrease *ad infinitum*. The polygon *ABCDEF* will then change to a curved line, and the centripetal force will thus act continuously upon the body and perpetually cause it to deflect from the tangent of the curve. Each area described (*SADS, SAFS*) is therefore proportional to the time elapsed.[104] Once again, Newton does not complete the proof for the first part of the proposition, although it follows directly from what has gone before.

Cohen interprets this transition from forces acting instantaneously to forces acting continuously as a *transformation* of the force of impact into a continuous

---

[100] Newton, *Principia mathematica*, 38r.29-31, Liber I, Sectio II, Propositio I: "Areas, quas corpora in gyros acta radiis ad immobile centrum virium ductis describunt, & in planis immobilibus consistere, & esse temporibus proportionales."

[101] *Ibidem*, "Dividatur tempus in partes æquales; & primâ temporis parte describat corpus, vi insitâ, rectam AB."

[102] *Ibidem*, "Idem secundâ temporis parte, si nil impediret, rectâ pergeret ad *c* (per Leg. I.) describens lineam Bc æqualem ipsi AB ; adeo ut radiis AS, BS, *c*S ad centrum actis, confectæ forent æquales areæ ASB, BS*c*."

[103] *Ibidem*, 39r.5-7: "Verùm ubi corpus venit ad B, agat vis centripeta impulsu unico sed magno ; efficiatque, ut corpus de rectâ Bc declinet, & pergat in rectâ BC."

[104] *Ibidem*, "Augeatur jam numerus & minuatur latitudo triangulorum in infinitum ; & eorum ultima perimeter ADF, (per corollarium Lemmatis tertii) erit linea curva: ideoque vis centripeta, quâ corpus à tangente hujus Curvæ perpetuò retrahitur, aget indefinenter; Areæ verò quævis descriptæ, SADS, SAFS, temporibus descriptionum semper proportionales, erunt iisdem temporibus in hoc casu proportionales."

force. More precisely, he writes: "Newton proceeds to the limit, whereupon the sequence of impulses becomes a continuous force and the series of line segments becomes a smooth curve."[105]

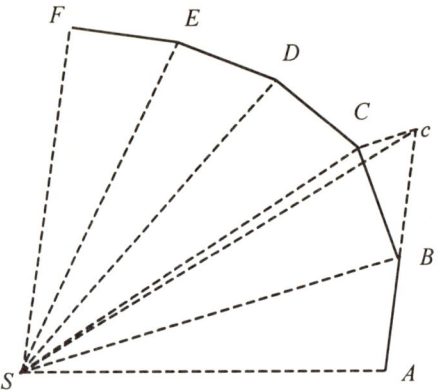

Figure 2.6 Illustration of Newton's proof of Kepler's law

From this, Cohen concludes that "Newton conceived the second law for continuous forces as obtainable in the limit from the Second Law as stated in the *Principia*."[106] In this way Newton's formulation of the second axiom appears to be a solution—albeit not perfect in the eyes of Newton himself—"to generalize his physics from the phenomenologically based dynamics of collisions and blows to the debatable realm of central forces, of gravitational attractions, and hence of continuous forces generally."[107]

---

[105] I.B. Cohen, "Newton's Second Law," 142 (version 1967) or 158 (version 1970).
[106] *Ibidem*, 142 (version 1967).
[107] *Ibidem*, 143 (version 1967) or 160 (version 1967). Cf. also —, *The Newtonian Revolution*, 176. Later Cohen added two appendices to his analysis, in which he defended the idea that the distinction between a classical-geometrical approach and an infinitesimal one, and the use of the expression 'within a given time' ("dato tempore") were all of vital importance in this transition (*ibidem*, 178-185 (version 1970)). However, Cohen was not able to continue along this potentially viable path because he made himself dependent on the mathematician Whiteside. Whiteside convinced him in personal conversations of the dubious idea that Newton would have used second order differentials. Cohen consequently applied this idea to Newton's reformulations of the second law in a revised version of his article, to explain Newton's sketches of curves. For example, Cohen says that Newton would speak of an "infinitesimal force-impulse acting in an infinitesimal time unit dt," which can then be divided up into second order infinitesimal units of time (*ibidem*, 166 (version 1970)). However, nothing in the relevant situations shows that Newton would not have talked about *finite* force-impulses. Cohen even goes so far as to state that the "continuous force could be conceived as a sequence of second-order infinitesimal force-impulses equivalent to a first-order infinitesimal continuous force: a micro-example of the physics of visible bodies" (*ibidem*, 168 (version 1970); see also 167, 158n.19, appendices I and II). This led Pierson to sigh: "The trouble is that Newton was being a little less than magisterial about this matter. Cohen seems always to rescue him from his lapses" (Pierson, "*Corpore Cadente*," 642).

Of course, this interpretation makes one wonder what caused Newton to retain a formulation of the second law for *instantaneous* forces, even though he knew and applied the law in its *continuous* form. A plausible explanation is given by Westfall. In his view, formulated cryptically, "[Newton's] concept of force was unable to emancipate itself from its own history." This 'history' was the geometric representation that Newton had developed in *De Motu* (1684). This work showed that the parallelogram is an outstanding aid in the mathematization of forces of impact. It linked the polygonal approach of the distance covered with the geometrical representation of impressed forces. The convenience of using the parallelogram subsequently tempted Newton to employ it for continuous forces as well. Here, however, he ran up against the problem that distance and force could not both be represented at the same time in one parallelogram. Thus the use of the parallelogram concealed the ambiguity in the concept of force.[108]

2.6.3 Force like Water

Cohen's interpretation, and Westfall's explanation, of Newton's dilemma are certainly valuable, but they do not reveal Newton's reason to reduce all forces to forces of impact. Although both writers say that, for Newton, both forces were in a certain way identical, they explain this identity only by referring to the fact that he believed continuous force was built up physically from instantaneous forces.[109] This leaves the question of how Newton looked at forces so that he could identify them, although, seen in a modern way, they are dimensionally incompatible.

This question does not contradict what was said in section 2.3 above. Even though one needs to refrain from dimensional analysis if one is to understand Newton thoroughly, this does not relieve us of the necessity of finding a logical basis for

---

[108] Westfall, *Force in Newton's Physics*, 476-477.
[109] I.B. Cohen *(The Newtonian Revolution,* 173-174) says that, as a consequence of Newton's use of the expression 'within a give time', the second law permits both the formulas $d(mv) \propto F$ and $d(mv) \propto Fdt$, as well as $d(mv) \propto \frac{1}{2}Fdt^2$. Westfall *(Force in Newton's Physics,* 474) states explicitly that "there can (...) be no doubt that Newton continued to consider Δmv and ma, as the measures of force, to be identical," and that "[w]here the distinction of force and total force (F and $\int Fdt$) implies dimensional incompatibility to us, the 'force' common to both implied identity to Newton." In an appendix he also says: "[Newton] was prepared to speak of motion being impressed as well as force," and adds that "it might be possible to find a common quantity (of impressed force or of motion generated by it) in all the above passages" *(ibidem,* 546). Westfall argues that continuous force might have a discontinuous origin, referring to Newton's treatment of resistance of a moving body in a medium *(ibidem,* 475-6). Cohen, on the other hand, considers this kind of construction as a kind of "physical hunch" (Cohen, *The Newtonian Revolution,* 176).

In 1983, Werner Kutschmann provided an excellent overview of the way Newton dealt with the concept of force, and how impetus, momentum, impact force, and continuous force were interrelated for Newton. He agrees with Cohen that continuous force is seen by Newton as a series of blows, like rhythmical hammer blows. However, he is concerned not so much with the underlying unity of the varying types of force, as with the process of their mathematical unification (Kutschmann, *Die Newtonsche Kraft; Metamorphose eines wissenschaftlichen Begriffs* (1983)).

his doings. Supposing that the identification of forces was not a matter of irrationality, we must make explicit the identity that Newton perceived in them.

The explanatory note to the second axiom clearly shows the identity of continuous forces and forces of impact. There Newton states that a multiple of the force will produce a multiple of the quantity of motion: "(...) whether [the force] is impressed at once and in its entirety, or gradually and successively."[110]

The identity Newton implicitly postulates here makes it possible to bunch a force together and stretch it out, just as clay can be pressed into a ball or pulled into a bar. The bunched-together form is the force of impact; the stretched-out form is the continuous force.[111] The scholium after the third axiom also expresses this identity:

> If a body falls, the uniform gravity, in order to act uniformly, impresses equal forces, in small, equal parts of time, upon that body, and generates equal velocities, and during the whole time it impresses the whole force, and generates the whole velocity proportionally to [that] time.[112]

The uniform, continuous force, whose action is an acceleration, can be broken down into forces of impact, whose actions are changes in the quantity of motion. In dealing with problems concerning continuous forces (such as the derivation of Kepler's law of areas, shown above), Newton approximates to a limit in the opposite direction, i.e., from forces of impact to continuous forces. However, both movements, one an analysis and the other a synthesis of continuous force, entail a qualitative transition between acceleration and change of velocity. Change of velocity is the foundation, it provides the building blocks of acceleration. From the perspective of modern mechanics, Newton identifies quantities of *differing dimension* here, since the dimension of the uniform force of gravity is kg·m·s^{-2}, while the dimension

---

[110] Newton, *Principia mathematica*, 13r.19-20, Liber I, Lex II: "(...) sive simul & semel, sive gradatim & successive impressa fuerit."

[111] See also McGuire's 1970 commentary on I.B. Cohen's "Newton's Second Law," in which McGuire says that the real distinction in Newton is not that between forces of impact and continuous forces, but between forces of impact and continuously *acting* forces (McGuire, [Commentary on I.B. Cohen], 187). Although forces of impact were of prime importance for Newton, it is certainly difficult to see why forces of impact and continuous forces have to be transformed into each other if *both* are fundamental.

[112] Newton, *Principia mathematica*, 21r.22-25, Liber I, Leges motus, Scholium: "Corpore cadente gravitas uniformis, singulis temporis particulis æqualibus æqualiter agendo imprimit vires æquales in corpus illud, & velocitates æquales generat: & tempore toto vim totam imprimit & velocitatem totam generat tempori proportionalem." Newton only added this passage in the third printing. One can see from his previous outlines that, at first, he had the following formulation in mind: "In falling bodies, the uniform gravity impresses forces and produces velocities that are proportional to the times" ("Corporibus cadentibus gravitas uniformis, vires imprimit et velocitates generat temporibus proportionales, ut supra") (I.B. Cohen & Koyré, [Notes on Isaac Newton's *Principia mathematica* (1972), 64-65). This notion corresponds with Sagredo's idea of the accelerated fall as a step by step increase in velocity (Galileo, *Discorsi*, Third Day, 199). The fact that Newton only added this kind of speculation about gravity, force and motion in the third impression is a solid argument for suspecting that his formulation was not a mistake, but, on the contrary, should be assigned an important role in his thinking about the concept of force.

of the *action* of the same is either kg·m·s^{-1} (integrated over time: momentum) or kg·m^2·s^{-2} (integrated over space: kinetic energy).

Newton makes the transition from the finite to the infinite without giving any indication that he realized that he had also made a qualitative transition. This is not in accordance with the calculus of those days, which he himself had developed twenty years previously: given that, during the seventeenth and eighteenth centuries, calculus was used rather intuitively, and by no means strictly, from the mathematical point of view, one could not expect the physical interpretation of its results to reflect upon the exact nature of the infinite. Moreover, as I have shown in section 2.2.3, there was no awareness of dimensional considerations, and this also prevented qualitative differences from being discussed. On the contrary, qualitatively different quantities were thought to be comparable by means of proportionalities very well.[113] Newton could therefore conceive of acceleration as the sum total of the actions of forces of impact. To put this another way: a continuous force is essentially identical to a force of impact, the only difference being that it is the summation of an infinite number of such forces. Thus for Newton, the action of a continuous force was analogous to that of a force of impact; it could be deduced from it mathematically, i.e., by means of calculus.

At this point the basis for this reduction comes into view. At the end of the previous section, I pointed out that Newton's use of the term of innate force was a remnant of his idea of 1684, that the innate force is the sum of all the impressed forces. Now we can see how this idea was carried over into more than simply the term 'force'; for Newton, continuous forces could be seen as a summation of forces of impact, although Newton himself no longer explicitly advocated the idea. The force of impact is proportional to the quantity of motion, which is in its turn proportional to the old concept of innate force, viewed as momentum.

In this way, Newton was trying to retain the continuity of cause and effect, of impressed force and innate force, in his dynamics. This continuity is no different to that between 'force' and 'impetus' in scholasticism. As in that school of thought, the cause of the *appearance* of the motion has given way to the cause of its *continuation*. Although Newton abandoned the strict identity, he replaced it with a transformation relationship, in which the external force, as far as its action is concerned, can be traced in the moving body.

This foundation for the various forces—continuous forces, forces of impact and of inertia—also reveals the meaning of the somewhat vague expression 'in a given

---

[113] Newton, *Principia mathematica,* 34r.18-22, Liber I, Sectio I, Lemma X, Scholium: "Si quantitates indeterminatae diversorum generum conferantur inter se, & earum aliqua dicatur esse ut est alia quaevis directe vel inverse: sensus est, quod prior augetur vel diminuitur in eadem ratione cum posteriore, vel cum ejus reciproca." ("If indeterminate quantities of different kinds are compared to one another, and if it is said that one of them is proportionate to [literally: is as] the other, whether directly or inversely, then this means that the first is increased or decreased in the same ratio as the latter, or as its inverse"). This scholium was added only in the third edition (I.B. Cohen & Koyré [Notes on Isaac Newton's *Principia mathematica,* 83]).

time' in the definition of the quantity of a centripetal force: a force that is transferred *entirely and immediately* produces a certain velocity (or quantity of motion) *instantaneously*.

This last observation has two implications. First, the action of the force is not seen as something mediated by space and time; i.e., a force is transferred into the quantity of motion of a material body, without any influence of the distance covered or the time elapsed. Second, the action of a force on the state of motion of a body is viewed as a kind of transfer in which, as in the transformation of energy a hundred and fifty years later, the motion that results also represents a certain force.

It is true that Newton, explaining his definition of the impressed force, explicitly rejected the idea that this might be transferred to the body on which it acts.[114] In fact, however, it still functioned in this way for him. From a modern point of view, this conception of force and its action is not so much physics as metaphysics.

Thus Newton's dilemma in bringing forces of impact and continuous forces together within a single approach becomes clear, if we regard it as the continuation of a traditional metaphysical view of the action of forces, i.e. the identification of the action of a force with the force itself. That which is transferred is at the same time the cause of its own transference. At the risk of being blasphemous, this could be expressed as: the action is its own cause.

The transfer of force is not mediated by space and time but can, instead, be compared to the stream of water flowing out of a hole in a barrel (see Figure 2.7). Here, the exit pressure is proportional to the head of water, and thus to the quantity of water. The exit pressure is also related to the exit velocity. As a result, the water in the barrel is simultaneously the cause of the outflow of water, and that which flows out of the barrel: the *why* and the *what*.[115]

Thus the *action* of the impressed force is equal to the impressed force itself. Strictly speaking, of all the external forces, only the forces of impact conform entirely with this, so it seemed evident to Newton that he should retain the force of impact in his mathematization of continuous forces.

2.7 Conclusion

Our archeological journey into the historical basis of the unity of dead and living force has led us, via Galileo and Descartes, back to the Paris nominalists, where we encountered the impetus theory as a solution for a problem arising from the Aristo-

---

[114] Newton, *Principia mathematica*, 2r.22-25, Definitio IV: "Constitit hæc vis in actione sola, neque post actionem permanet in corpore. Perseverat enim corpus in statu omni novo per solam vim inertiæ" ("This force exists solely in the action itself, and does not remain in the body after the action [is completed]. For a body persists in each new state solely by its force of inertia").

[115] Strictly speaking, the analogy makes use of a metonymy: it is not the water as a whole, but its *weight* that causes the outflow. In can be deduced from Bernoulli's law that the head of the water is proportionate to the square of the exit velocity: $gh = \frac{1}{2}v^2$, in which g is the gravitational acceleration.

telian theory of the projectile motion. It appeared that impetus, the force of a body in motion, ought to be viewed as the internalized motive force. In this way, impetus constituted the link between 'material resistance' and 'external force'. It was to be expected, therefore, that the unity of the concept of force would not only be found in the development of the relation between 'impetus' and 'external force', but also in that between 'impetus' and 'material resistance'. For this purpose, we have examined several stages of the development of the concept of inertia, ultimately linking these with a basic problem in Newton's *Principia*.

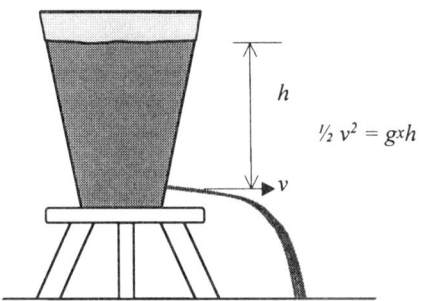

Figure 2.7 Metaphor of the action of force with water streaming out of a barrel

In Newton's thought, the premise of the unity of force is evident both in the relationship between innate force and impressed force, and in his basing of continuous forces upon forces of impact. This conclusion nuances our view of the historical relationship between Newton's concept of force and the scholastical one, and brings them closer together, in the sense that it points to common metaphysical ground without holding on to a univocal development of the scholastic concepts relating to force. Force is a cause that can be transferred from one substance to another, and during that process manifests itself in different forms, such as quantity of motion, resistance, and weight.

Chapter 3 complements the archeological mission of the present chapter with a discussion of the meaning and development of Leibniz's concept of force. Despite his distinction between dead and living force, or perhaps precisely because of it, the unity of force was preserved in Leibniz's thought, more than in Newton's. One important reason for this is the brilliant way in which he attempted to unite Aristotelian-scholastic metaphysics, which emphasizes substance, with mechanical-mathematical physics, which focuses on motion. This, in combination with his efforts to provide clear definitions of fundamental concepts and axioms, led to a dynamics that highlighted one thing much more clearly than in Newton's dynamics (where it was buried under mathematical instrumentation): the unity of force as the leading principle in its mathematization.

CHAPTER 3

LEIBNIZ: FORCE AS THE ESSENCE OF SUBSTANCE

> *The creation of modern dynamics was a matter, not of experimentation and new discovery, but of drawing consequences from accepted conclusions, of clarifying ambiguities and resolving conceptual tangles, above all of breaking through the rigidity of received intuitive conceptions of dynamic action.*
> Richard S. Westfall[1]

3.1 INTRODUCTION

3.1.1 Leibniz's Influence on the Eighteenth Century

Everyone who addresses the controversy surrounding the concept of living force discusses Gottfried Wilhelm Leibniz, or rather, his legacy. In fact we owe both the definition of the term and the beginning of the controversy to Leibniz.[2] One might consider a less emphatic view of his importance, if one took only the superficial and one-sided transmission of Leibniz's philosophy in the eighteenth century into account. Many of the physicists from that period, even those who were sympathetic to his ideas, had a caricatured picture of Leibniz's system.

For example, Nicolas Béguelin was a natural scientist of some importance at the Berlin Academy. In two essays dating from 1766 and 1769, Béguelin tried to combine Leibniz's metaphysics with Newton's physics.[3] In the first of these he interpreted gravitation as a low-level form of a universal tendency to associate with equals ("similis simili gaudet"), with the level being determined by the clarity of the "sentiment," i.e., by the degree of spirituality of the being. The most material being

---

[1] Westfall, *Force in Newton's Physics. The Science of Dynamics in the Seventeenth Century* (1971), 289. Westfall presents this proposition in the course of identifying Leibniz's primary significance for the development of classical mechanics.
[2] In Leibniz, "Specimen dynamicum pro admirandis naturae legibus (...)" (1695) and "Brevis demonstratio erroris mirabilis Cartesii (...)" (1686), respectively.
[3] Béguelin, "Essai d'une conciliation de la métaphysique de Leibnitz avec la physique de Newton" (1766) and "Conciliation des idées de Newton et de Leibnitz sur l'espace et le vuide" (1769).

seeks a general and indiscriminate association and "embraces all classes of beings."[4] The inclination of the most spiritual being, in contrast, is limited to a very exceptional class. Thus, twenty years before the French Revolution, Leibniz dons the uniform of a footman to the aristocrats, although he always argued against the existence of 'attractive tendencies'!

One of the first competition questions set by the Berlin Academy, the question for 1747, concerned Leibnizian monadology. The way in which it was answered again revealed a stark contrast between Leibniz's philosophy and its transmission.[5] D'Alembert, Euler and Maupertuis, likewise, had only a superficial picture of Leibniz, as we will see in Chapters 4 and 5.

This superficial transmission can in part be attributed to Leibniz's working method. Because he was mainly responding to other people's opinions, he wrote little that could be published independently. Another reason, of at least equal importance, is the gradual shift in his interest from the synthesis of mechanics and metaphysics to pure metaphysics and natural theology, resulting in the *Essais de théodicée* (1710), the famous correspondence with Samuel Clarke (1717), and the shorter popularising works *Principes de la nature et de la grâce, fondés en raison* (1718) and *La Monadologie* (1720). This shift may also be the reason why it was at first mainly theologians who concerned themselves with editing of Leibniz's work. The first collections of Leibniz's writings therefore focus on the portion of his work that is not immediately relevant to physics. One example is the *Des Freyherrn von Leibnitz kleinere philosophische Schriften*, published by C.-J. Huth in 1740. This contains the *Monadologie*, the defense of his idea of the pre-established harmony against the criticisms of Pierre Bayle, his correspondence with Clarke, and a number of pieces composed by the publisher.[6]

Despite enthusiastic attempts, it was a long time before a collected edition appeared containing both his unpublished manuscripts and correspondence and the articles that had been published separately and were difficult to obtain. In 1731 the Leibniz populisor Carl Gunther Ludovici attempted to republish all of Leibniz's published articles, but the plan was torpedoed by opposition from Elector Friedrich-August von Sachsen.[7] Most subsequent attempts at a more comprehensive and systematic edition also failed, with the exception of a few fragmentary editions of correspondence.[8] Diderot was quite right when, in 1765, he accused Germany of still

---

[4] "embrassera toutes les classes d'Etres" (Béguelin, "Essai d'une conciliation," 375).
[5] Harnack, *Geschichte der Königlich Preussischen Akademie der Wissenschaften zu Berlin* I (1900), 402-403; Bongie, [Introduction to Condillac's *Les Monades* (1747)] (1980), 21-35; Buschmann, "Der Monadenstreit um 'ein Prämium von 50 Ducaten' " (1987), " 'Die elendste Schrift, die je gekrönet worden' " (1988), and "Die philosophischen Preisfragen und Preisschriften der Berliner Akademie der Wissenschaften im 18. Jahrhundert." (1989), 182-186.
[6] C.J. Huth ed., *Des Freyherrn von Leibnitz kleinere philosophische Schriften*. (Revision and extension of Köhler's 1720 translation of Leibniz's *Monadologie*). Ravier 412. Jena: Mayerischen Buchhandlung 1740.
[7] Ravier, *Bibliographie des œuvres de Leibniz* (1937), 165.
[8] See Aiton, *Leibniz: A Biography* (1985), 351-352.

not having published what had flowed from the pen of this man, who alone brought Germany as much fame as Plato, Aristotle and Archimedes together brought to Greece.[9] The difficulty of obtaining Leibniz's own work was only partially compensated by the perpetuation of his thinking in the philosophy of Christian Wolff. Wolff, who in the middle of the eighteenth century won great respect for German speculative philosophy and so created the necessary conditions for the development of comprehensive systems such as those of Hegel and Schelling, derived his own philosophical system largely from Leibniz. While Leibniz was extensively popularized as a result—Wolff's system was declared to be the official 'School philosophy' in Germany—this popularization was coupled with simplification and materialization.[10]

Finally, in 1768, Louis Dutens brought out the first real collected edition. It was not complete, because Dutens was refused access to the archives of the library of Hanover![11] Because of this refusal, most of Leibniz's unpublished work and correspondence did not become generally available for evaluation until much later. The edition published by Dutens would remain the standard work until C.I. Gerhardt's very extensive edition in the second half of the nineteenth century.[12]

The superficial and faulty reception of Leibniz in the eighteenth century cannot alter the fact that his contribution to and criticism of mechanics has been historically important. Especially in the last two decades of the seventeenth century, the period in which his metaphysical system developed, Leibniz was closely involved in developments in mechanics. In those years, the development of his metaphysical system was closely linked to his attempt to provide foundations for mechanics. At that time metaphysics and mechanics for Leibniz were still equal, in contrast to the situation

---

[9] Diderot, "Leïbnitzianisme, ou Philosophie de Léïbnitz" (1765), 379.

[10] Until recently, Wolff's philosophy was regarded as a systematization and faithful elaboration of Leibniz's philosophy, as shown by references to the 'Leibniz-Wolffian School', a term which was in use even during Wolff's life. In recent decades more emphasis has been placed on the fundamental differences between the two. See Heinekamp ed., *Beiträge zur Wirkungs- und Rezeptionsgeschichte von Gottfried Wilhelm Leibniz* (1986), especially Jean École, "Des rapports de Wolff avec Leibniz dans le domain de la métaphysique," and Hans-Jürgen Engfer, "Teleology und Kausalität bei Leibniz und Wolff; Die Umkehr der Begründungspflicht." This distinction between Leibniz and Wolff also has implications for the study of Leibniz himself, for instance by enabling the relationship between Leibniz's dynamics and metaphysics to be considered anew. Although views of Leibniz have been immeasurably broadened by Gerhardt's publication of the sources (see below, in the text), disproportionate emphasis has continued to be given to his later documents and correspondence. This disproportionate emphasis could be an important factor in the one-sided interpretation of Leibniz's philosophy as a metaphysical and logical system (Daniel Garber, "Leibniz and the Foundation of Physics: The Middle Years" (1986), 27). In fact, it is in the later period of his thinking that Leibniz focuses mainly on the question of individual substance, and that the relationship of the concept of substance with the phenomenal reality of space and time becomes problematical.

[11] Ravier, *Bibliographie*, 176.

[12] In this book I refer mainly to Gerhardt's editions, referred to respectively as 'Gerh. *Math. Schr.*' (for *Leibnizens mathematische Schriften*) and 'Gerh. *Phil. Schr.*' (for *Die philosophische Schriften von G.W. Leibniz*).

after about the turn of the century. For example, the reality of matter was not yet in dispute; only the need for a metaphysical foundation for matter was being discussed. With perhaps a little exaggeration, one could say that Leibniz still saw matter at this time as an *abstraction*, whereas later it came to have the character of an *illusion*.

3.1.2 Leibniz's Significance for Mechanics

What then was Leibniz's significance for mechanics? Like Newton, but in a completely different way, Leibniz conceives of forces in mechanics within a metaphysics of substance, and assumes on that basis that force is a unity. Newton initially tried to mathematize all *external* forces ("vires impressae"), as external impact forces or as their derivatives, and thus also to explain continuously acting forces on the basis of instantaneously acting forces. Leibniz, in contrast, focuses on the *internal* force, which is the cause of the continuation of the motion of the body, and posits the conservation of force as a general principle for the determination of the laws of motion. The approaches of Newton and Leibniz, like the internal and external forces, can be regarded as complementary. While Newton mathematizes the continuously acting impressed forces, by reducing them to instantaneously acting forces, Leibniz tries to find a mathematical expression for the force of a body in motion, which he at first called the 'moving' and later the 'living' force. In his case, however, the old idea of the unity of cause and effect is much more evident than with Newton; in fact for him it has become a principle, as I will show in the following sections.

In addition to this complementary conceptualization of the concept of force, Leibniz's mechanics contributed an important conceptual clarification that resulted from his idiosyncratic philosophy. He opposed the definition of the essence of matter as extension in space, because it is impossible to explain the laws of motion on this basis. For his own explanation he goes back to the Aristotelian-scholastic metaphysics of substance, which he had learnt to know in his youth.[13] He thinks he is able to solve the problems that he observes in mechanical philosophy by means of a differentiation of the original homogenous reality into a metaphysical and a phenomenal level. In this differentiation the concept of force plays a connecting role

---

[13] At the end of the seventeenth century and the beginning of the eighteenth it was not unusual to make a connection between mechanical philosophy and Aristotelian-scholastic metaphysics, but it was generally made by Aristotelian Jesuits who included mechanical philosophy in their physics only as a *hypothesis*, or even as an *instrument*,. Heilbron writes that the Jesuits initially entirely rejected Descartes's teachings but later gradually assimilated his physics. The 15th congregation of 1706 permitted the defense of Descartes's physics as a *hypothesis*. One famous Cartesian Jesuit was Ignace Gaston Pardies (Heilbron, *Electricity in the 17th and 18th Centuries* (1979), 36-37). However, Leibniz, who had grown up with Aristotelian ideas but was converted in his youth to mechanical philosophy, develops his thought in a very different way to the Jesuits because he did not assimilate mechanical philosophy but rather *returned* to Aristotelian-scholastic metaphysics in order to be able to provide foundations for mechanics.

because, on the one hand, it constitutes the substances, and, on the other hand, it is the cause of material, phenomenal motion.

3.1.3 The Structure of this Chapter

I do not in any way pretend to summarize Leibniz's system in its entirety here. His theology, his moral philosophy and his logic, for example, are entirely beyond the scope of my theme, which is the contribution of his metaphysical system to the foundations of mechanics. In broad terms the remainder of this chapter follows two interrelated themes.

The first theme asks how Leibniz, implicitly, based his mechanical concept of force, on the phenomenal level, on a concept of transference. An important subsidiary question is how Leibniz related the existence of the mechanical action of force to his metaphysical idea that substances can not act on one another, but only on themselves. The second theme follows his attempt to seek the connection between the metaphysical and the phenomenal level in the concept of force.

My argument will proceed first from mechanics to metaphysics, and then return. The starting point is in 1686, when Leibniz introduced the theme of force in a critique of the Cartesian definition of the moving force ("vis motrix").[14] Force is here a concept that must be hypothesized as a fundamental of mechanics *in addition to* the concept of extension, in order to be able to explain the laws of motion. Around 1695, the year in which his mechanics and metaphysics were most closely related, Leibniz extended the scope of the concept of force.[15] From then on, force is also the foundation of substance. The concept of force then assumes a complicated structure, with fundamental and derivative forces, each of which exists in both passive and active forms. What Leibniz referred to in 1686 simply as 'force' is now called 'derivative active force'. The question then is, what problems this structure produces for the foundation of mechanics, and in particular for the concept of mechanical force.

---

[14] Leibniz, "Brevis demonstratio" and the posthumously published "Discours de métaphysique" (both from 1686). Leibniz's earlier eclectic views, in which he still accepted the tradition dogmatically, are less important here. Loemker draws the demarcation line between his early and mature periods in 1690, because only then were the fundamental concepts of individuality, grades of substance, force, and the monad introduced (Loemker, [Commentary on *Philosophical Papers and Letters of Leibniz*] (1956), 355). Although I agree with Loemker as regards the formation of Leibniz's system as a whole, the "Brevis demonstratio" and the "Discours de métaphysique," both written in 1686 although the latter was only published in 1846, cannot be excluded from this study because of their significance for the inception of the *vis viva* controversy and for a number of later problems relevant to the relationship between metaphysics and dynamics.

[15] Leibniz, "Specimen dynamicum" and "Système nouveau de la nature et de la communication des substances (...)" (both from 1695).

3.2 LEIBNIZ'S EARLY CONCEPT OF FORCE

3.2.1 Leibniz's Criticism of Descartes

In 1686, Leibniz began an attack on the Cartesian concepts of matter and force that was to continue throughout his life. The attack began with a short proof, "Brevis demonstratio erroris mirabilis Cartesii," published in the recently established journal *Acta eruditorum*, and with a passage in the "Discours de métaphysique" that was not published until after his death. Descartes is said to have incorrectly claimed that God conserves the quantity of motion ("quantitas motus") in His creation.[16] Leibniz thought that this error was based on the assumption that moving force and quantity of motion are equivalent.[17] Although the argument is short and concisely formulated, it contains a wealth of meaning, which makes it difficult to understand. Moreover,

---

[16] Descartes, *Principia philosophiae* (1644), II.36.

[17] It is comical that Descartes does not say, in his *Principia*, that the moving force or any other force is equal to the quantity of motion, and thus also does not say that this identity is the basis of the conservation of the quantity of motion. In the *Principia*, Descartes talks about 'force' only as a phenomenon, consisting of the determination that each body has to continue in the state in which it then finds itself (Descartes, *Principia philosophiae* II.43. See also above, section 2.5.3).

Thus it is at first sight remarkable that Leibniz attributes the idea of the conservation of force to Descartes, when Descartes only explains the conservation of the quantity of motion, and moreover never uses the term 'force' in this explanation, seeking support only in the creation of matter and motion by God (*ibidem*, II.36). The fact that none of the Cartesians who responded to Leibniz's article, such as Catelan and Arnauld, made any protest to this, only indicates that between 1644 and 1686 this fundamental reshaping of Descartes's thoughts—because that is what it is—must have become a commonplace.

However, sometimes Leibniz appears to realize that the attribution is not correct. For instance, in 1692 in his commentary on Descartes's *Principia philosophiae* he says that Descartes's mistake was that he chose to explain the conservation of the wrong quantity. After all, Descartes goes immediately from the perfection of God to the conservation of the quantity of motion. But, Leibniz says, the premise of the perfection of God only shows *that* something is conserved, but not *what* is conserved. It remains to consider whether it is the quantity of motion or something else, such as the quantity of force (Leibniz, "Animadversiones in partem generalem Principiorum Cartesianorum" (1692), 370).

Nevertheless, in his "Specimen dynamicum" Leibniz returns to the incorrect attribution of the identification of force and quantity of motion to Descartes: "[Cartesiani] vires creduntur esse in ratione composita corporum et celeritatum" ("[The Cartesians] consider that forces are proportional to the velocity and the bodies [i.e., the quantity of matter, or mass, JCB]") (Leibniz, "Specimen dynamicum," 245). Although strictly speaking Leibniz is talking about Cartesians and not about Descartes himself, it is clear that he does not make any distinction between the leader and his followers. See also Leibniz ("Brevis demonstratio," 119): "Cartesiani autem non pauci vereor ne paulatim Peripateticos complures imitari incipiant, quos irridenti, hoc est ne pro recta ratione et natura rerum, consulendis magistri libris assuefiant" ("I fear that there are not a few among the Cartesians who gradually begin to imitate various Peripatetic philosophers, although they ridicule them, and this is because they have become accustomed to consulting the books of the master instead of true reason and the essence of things").

The reason for this confusion could be that Leibniz based himself on the published letters of Descartes concerning mechanics. Arnauld had drawn his attention to these in a letter of 28 September 1686, and it would appear from Leibniz's answer (28 Nov./ 8 Dec. 1686) that he had indeed studied them. However, it is not known whether he had done so before writing the "Brevis demonstratio" (Gerh. *Phil. Schr.* II, 67-68, and 80, respectively).

any explanation runs the risk of overlooking something important or creating a misunderstanding. Nevertheless, I shall make an attempt.

Leibniz's argument aims simultaneously to refute the proposition of the conservation of the quantity of motion and to provide an alternative path to the quantitative determination of force. He says that he agrees with the premises that there is always the same quantity of force conserved in the universe, and that the force of a body can only be reduced by transference to another body with which it is in contact.[18] It is clear that, although Leibniz does not yet explicitly use the *term* 'living force' here—he was not to make the distinction between dead and living force until a letter to Arnauld later in that year[19]—he does have the *concept* in mind, i.e., he is thinking of an *internal cause of motion*.

What would be a good way to quantify this force and so prove that Descartes was wrong? Leibniz bases himself on two premises. The first is that a body that falls downwards from a certain height acquires just enough force to enable it to rise to the same height, if the 'direction' of the force is reversed and if there are no external hindrances. An example of this is the motion of a pendulum which, if we disregard air resistance, always reaches the same height.[20] The second premise is that force is proportional to the product of the weight and the maximum increase in height.[21]

---

[18] "Itaque cum rationi consentaneum sit, eandem motricis potentiae summam in natura conservari, et neque imminui, quoniam videmus nullam vim ab uno corpore amitti, quin in aliud transferatur, neque augeri, quia vel ideo motus perpetuus mechanicus nuspiam succedit, quod nulla machina ac proinde ne integer quidem mundus suam vim intendere potest sine novo externo impulsu (...)" (Leibniz, "Brevis demonstratio," 117; see also Leibniz, "Discours de métaphysique," 442 (section 17)).

[19] Letter from Leibniz to Arnauld, 28 Nov./8 Dec. 1686 (Gerh. *Phil. Schr.* II, 80). See note 35 below.

[20] Leibniz, "Brevis demonstratio," 118. This axiom, including the illustration from a pendulum, was formulated in 1669 by Huygens in a *kinematical* form, that is: in terms of acquired *velocity* rather than acquired *force*. See Dijksterhuis, *The Mechanization of the World Picture*, 370-373 (section IV.141-142).

[21] Leibniz, "Brevis demonstratio," 119. Descartes himself made the same assumption: he defined 'action' as the force that is required to raise a weight through a certain vertical distance (see above, section 2.3.2). In "Brevis demonstratio" Leibniz does not give any further explanation of the basis of his method of estimating the quantity of force. He gives the requirements for this only in his retrospective summary in "Specimen dynamicum" (243). He says there that the quantity of force can be determined *a posteriori* from the effect that it produces when it is consumed ("ab effectu quem producit se consumendo"). There are two requirements for the effect: it must be "violentum," that is, force-consuming, and "as far as possible, able to be homogeneously divided into parts of the same shape and size" ("maxime capax est homogenei seu divisionis in partes similes et aequales"). He adds to this that the period over which this effect is produced is unimportant. Here Leibniz is responding to the Cartesian Abbé de Catelan's critique of his "Brevis demonstratio" (Catelan, "Courte remarque de M. l'abbé D.C." (1686)), which claimed that force should be estimated not from the resulting height itself, but from the increase in height *per unit of time*. Catelan appealed to a metaphor that shows very acutely how strongly the concept of force was linked to everyday experience. It is clear, he argues, that if an adult performs a quantity of work in three hours, and a child does the same work in eight hours, one could not say that they had the same power, or the same force. It is not only the work done, but also the time required, that constitutes a measure. That is, *the ratio* between work performed and time required is the true measure of a person's capacity for work. In mechanics, in the same way, it is not the resistance overcome—for example the in-

continued on next page

To quantify the force, Leibniz now proposes the following thought experiment: a body of 1 pound falls from a height of 4 yards, and in the same way a body of 4 pounds falls from a height of 1 yard.[22] According to his first premise, they both acquire exactly the force required to return to the same height. According to the second premise, at the lowest point the forces of the two bodies are equal. If the forces were proportional to the product of weight and velocity, as Descartes had said, the velocities would at that moment have to be in the proportion of 1 : 4. However, Galileo had proven that in a fall the decrease in height is proportional to the square of the resulting velocity, so at the lowest point the ratio between the velocities must be 1 : 2! Descartes's measure is therefore incorrect.[23]

Leibniz considers that this thought experiment confirms that the quantitative determination of force must be based on the *effect* that it can cause, i.e., the increase in height, and not on the velocity, which is in fact only linked with the force "per accidens." It is only in particular situations that the force of a body is indeed proportional to the product of weight and velocity, as in the case of the five simple machines: the lever, windlass, pulley, wedge and screw. If in these machines the quantities of motion are equal, the products of weight and height difference are also equal, and thus also the forces.[24] However, this is not generally true, as can be seen from the motion of free fall, in which the product of the weight and the *square* of the velocity is in accordance with the effect.[25]

3.2.2   The Measure of Force

Leibniz's "Brevis demonstratio" is usually interpreted in such a way that Leibniz proposes his own measure in the same breath as he rejects Descartes's measure. However, on closer examination this interpretation appears to be based only on Leibniz's later development and on the fact that Leibniz does not expressly reject the quantity mv^2 as the general measure, as he does with mv. However, from Leibniz's

---

continuation of former page
crease in height—that is the true measure of force, but rather the ratio between this and the time required. Now, if a body with mass m is thrown upwards with velocity v, it reaches a maximum height h according to the relation $h/t \sim mv$, in which t is the time required. The quantity of force is thus equal to the quantity of motion. See also Papineau, "The *Vis Viva* Controversy" (1977), 136-138.

[22]   What is important here, naturally, is the relative and not the absolute quantities. In the spirit of the times, I have chosen to use older units here.

[23]   Leibniz, "Brevis demonstratio," 118. Later, in "Essay de dynamique" (1692), and in "Specimen dynamicum" (1695), 245-246, he reduces the Cartesian equation of force and the quantity of motion to absurdity. In "Essay de dynamique" he deduces from his thought experiment that, if Descartes were right, perpetual mechanical motion would be possible ("Essay de dynamique" (1692), 120 ("Proposition 5")). That is: it would be possible to produce a mechanical effect or outcome without any external input (*ibidem,* 110 ("3. Definition")). This contradicts his axiom that "l'effet entier est égal à la cause totale" ("The whole effect is equal to the total cause") (*ibidem,* 112 ("Axiome 1")).

[24]   See page 98 below.

[25]   "Itaque per accidens ibi contingit, ut vis a motus quantitate possit aestimari. Alii vero casus dantur, qualis is est, qeum supra attulimus, ubi non coincidunt" (Leibniz, "Brevis demonstratio," 119).

own argument, in both the "Brevis demonstratio" and the "Discours de métaphysique," it appears that, at least in 1686, he was not at all intending to propose a quantity such as mv^2 as a measure for force. On the contrary, velocity is only incidentally linked to force, in one instance as mv (in the case of the five simple machines), and in another instance as mv^2 (in the case of free fall).[26]

In the "Discours de métaphysique" he is clearer in rejecting the determination of force from velocities. Both quantities, mv and mv^2, relate force to motion, although, as he indicates in section 18, motion is something that formally speaking does not really exist. After all, rest and motion are only relative concepts.[27] Leibniz's main concern is to argue against the Cartesian concept of matter as extension. His refutation of the identity of force and the quantity of motion in section 17 is in the first place intended to supplement the Cartesian mathematical approach with a metaphysical approach. Force is not one of the modalities of extension—such as motion, magnitude and shape—and therefore requires a metaphysical approach.[28]

In "Brevis demonstratio" Leibniz barely touches on the ontological distinction between force and motion.[29] Force is not disassociated from the changes that it can produce. His chief concern is to defend the correct method of quantitative determination. Quantification must begin with the magnitude of the effect produced. This effect may consist of an increase in height, but cannot consist of velocity changes in collisions.[30] The emphasis is on determining the *positive* relationship between the concept of force and the level of extension and its modalities, rather than on determining the *negative* relationship. For while force proper is real, and thus cannot fall in the sphere of mathematics, it is also the direct cause of motion. Forces can be attributed to bodies and they can even be compared to one another through the motions they produce. So the aim of his proof in this instance was only to refute a bad argument and propose a correct method of quantifying force.

Thus in both the "Discours de métaphysique" and the "Brevis demonstratio," Leibniz mentioned the quantity of mv^2 only as an *incidental* measure of the effect of force. The proportionality of force to mv^2 is not highlighted as forcefully as the

---

[26] Leibniz's proof is therefore entirely different in character to Huygens's proof of the conservation of mv^2. In the first place, Leibniz does not in any way say that the quantity mv^2 is conserved, and in the second place his purpose is rather to demonstrate that 'conservation' occurs at the level of force itself, and not at the level of relative quantities that are based on the concept of motion, and mv^2 is such a quantity.

[27] Leibniz is referring here to Descartes's definition of motion (Descartes, *Principia philosophiae*, II.25).

[28] "Or cette force est quelque chose de different de la grandeur, de la figure et du mouvement, et on peut juger par là que tout ce qui est conçû dans les corps ne consiste pas uniquement dans l'étendue et dans ses modifications, comme nos modernes se le persuadent" (Leibniz, "Discours de métaphysique," 444).

[29] It is notable only in his conclusion, where he says that the velocities are not produced in actuality (Leibniz, "Brevis demonstratio," 119).

[30] "Ex his apparet, quomodo vis aestimanda sit a quantitate effectus, quem producere potest, exempli gratia ab altitudine, ad quam ipsa corpus grave datae magnitudinis et speciei potest elevare, non vero a celeritate quam corpori potest imprimere" (*ibidem*, 118).

conservation of mv is rejected. This should be attributed to an internal tension in Leibniz's metaphysical interpretation of motion. On the one hand, Leibniz used the relativity of motion as an argument against Descartes's law of conservation.[31] This can be found in his "Discours de métaphysique," as we have seen, but also, although less forcefully, in "Brevis demonstratio." On the other hand, to be able to propose mv^2 as the measure of force, Leibniz had to concede that there is a positive relationship between motion and the concept of force: the force of a body is linked to its velocity. He tries to escape this dilemma by treating velocity as secondary to the increase in height, and focusing his conclusion on this height increase: "forces are composed in proportion to the bodies (...) and the height that can be produced from the velocity."[32]

Some years later, the formula mv^2 wás used as the measure of force. In his "Essay de dynamique," written in 1692 but not published until the nineteenth century, Leibniz gives his argument from the "Brevis demonstratio" an axiomatic deductive form. The lack of clarity in the meaning of the proportionality of force to mv^2 has largely disappeared by this time. In a scholium under the sixth and seventh propositions, he states explicitly that: "the forces of bodies are directly proportional to their mass and the square of their velocity."[33] However, there is still a tension with the phenomenality of motion. The source of tension now is not the Cartesian relativity of rest and motion, but the lack of a Leibnizian unity in motion. At the end of his essay he says: "motion is a transient thing that never, strictly speaking, exists, since its parts are never all together."[34] In the end it is only force itself that is real, and the only effect by which it can be measured is the increase in height or something analogous; the magnitude mv^2 is no more than a *derivative*.

Leibniz never accorded the magnitude mv^2 any higher status than as a derivative measure of the effect of force. However, the tension between the phenomenality of motion and the reality of force took on a new form when Leibniz differentiated the concept of force. This is the task he attempted in his "Specimen dynamicum." I will discuss the nature of this differentiation, and its significance for the concept of mechanical force, below. First I want to answer another question, in relation to the early period of Leibniz's system: how is force conceived as the cause and the content of the transference of motion?

---

[31] See page 78 above.
[32] "(...) vires esse in composita ratione corporum (...) et altitudinem celeritatis productricium (...)" (*ibidem*, 119).
[33] "les forces des corps sont en raison composée de la simple de leurs masses et de la doublée de leur vitesse" (Leibniz, "Essay de dynamique" (1692), 122).
[34] "le mouvement est une chose passagère qui n'existe jamais à la rigueur puisque ses parties ne sont jamais ensemble" (*ibidem*, 130).

3.2.3 *Force as the Content and Cause of Transference*

In his first observations about the conservation of force and its effect, 'force' consistently means what Leibniz was later to call 'living force'.[35] The relationship of transferal can clearly be seen: force is both the cause of the change and what is transferred. That force is something that is transferred from one substance to the other can be seen from Leibniz's more detailed specification of the conservation of force. In "Brevis demonstratio" he says that "we never observe any force lost in a body unless it is transferred to another."[36] In "Discours de métaphysique" this is expressed in a more general way: "the force of a body is only diminished to the extent that it is given to some body with which it is in contact, or to parts of itself, to the extent that these move separately."[37] However, there is one problematic point in this, in that Leibniz can only illustrate this conservation, strictly speaking, for elastic collisions, whereas what he regards as the *effect* of force—an increase in height—is linked to the *consumption* of force. It is by no means clear—without making further assumptions about the underlying mechanisms—where the force is in the period that it cannot be observed temporarily because it is being consumed.

It is remarkable that in both the pieces written in 1686 and in the "Essay de dynamique" in 1692, Leibniz speaks about force almost exclusively as if it is a *quantity* of something that is needed for an increase in height, is obtained from the loss of height, can only be increased by a new momentum from outside, and can only be absorbed by resistance and other obstacles. The modern reader, knowing the concept of kinetic energy, could easily think here that Leibniz, when he refers to 'force', is thinking only of the *content* of what is transferred in a collision.

This is not the case. Leibniz clearly shows that he really does understand 'force' as a cause of change. In the first place Leibniz writes in the introduction of "Brevis demonstratio" that by 'force' he means the moving force.[38] It is the moving force that is conserved. Leibniz's force is thus in the first place the force that sustains the motion of a body. In the second place it can also cause changes in other bodies. In

---

[35] Leibniz does make incidental use of the distinction between dead and living force that was later to become so important. For example his letter to Arnauld of 28 Nov./8 Dec. 1686 (Gerh. *Phil. Schr.* II, 80) refers to "puissances mortes" and "force absolue." The "Essay de dynamique" (1692) speaks of "force morte" and "force vive." The distinction serves only as a means of distinguishing equilibrium situations from dynamic situations, because the *dead* forces of statics, at the first moment and for as long as there is no motion, are proportional to the *velocity*, while the *living* forces can only be said to be proportional to the *effect*, that is, to the possible increase in height. But the transition from dead to living force—i.e., the generation of the living force from the continual effect of dead forces—will not be discussed at this point. I will return to this question in section 3.4 below.

[36] "videmus nullam vim ab uno corpore amitti, quin in aliud transferatur" (Leibniz, "Brevis demonstratio," 117).

[37] "la force d'un corps n'est pas diminuée qu'à mesure qu'il en donne à quelques corps contigus ou à ses propres parties en tant qu'elles ont un mouvement à part" (Leibniz, "Discours de métaphysique," 442).

[38] Leibniz, "Brevis demonstratio," 117.

"Discours de métaphysique" Leibniz calls it the "immediate cause of (...) changes" in the relative positions of various bodies.[39] In the third place, with respect to the relationship between force and its effect, Leibniz says that force *produces* the effect.[40] But to produce, literally 'to lead forth', is to cause.

We can therefore conclude that Leibniz, in his criticism of Descartes's principle of conservation, based himself on a concept of force that united in itself the *cause of change* and *the contents of the transference of motion in a collision*. This identity means that the action of a force entails the transference of all or part of that force from one substance to another. Leibniz's thesis of the conservation of force thus takes the scholastic idea of force to its natural conclusion. Although substances become increasingly isolated from one another in the further development of his dynamics, the same metaphysical principle is preserved in Leibniz's later mathematization of the concept of force.

3.3 LEIBNIZ'S DIFFERENTIATION OF THE CONCEPT OF FORCE

3.3.1 *Aristotelianism and Atomism*

Between 1686 and 1695 there is an enormous development in Leibniz's thought regarding force. In the course of this development he makes force the ground of both metaphysical substance and phenomenal matter. In 1695 he published "Specimen dynamicum" and "Système nouveau." These publications play complementary roles within his 'system of pre-established harmony'—as he began to call his philosophy from that year.[41] On the one hand one has the philosophy of matter, on the other hand the philosophy of the harmony of substances, both founded in a philosophical approach that seeks to link substance and phenomenon through the concept of force.

Leibniz's aim is ambitious. His goal is the unification of all philosophical positions, and not as an eclectic structure, the mere sum of seminal ideas, but as a synthesis in which all the different points of view are united in one framework, and the world is shown to be reasonable and understandable.[42] As Erdmann said in a

---

[39] "cause prochaine de (...) changemens" (Leibniz, "Discours de métaphysique," 444).
[40] Leibniz, "Brevis demonstratio," 119 ("productricium"); "Discours de métaphysique," 443 ("produire").
[41] Finster, *Gottfried Wilhelm Leibniz* (1992), 65. As Karskens explains, Leibniz's system is not at all a comprehensive system, it is *the idea of such a system* (Karskens, [Introduction to Leibniz's *Metafysische verhandeling*] (1981), 36).
[42] See also the conclusion to his published answer to Bayle's criticism (Leibniz, "Éclaircissement des difficultés que Monsieur Bayle a trouvées dans le système nouveau de l'union de l'âme et du corps" (1698), 523-524): "A consideration of this system will also prove that, when one gets down to the roots of things there is more reason not to believe in most schools of philosophy belief. The scant substantial reality that the Skeptics allow to sensible things; the reduction of everything to harmonies or numbers, ideas and perceptions in the Pythagorean and Platonic systems; (...) the forms and entelechies of Aristotle and the Scholastics; and even the mechanical explanation of every specific phenomenon after the manner of Democritus and the moderns etc., *all are united in a single gaze whose object exhibits its regularity and the conformity of its parts (...)*" [my italics].

commemorative lecture in 1916, Leibniz shows us that it is not struggle which fathers all things, but mutual understanding and the reconciliation of inevitable differences.

Naturally Leibniz's pursuit of a synthesis does not entirely determine the contents of his 'system'. This system only becomes transparent if we perceive the basis of his rejections of Aristotelianism and of atomism. In "Système nouveau" he outlines how he was converted from his original training in Aristotelianism to Gassendi's atomism and finally arrived at his own philosophy:

> At first, when I had freed myself from the yoke of Aristotle, I fell in with the void and the Atoms, because this is what best satisfies the imagination. But thinking again about this, after much meditation, I realized that it is impossible to find the principles of a true Unity in matter alone, or in that which is merely passive, because everything is but a collection or accumulation of parts ad infinitum. Now, the reality of a multiplicity must reside in true unities that come from elsewhere. These are something quite different to points, as it is obvious that something continuous cannot be composed of points. So, in order to get to these true unities I had to have recourse to a formal atom, since a material thing cannot simultaneously be material and perfectly indivisible, i.e., possessed of a genuine unity.[43]

What Leibniz says here, briefly, is that Aristotle, and to an even greater extent the scholastic philosophers, have certainly tried to express the unity of things, but did this with unclear and confused concepts. The atomists, on the other hand, have described the changes in nature, and have presented their conceptualizations of nature, but could not identify the true unity of nature. Thus in Leibniz's own sketch, Aristotelian-scholastic metaphysics and atomist-mechanistic physics were the two preparatory links leading to his own system. In this system, Leibniz tries to do justice to both approaches. Given this starting point, it is extremely unlikely that Leibniz would have been seeking to explain the world of space and time as an illusion. Even in 1705, when his doctrine of substantial forms had already been narrowed to the famous doctrine of monads, with the result that mechanics had been moved from the center of his thinking to the periphery, he could still write: "In fact I do not eliminate the physical, but I reduce it to what it is."[44]

---

[43] "Au commencement, lorsque je m'estois affranchi du joug d'Aristote, j'avois donné dans le vuide et dans les Atomes, car c'est ce qui remplit le mieux l'imagination. Mais en estant revenu, après bien des meditations, je m'apperceus, qu'il est impossible de trouver les principes d'une veritable Unité dans la matiere seule ou dans ce qui n'est que passif, puisque tout n'y est que collection ou amas de parties jusqu'à l'infini. Or la multitude ne pouvant avoir sa realité que des unités veritables qui viennent d'ailleurs et sont tout autre chose que les points dont il est constant que le continu ne sçauroit estre composé; donc pour trouver ces unités reelles, je fus contraint de recourir à un atome formel, puisqu'un estre materiel ne sçauroit estre en même temps materiel et parfaitement indivisible, ou doué d'une veritable unité" (Leibniz, "Système nouveau," 478-479). The last sentence quoted is given in a footnote by Gerhardt, and comes from the *printed* edition. Leibniz later expanded and changed it.

[44] "Ego vero non tollo corpus, sed ad id quod est revoco" (letter from Leibniz to De Volder, undated, probably written between 14 November 1704 and 11 October 1705, Gerh. *Phil. Schr.* II, 275).

The concept of force is at the heart of Leibniz's synthesis. It constitutes the substances and underlies the laws of motion. Although this means that force is the connection between substance and phenomenon in Leibniz's system, this connection is only possible because force itself has been differentiated into fundamental and derivative forces. Force has a remarkable double nature, like the double nature of the reality of which it is the foundation.

Is such a foundation necessary? Can mechanics not proceed using forces that do not in turn assume more fundamental forces? I do not intend to comment here on the question of whether such a foundation is necessary or not, but I do think that, in a historical perspective, Leibniz's dynamics is the most explicit elaboration of the old metaphysical approach to reality, in which the structure of reality is based on the substances that comprise its elements, and from which it is derived in an ontological sense. At the same time, Leibniz is the person who has dealt in the most consistent way with· the problems entailed by a dynamic approach. Naturally Leibniz's dynamics cannot be regarded as the only possible synthesis of scholastic metaphysics and mathematical mechanics, but it is certainly a first-rate example.

The reader might well be saying "Enough abstractions! Get to the point!" in response to all this abracadabra from the Leibnizian alchemy of force and substance. "Willingly!" is my answer, "but have patience for a few more pages before I return to the problems of Leibniz's foundation."

3.3.2   From Moving Force to True Unity

In "Specimen dynamicum" Leibniz begins his argument by repeating what he had presented in his "Brevis demonstratio" and elsewhere as the central point of his dynamics, namely the proposition that material things must contain natural forces in addition to having the characteristic of extension. These forces do not consist only of a capacity ("facultas"), they also possess a drive or urge ("conatus" or "nisus").[45] Everything that is more than pure extension is founded in these forces. Thus motion—conceived as a displacement per unit of time, i.e., as the ratio between two *finite* quantities—is not real *as a whole,* because the *parts* of the motion do not exist *simultaneously* but rather *successively.*[46] Only momentaneous velocity—what we would now call velocity—exists in a proper sense, and it consists of a *force seeking change.*[47]

Although the concept of force appears thus far to be an extension of the concept of the moving force, in his introduction Leibniz links the concept of force with two new elements that change it radically. First, he says that force is the "deepest nature

---

[45] Leibniz, "Specimen dynamicum," 235.
[46] The premise here is: "nunquam totus existit, quando partes coexistentes non habet" ("no whole can exist that does not have coexistents parts") (*ibidem*).
[47] "(...) quod in vi ad mutationem nitente constitui debet" (*ibidem*).

of bodies," since it is a "characteristic of substances to act."[48] Then he announces that he wants to reduce the forms or entelechies from the philosophy of Aristotle, which he says are unclear, to forces and so turn them into understandable concepts.[49]

Here Leibniz implicitly alludes to the question that underlies the outline of his system of forces which follows this passage. This is the question, or the quest, of the *true unity*, the "veritable Unité." In his complementary article, "Système nouveau," Leibniz addresses this question explicitly. I will therefore turn briefly to this work, before going further with the distinctions Leibniz makes within the concept of force. In the passage cited above in which Leibniz outlined his own development, he stated that mechanical philosophy could never yield the "principles of a true unity."[50] He, however, was able to achieve this by resorting to the 'formal atom'. What does he mean by 'true unity' here, and what are its principles?

Without saying so, Leibniz is operating in two different domains: the domain of Aristotelian tradition and that of the atomists and 'extensionalists', each maintaining the truth of what the other labels an illusion. This can perhaps be seen most clearly from the way each looks at the living organism. For Aristotle, the unity of an organism, and in general of every natural being, is in its *form*. A substance has both a formal and a material principle. The form, for example the form 'horse', constitutes the unity, while matter is that which receives the form: the form 'informs' the matter. But form and matter cannot exist independently of one another: they are *principles*, that is, the preconditions for the existence of a substance. Atomists in contrast saw organisms as being compounded of smaller building blocks. For them, therefore, the unity of the organism could only come about through external manipulation. The organism in itself is no more than a large number of elementary building blocks, and it is these that are the true unities.

Leibniz uses the term 'true unity' at some points as it is derived from the Aristotelian tradition, and at other points from the atomistic tradition. Sometimes it means organic unity, at other times it means the elementary building blocks. This can be seen as early as the third section of "Système nouveau." When he says that it is impossible to find the principles of a true unity in passive matter alone, the 'true unity' is *organic* unity, and the 'principles' are the *building blocks*; but when, in the following sentence, he writes that "the reality of a multiplicity must reside in true unities" the 'true unities' are the building blocks.

This double derivation is not reflected unambiguously in Leibniz's understanding of unity. After all, Leibnizian forms are not Aristotelian forms, any more than the building blocks of Leibniz are the building blocks of the atomists. His characterization of the true unities combines both origins.

---

[48] "(...) ut intimam corporum naturam constituat, quando agere est character substantiarum, (...)" (*ibidem*). The term 'character' was a foreign word in Latin. Leibniz has coined it from the Greek word 'χαρακτήρ'.
[49] *Ibidem.*
[50] See above, page 82.

However, the path Leibniz travels to get to his concept of true unities begins with only one of these origins, the atomistic. Leibniz criticizes atomism, because its building blocks, the physical atoms, are not at all true unities. After all, all matter, including a physical atom, is extensive and thus endlessly divisible. But how can a true unity be divisible? This definition of atoms, as indivisible pieces of extension, is therefore a contradiction in terms. In the same way, he criticizes the attempt to seek the true unities in mathematical points through a notional continuation of subdivision into infinity. Mathematical points may be indivisible, but they are only the extremities of extended things, mere modalities. Therefore a continuum, such as physical matter, cannot be composed of mathematical points.[51]

It is striking that Leibniz's criticism relies on an analogy between the physical continuum and the geometric continuum. Therefore it is perhaps not so strange that, on the basis of the "Système nouveau," Foucher—who was a Canon of Dijon and one of Leibniz's correspondents—considered that Leibniz's real concern was the "essential principles of extension."[52] But while Leibniz treated the mathematical continuum as analogous to the physical continuum, this does not mean that it was the same. He is not concerned with the principles of extension as such, but with the principles of *physical* extension. That is, he is seeking the principles of "effective extension, or of corporeal matter."[53]

Thus Leibniz has shown that an abstract geometric idea of matter entails an absurdity, because if it is pushed to its logical conclusion it proves to be fundamentally groundless. An atomistic idea produces a contradiction, while composition from mathematical points is impossible. Leibniz concludes from this that it is necessary to reintroduce the unpopular substantial forms.[54]

A few years later, Leibniz comes to the same conclusion via a different route, this time by examining the preconditions for the existence of laws of nature. He published the arguments for this in his "De ipsa natura sive de vi insita actionibusque creaturum, pro Dynamicis suis confirmandis illustrandisque" (1698). The last part of the title—'to strengthen and clarify its dynamics'—indicates that the article was intended as a supplement to his dynamics. It was at the same time an answer to the Cartesian Johann Christoph Sturm.

Following Robert Boyle, Sturm defended the position that only God can be the source of material movement and of the laws of motion. It is not possible for created

---

[51] In Cantor's theory of infinity, this distinction is expressed by means of various degrees of infinity: the set of points in a line segment has a degree of infinity referred to as c, that is the cardinal number of the power set of the set of all natural numbers. According to Cantor's continuum hypothesis, this is equal to Aleph1. L.E.J. Brouwer, however, criticized the identification of a line segment with the set of its points on the basis of arguments similar to those of Leibniz (L.E.J. Brouwer, *Over de Grondslagen van de Wiskunde*, Amsterdam 1907, reprint: Amsterdam 1981). I am grateful to Louk Fleischhacker for his addition to and correction of the original Dutch footnote.

[52] "principes essentiels de l'étendue" (letter from Foucher to Leibniz, 12 September 1695 in: Gerh. *Phil. Schr.* IV, 487-490, this passage at 487).

[53] "étendue effectif, ou de la masse corporelle" (Leibniz, "Éclaircissement du nouveau système," 494).

[54] "Il fallut donc rappeler et comme rehabiliter les *formes substantielles* (...)" (Leibniz, "Système nouveau," 478-479).

reality to have a force within itself.[55] In his answer, Leibniz focuses on the basis of the regularity of material motions, and especially on the basis of the principles of conservation.[56] He cites with approval Sturm's idea that the laws of motion are eternal and that they are promulgated by God as a commandment.[57] But, says Leibniz, such a commandment, promulgated in the distant past, can only have effect in the here and now if there is something that maintains it. This means either that God continually intervenes from outside, as is argued in occasionalism, or that there is an inherent law ("lex insita"), effectiveness ("efficacia"), form or force in things which ensures that the laws of nature are in accordance with what God ordained in the beginning. Leibniz argues that it would contradict God's omnipotence and absolute will if He could not perpetuate eternally what He wills, so the latter possibility is closest to the truth.

Leibniz has justified the necessity of the existence of true unities in two ways, first by a critique of the mechanistic concept of matter, and later through a critique of the foundation of the laws of motion. The common element in the two is the interaction of particles, i.e., force. Thus the consideration of matter leads Leibniz to recognize something in matter in addition to extension, that most closely resembles the soul, because it entails endeavor and intent.

Leibniz does not derive the resemblance with the soul from his critique, but rather from his conviction of the analogy between all things. In "De ipsa natura" this is quite clear. He says there, that if our spirit can produce immanent effects, there is no objection to the same force also being present in the "anima," "forma" or "natura" of other substances. In a letter to De Volder of 30 June 1704, he explains that the analogy between the various substances is universal. In reality, the nature of things is uniform, and human nature can therefore not be entirely different from the other simple substances that comprise the universe. This means that there is no reason why there should be perception or appetite in one substance and not in the other.[58]

---

[55] Johann Christoph Sturm, "De natura sibi incassum vindicata" (Altdorf 1698). See Loemker, [Commentary], 498.
[56] Leibniz, "De ipsa natura," 505-506. He identifies two such principles, the principles of the conservation of "potentia actrix," and of the conservation of "actio motrix." The name of the first of these quantities may be due to a printer's error, and should presumably read 'potentia activa', as Loemker has translated it (Leibniz, *Philosophical Papers and Letters*, 499). The "actio motrix" refers to mv^2.
[57] "(...) motus qui nunc fiunt, consequi *vi aeternae legis* semel a Deo latae, quam legem mox vocat volitionem et *jussum*" ("(...) that the motions that take place now, are in accordance with *an eternal law* which was once promulgated by God, and this law He then calls a will and *command*") (Leibniz, "De ipsa natura," 506).
[58] Letter from Leibniz to De Volder, 30 June 1704 (Gerh. *Phil. Schr.* II, 270-271).

3.3.3 A Further Analysis of the True Unities

It should now be clear why it was not possible to discuss Leibniz's differentiation of the concept of force without first introducing the underlying metaphysical endeavor: force is the essence of true unity. We will therefore deal with the way Leibniz differentiated the concept of force together with the characterization and analysis of his true unities. The best example for the purpose, and a favorite example for Leibniz, is a plant or animal organism. In fact Leibniz's true unity refers to both organic unity and the elementary building blocks. It is clear that an organism is something more than a composite of elements: it has its own unity. This unity constitutes the qualitative difference between a natural organism and a technical artifact.[59] It corresponds to the human mind or soul, because it has the characteristics of immateriality and of a certain 'perception'. However, there is an important difference with respect to dependence on matter. The soul, while it is linked to a particular quantity of matter, is not dependent on matter. Organic unity, on the other hand, cannot exist without matter: it is "sunk in matter."[60]

Thus an organism consists of a form, which determines the unity, and matter, which is composed of elements. Nevertheless—and this is an important insight if we are to understand Leibniz's true unity—this is a relative contrast, because both are true substances. The form of the organism itself is a substantial atom ("atome de substance"), as are the substances that constitute matter.[61] The difference between the two substances is only that the 'form' substance is the organizing principle of the 'matter' substances. In qualitative terms there is no distinction; both are characterized in the same way by active and passive forces.[62] However, the term 'matter', in

---

[59] Leibniz, "Système nouveau," 482. Leibniz sees organisms as "machines de la nature," but in contrast to "machines de l'art" they can be divided indefinitely without ceasing to be machines. Moreover, they have a "true unity that corresponds to what we call 'I' in ourselves" ("veritable unité qui repond à ce qu'on appelle *moy* en nous"). Leibniz therefore says, with respect to organisms, that "a natural machine remains a machine in its smallest parts" ("[u]ne machine naturelle demeure encor machine dans ses moindres parties"). This idea of matter explains how he can suppose that extension is nothing more then the "continuatio" or "diffusio" of a substance that is already striving or resisting. Leibniz's thinking about genesis seems to imply that the hierarchy of material substances—with the animals at the top and the building blocks at the bottom—is dynamic. There is a circular process in which the monads at one moment manifest themselves in higher organisms and at another are reduced to mere building blocks, for the benefit of another organism.

[60] Leibniz, "Système nouveau," 479: "enfoncée dans la matière."

[61] *Ibidem*, 482.

[62] Another interpretation is given by Daniel Garber. In broad terms, he supposes that the 'form' substance is an immaterial substance, i.e., a substantial form that constitutes a substance without any material principle. Garber's interpretation of Leibniz's philosophy of the 1680s and 1690s is intended to show that the mechanical laws govern a "real world of quasi-Aristotelian substances" (Garber, "Leibniz and the Foundation of Physics: The Middle Years" (1986), 28). However, he did not need to resort to this rather forced concept of substance. In my interpretation the forces in mechanics are also not purely phenomenal, but are based on the intrinsic nature of the substances themselves. Nevertheless, I agree with Garber where he says that the usual interpretation of Leibniz's system makes an excessively radical distinction between the phenomenal and real world, so that mechanical forces are incorrectly denied any reality. It is important to understand that the

continued on next page

the second case, is not arbitrary. In this case the substances are considered under the aspect of their external relationships in space and time, so that we can bring in concepts such as mass, extension, derivative forces (living and dead force), inertia, etc. Strictly speaking, a substance is locked up in itself, and cannot therefore have external relationships to other substances. For simplicity's sake—and I will explain below why this is possible—I will speak as if external relationships are possible. The interactions between substances give rise to the impression of matter. But this is *secondary* matter, which, in contrast to *primary* matter, is not real in the metaphysical sense. Just as an organism consists of form and matter, the individual substance also has form and matter, in this case *substantial form* and *primary matter*. These are the principles of substance.

Leibniz distinguishes three grades of substance, each associated with its own forces and laws. The substantial form is paired with the material substance; the "anima," "ame brute" or "ame materielle" with animals; and the rational soul with human beings. Of these the rational soul is the only one that is not linked to matter and which constitutes a substance in itself. This means that it is subject to *moral* laws. The other "animae," or "formae substantiales," are analogous to the human soul, but differ from it because they are "sunk in matter." Living organisms, plants and animals, like material substances, are therefore subject only to the laws of *nature*.

It will now be clear that there are two levels of reality, both built up of matter and form—or even three, if the phenomenal reality of mechanics is included. But what is complicated about Leibniz's system is that, while form and matter are the metaphysical principles of (elementary) *substances* and as such are the same as primary forces, they are only *component elements* of *organisms*. The concept of matter thus has a stratified quality, consisting of a primary level without form, where the principles can be found, and a secondary level that includes form, and where the unity of matter is constituted. These levels correspond to the 'primary matter' and 'secondary matter' ("materia prima" and "materia secunda") respectively.[63] A figure can clarify the relationship (see Figure 3.1).

By thus reintroducing Aristotelian form to mechanics, Leibniz hopes to rectify the mechanistic division of reality into mind and matter. Even the smallest particles of matter must consist of both matter and form if they are to be real. This is premised on the analogy between substantial atoms and organisms. Substantial atoms are analogous to the material souls of animals and plants, while these in turn collectively correspond to, but are lower than, the human rational soul.

---

continuation of former page
mechanical forces are in fact the connection between that invariable, isolated world of substances and the mere mutable, relational world of space and time.

[63] According to Jespers, the intermediate level of corporeal substances is postulated as a result of Leibniz's correspondence with Arnauld (Jespers, *De kracht in alles; Het mechanistisch en metafysisch systeem van Leibniz* (1997), 43).

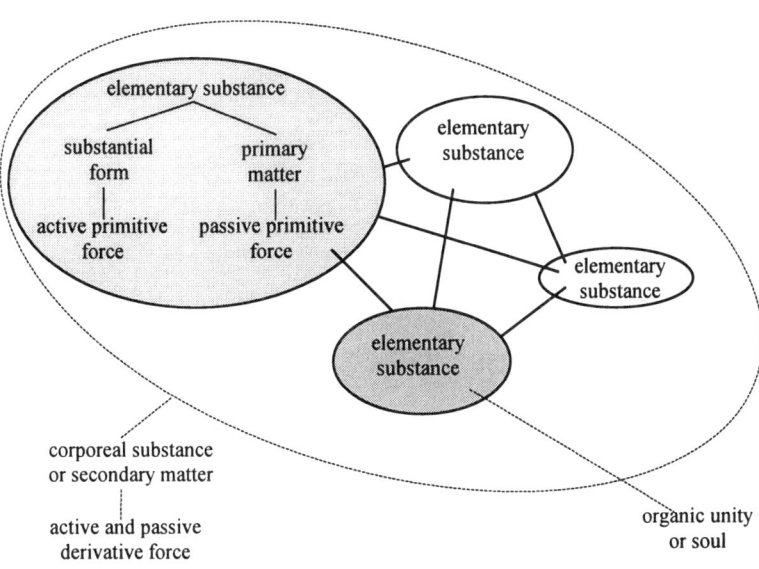

Figure 3.1 Schematic presentation of Leibniz's structure of substances

3.3.4 Force as the Essence of True Unity

As has been said, force is the essence of true unity, and thus of substance. Leibniz expresses the fundamental nature of force in the famous proposition: "agere est character substantiarum."[64] In "Système nouveau" he does not consider these forces in detail. Therefore I return to the "Specimen dynamicum" for a more precise characterization.

In the first place one has the forces of the individual substance. These forces are fundamental forces ("vires primitivae"), that is, they are "present in every physical substance on the basis of the substance itself."[65] There is an active and a passive force, these being the substantial form and the primary matter. The active force, which is the form, relates to general causes—to the sources of the things—and cannot be used to deduce specific phenomena. Leibniz makes the passive force, or primary matter, equal to the capacity to suffer (in the sense of being affected by something) and to resist.[66] It is the cause both of impenetrability, and of inertia ("ignavia") and resistance to motion ("ad motum repugnatio").[67] Together with the

---

[64] Leibniz, "Specimen dynamicum," 235. See also note 48 above.
[65] "(...) quae in omni substantia corporea per se inest (...)" (*ibidem*, 236):
[66] "patiendi seu resistendi" (*ibidem*).
[67] Here Leibniz is employing a concept of inertia according to Kepler's definition, as resistance to motion *as such* (Dijksterhuis, *Mechanization*, 314 (section IV.46)). This should not be confused with the modern concept of inertia! (See also section 2.5 above). Later, in "De ipsa naturae" (1698),

continued on next page

substantial form, serving as the active principle, they constitute a true and complete substance, or monad. Thus an individual substance always acts through its form, and suffers and resists through its matter.[68]

In an earlier article, "De primae philosophiae emendatione et de notione substantiae" (1694), Leibniz contrasted his concept of force, or at least of the active force, to the scholastic concept of force. The scholastic "potentia activa" is only the immediate possibility of action ("propinqua agendi possibilitas"), needing an external stimulus ("aliena excitatio") to result into an action ("actus").[69] The active force, in contrast, is self-activated ("per se ipsam in operationem fertur"). He claims that this makes his concept of force rather similar to Aristotle's "εντελέχεια ἡ πρώτη."[70] Thus the active force is an intermediary between capacity and action, and contains a purposeful drive.[71] Here we can see that Leibniz's concept of force refers to something that is, on the one hand, inherent to substance and, on the other hand, the cause of its action.[72]

The terms 'to act', 'to suffer' and 'to resist' are in fact misleading. As I have already mentioned, in "Système nouveau" Leibniz emphasizes that the unitary nature of the substance means that there can be no real transfer between substances: "the action of one substance on another is not an emission or the transfer of an entity."[73] This belief is expressed again in the second part of "Specimen dynamicum," where Leibniz illustrates the principle that "everything a body suffers either occurs spontaneously or is due to an internal force, even if it is on the basis of

---

continuation of former page
Leibniz did use the modern concept of inertia, yet he still claims that he has derived the concept from Kepler (see below, note 87).

[68] "(...) ob formam corpus omne semper agere et ob materiam corpus omne semper pati ac resistere" (Leibniz, "Specimen dynamicum," 237).
[69] Compare this to the metaphor with the workman's axe that Thomas Aquinas used, frequently repeated by later scholastic philosophers: "as art can give the axe its sharpness as a permanent form (...), so it is possible for a natural thing to be given its own proper power as a permanent form within it (...) [Yet as] it clearly cannot be given to the [axe] to work unless it be moved by [the craftsman,] so neither can it be given to a natural thing to operate without the divine operation (...)" (Aquino, *Quaestiones disputatae de potentia Dei* Q.3, section 7, quoted in: Keith Hutchison, "Dormitive Virtues, Scholastic Qualities, and the New Philosophies" (1991), 267).
[70] Leibniz, "De primae philosophiae emendatione," 469. The clarifying "ἡ πρώτη" had not yet been inserted; this dates from 1695 in the "Specimen dynamicum," 236.
[71] *Ibidem*, 469: "(...) inter facultatem agendi actionemque ipsam media est, et conatum involvit."
[72] Leibniz says nothing about the passive force, and it would not appear to be correct to make the same assumption as for the active force. That would mean that an action can be produced from something passive, which entails a contradiction. See Leibniz's letter to Johann I Bernoulli of 18 November 1698: "Nempe ejus, quod essentia sua mere passivum est, nullae possunt esse modificationes activae, quoniam modificationes limitant magis, quam augent vel addunt (...)" ("In fact, there can not be any active things that are modifications of something that has a purely passive essence, since modifications limit, rather than supplement or add; (...)") (Gerh. *Math. Schr.* III, 552).
[73] "l'action d'une substance sur l'autre n'est pas une emission ny une transplantation d'une entité" (Leibniz, "Système nouveau," 486). See above, page 88.

an external occasion."[74] Real transfer would contradict the indivisibility of the substance. Thus the force of the substance can only work on the substance itself.

This directly contradicts everyday experience, which shows us continual mutual influences between the things around us. The laws of mechanics also suggest changes from outside. The problem Leibniz is facing, therefore, is how these phenomena are to be saved.

Leibniz does not resort to the Spinozan solution of supposing that there is only one substance and that all other phenomena such as thinking, perception and extension are its attributes. This is not required, so long as there is an explanation for the appearance of interaction. Such an explanation is possible, if we assume that the forces that a substance exerts in itself are at any moment exactly equal to the forces that all other substances *appear* to exert on the first. This is Leibniz's "hypothèse des accords," the theory of a harmony between spontaneous action and the external world.

Nevertheless, it is a solution that solves the metaphysical problem by internalizing the external world. What we think is happening outside of us, is in reality taking place separately within each substance, each from its own point of view,[75] and can therefore also be explained from the passive and active forces that constitute the substances. Since motion ("motus localis") provides the explanation for all other material changes,[76] Leibniz concentrates first on the laws of motion and the forces relating to it, that is, on mechanics. Nevertheless, it is important to bear in mind that force does not only produce motion.

In the next section I will address what Leibniz's metaphysics means for mechanics, and especially for the principle of conservation of force.

3.4 LEIBNIZ'S CONCEPT OF MECHANICAL FORCE

3.4.1 Fundamental and Derivative Forces

How far have we now come? In mechanical philosophy, the essence of matter was considered to be extension and nothing else. All else stemmed from extension. Leibniz's fundamental criticism of this led him to reinstate force as the foundation of mechanics and of matter (section 3.2). But he went further, and united the concept of force with his concept of substance. Given this foundation, the whole of nature, including mind, can be understood and explained by analogy (section 3.3). On this basis, he now seeks to refound mechanics in such a way that explanations in mechanics could be based entirely on mechanical efficient causes, and do not need

---

[74] "(...) omnis corporis passio sit spontanea seu oriatur a vi interna licet occasione externi" (Leibniz, "Specimen dynamicum," 251).
[75] Leibniz, "Système nouveau," 483.
[76] "Nam per motum localem caetera phaenomena materialia explicari posse agnoscimus" ("Because we recognize that we can explain other phenomena as changes in position") (Leibniz, "Specimen dynamicum," 237).

metaphysical forms or entelechies.[77] But we should first see how Leibniz regards mechanics.

Space and time are not characteristics of metaphysical reality; rather, they should be regarded as the structures of phenomenal reality. This means that space and time are not absolutes, but are the way in which real substances relate to one another and to themselves. In Leibniz's answer to Bayle, he calls extension "an ordering among coexistents," and he calls time "an ordering of changes."[78] But it would not be correct to say that space and time are illusory for Leibniz. Although he frequently characterizes them as 'phenomenal', this refers, at any rate in the documents of the 1680s and 1690s, mainly to their lack of any basis in themselves. Space and time need a true unity as a foundation.[79] There is something real in the world of space and time, or rather: there is a real basis. This reality is expressed in the *momentaneous motion*, which simultaneously comprises the state at that moment and the germ of the state at the following moment; it is therefore the same as a *force*.[80]

The greatest problem now, of course, was the relationship between fundamental and mechanical forces. In mechanics we do not deal directly with these fundamental forces, but with *derivative* forces. It is through derivative forces that bodies actually act on other bodies and suffer the actions of other bodies.[81] The laws of motion therefore result from derivative forces. Leibniz explains the existence of these derivative forces in "Specimen dynamicum," but the explanation is not very clear: "[the derivative force] is produced in various ways, by a curtailment of the fundamental [force] as a result of collisions between particles."[82] Ten years later, Leibniz introduced two analogies that greatly clarify the relationship: derivative forces are related to fundamental forces as the separate elements in a series are related to its general formula,[83] or as a geometrical figure is related to extension itself.[84]

---

[77] *Ibidem*, 242-243.
[78] "un ordre dans les coëxistences," and "un ordre dans les changemens," respectively (Leibniz, "Éclaircissement des difficultés," 523).
[79] For example Leibniz, "Specimen dynamicum," 235: "In rebus corporeis aliquid esse praeter extensionem, imo extensione prius (...)" ("that there is something beyond the extension, and even prior to extension, in physical things (...)") Such a formulation does not detract from the real character of space and time, but does give it a metaphysical basis.
[80] *Ibidem*.
[81] "[Vis], qua scilicet corpora actu in se invicem agunt aut a se invicem patiuntur" (*ibidem*, 237). The adverb "actu" ('actually') is an indication of the ambivalence in Leibniz's thought.
[82] "(...) quae [vis] primitivae velut limitatione, per corporum inter se conflictus resultans, varie exercetur" (*ibidem*, 236). Literally, Leibniz is referring here only to *active* forces; but we can assume that the same applies to *passive* forces, since he says that the fundamental and derivative forms have the same relationship ("Similiter vis quoque passiva duplex est, vel primitiva vel derivativa," *ibidem*). However, his explanation is circular: the limitation of the fundamental forces is the result of the collisions of particles, while those collisions are supposed to be explained by the limited forces. In this way, what was to be explained has itself become part of the explanation. The circular nature of the explanation can be broken by distinguishing between the occurrence of collisions and the course of a collision. The occurrence of collisions leads to the curtailment of the fundamental forces, while the resulting derivative forces determine the course of the impact.
[83] Letter from Leibniz to De Volder, 21 January 1704 (Gerh. *Phil. Schr.* II, 262).

3.4.2 Active and Passive Forces

In mechanics, as on the metaphysical level, a distinction must be made between active and passive forces. Dead and living force are active forces; inertia is a passive force. On the phenomenal level, as on the fundamental level, passive forces have an unwarranted subordinate position, as compared to the active forces, in Leibniz's presentation. Leibniz used them mainly to provide a transition from the contemporaneous mechanistic approach to his own metaphysical approach.

Although Leibniz touches on the subject of passive force in "Specimen dynamicum" he does so very generally. We know that the primary passive force is the cause of impenetrability, laziness ("ignavia"), and resistance to motion ("ad motum repugnatio"). In the same way, the derivative passive force expresses itself in two ways, when a material body suffers ("patitur") or resists ("resistit") a change.[85]

We are told little about the relationship between passive and active forces on either the fundamental or phenomenal level. It is however clear that the individual law that embodies all the changes that the individual substance goes through must be attributed not only to the substantial form, but also to the primary matter. After all, primary matter is also a force, and consequently also results in a number of laws such as the law of inertia.[86] However, it is not clear here where the transition from fundamental to derivative forces lies. This is because it is not clear how primary matter could add something to substantial form. In mechanics, primary matter is simply equated with phenomenal matter, with mass and inertia. Consequently, it is hardly possible to differentiate the primary passive force from the derivative passive force.

In "De ipsa natura" (1698) Leibniz no longer speaks about the difference between fundamental and derivative passive forces; he now relates the passive force to the moving force. How does he do this?

As we have seen in Chapter 2, Galileo denied that a material body has resistance to motion. For him, matter was indifferent to motion or rest. Leibniz admits that matter cannot move or change its motion or the direction of its motion of its own accord, but rejects the idea that these characteristics entail matter's *indifference* to motion or rest. In fact, matter does have a resistance to *changes* in motion.

Leibniz provides two interpretations for this resistance. In the first place, as "inertia naturalis,"[87] it is one of the effects of the passive force of primary matter or

---

continuation of former page
[84] Letter from Leibniz to De Volder, 30 June 1704 (Gerh. *Phil. Schr.* II, 262).
[85] Leibniz, "Specimen dynamicum," 236-137.
[86] *Ibidem.*
[87] Leibniz claims to have derived this concept from Kepler. But by 'inertia', Kepler meant resistance to movement *as such*! See above, note 67.

"molis" (not of secondary matter!).[88] In the second place, it supposes perseverance in motion, that is, it supposes an *active* force.[89]

Logically, the active and passive forces are identical: the *resistance* to a *change* in motion is equal to *perseverance* in the motion itself. In view of his systematic plan, it is understandable that Leibniz presents them at this stage as two different forces, but it also shows how difficult it is, at both the fundamental and phenomenal levels, to relate the active and passive forces to one another. Sooner or later this relationship would become a problem in any consistent elaboration of Leibniz's system. Because Leibniz, in his critique of the mechanical philosophy, is focusing mainly on the active force, he naturally tries only to find a mathematical expression for the active force at this point. This means that he does not immediately have to deal with the problem of the relationship between active and passive forces.

Returning again to the "Specimen dynamicum," we see that the active force must be sought in the momentary drive towards change and that it is proportional to the mass. In mechanics it is called the moving force ("vis motrix"). It is present in two forms, as living and dead force. The living force ("vis viva") is linked to actual motion. The dead force ("vis mortua") only has the drive ("sollicitatio") to motion, without actually being accompanied by motion. Leibniz uses the centrifugal force that drives a stone in a whirling sling outwards as an example of the latter. Gravity and the elastic force of a stretched spring are also dead forces. The living force for him is the ordinary force, that is, the force that is effective in an impact.[90] This is the same force that Leibniz quantified in his "Brevis demonstratio" by assuming it to be proportional to the product of the mass and the maximum increase in height. This means that what I concluded above regarding moving force, that it is a cause of change, applies equally to Leibniz's concept of living force.[91]

3.4.3   Conservation of Living Force

Now that Leibniz has redefined moving force as a living force and therefore a derivative active force, a new problem arises. How can the conservation of living force be derived from the fundamental forces? The supposition of the "Brevis demonstratio," that living force is transferred from one body to the other, no longer applies. Leibniz's metaphysical views have now made this impossible. Nevertheless, in *mathematical* terms, it must still comply with the same transference relationships.

This dilemma is discussed a few years later in Leibniz's correspondence with the Cartesian De Volder. The problem here is that the fundamental force acts only on

---

[88]   However, Leibniz does not attempt a quantitative determination of inertia. On the basis of what he himself says, one would not be justified in saying anything more than that inertia is proportional to the mass, and thus to the size, of the body.
[89]   Leibniz, "De ipsa natura," 510-511.
[90]   Leibniz, "Specimen dynamicum," 238-239.
[91]   See section 3.2, page 81.

the substance itself, but the derivative active force appears to extend outside the substance. This force acts not only on the material particle of which it is a part—where it serves to maintain its motion—but also on other particles, by changing their motion in an impact. Leibniz now explains that the living force in each particle really is conserved in an impact, at least if we consider the motions in comparison to the center of mass of the colliding particles.[92] This is because, in relation to the center of mass, the quantities of motion are of equal but opposite magnitudes and the velocities before an impact are equal and opposite to those after the impact.

Thus in mathematical terms Leibniz has a good argument, but it does not really solve his problem. There are two points that have not been adequately explained. In the first place, a consequence of his solution is that phenomenal space must contain a subset of reference spaces, which are more real than other spaces. In every collision there is one moving space that is as it were the point of view from which the substantial reality is mathematically visible. His 'true measure of living force' should therefore correlate to a true mathematical space, that is, to true motions on the part of the particles. However, this contradicts his idea of time and space as relative orderings that do not have any reality in themselves. Thus Leibniz's derivation of the conservation of living force threatens to undermine the basis of his distinction between phenomenal and substantial reality.

In the second place, Leibniz has always taken the conservation of living force primarily in the sense that the *sum* of the living forces is conserved within any reference system. In a reference system, chosen at random, the living forces of the separate substances do usually change. The constancy of the *sum* of the living forces follows from the conservation of living force in each substance only in a *mathematical* sense. However, he does not discuss how it should be made consonant with the unchangeability of the fundamental forces in a *metaphysical* sense, for example by means of pre-established harmony.

The above problems originate in the conflict between a particular form of the metaphysics of substance and the mathematical regularity in mechanics. The most important problem as regards Leibniz's connection of force and substance is the isolation of the substance in itself. The principle of the conservation of living force cannot be rooted in individual substances, but must be sought in a higher unity that links the different reference spaces.

3.4.4 *The Integration of Dead Force*

Leibniz's metaphysical idea of force as the essence of substance also has implications for his idea of the relationships between derivative forces. These implications may be less striking than those discussed above, but they are even more important for the history of mechanics. Although Leibniz had already made a distinction

---

[92] Letter from Leibniz to De Volder, 20 June 1703 (Gerh. *Phil. Schr.* II, 248-253, see especially 251-252).

similar to that between living and dead force before he read Newton's *Principia mathematica*,[93] at that time he used this distinction only to distinguish the laws of static equilibrium from those of motions and impacts. Perhaps inspired by Newton's successful attempt to derive laws of motion from the action of external forces, Leibniz now adds a discussion that is intended to clarify the relationship between dead and living force. However, he does not mathematize this relationship, limiting himself to a half worked-out, qualitative formulation. This is even more striking if we realize that he saw the measuring of living forces as the gateway to metaphysics.[94]

How does Leibniz link living and dead forces? Living force "arises from infinitely many sequential impressions of dead force."[95] After a certain period of fall it is infinite, compared to the simple drive ("simplex nisus") of gravitation. This formulation certainly points to a relationship of integration. Nevertheless, Leibniz is somewhat vague about it at this point, causing later interpreters to read the modern idea into the way he expresses himself here. The modern idea would be that dead force must be integrated *over time or over distance* to obtain living force: $F_{living} = \int F_{dead} ds$ (or dt). But if Leibniz had actually intended this then he would certainly have indicated whether dead force should be integrated over time or distance. Integration over time results in a quantity that is in fact proportional to the change in the quantity of motion (mv), while the result of integration over distance is proportional to the change in living force (mv^2). Probably he would also have supplemented the qualification 'infinite' with a comparison, as in the "Système nouveau," for example by comparing it to the relationship between a point and a line, so as to indicate what sort of infinity this is.

Why did Leibniz not do this? Westfall has suggested that it might be because of Leibniz's metaphysical idea of derivative force. The derivative force is a momentary state in the law that controls the development of the substance. Integration of the momentary efforts over time expresses this law in a mathematical manner, and would therefore make sense, while integration over distance would not.[96] But this assumes greater consistency in Leibniz's system than it actually has. Did not Leibniz define the quantity of living force with reference to a *distance,* namely the maximum increase in height, without considering the time required? What would then be more natural than to define the action of the *dead* force in relation to distance as well?

There is however another reason: Leibniz imagined this integration as the aggregation of infinitely small units of the dead force, that is: $F_{living} = \int dF_{dead}$. The dead force results in a continual series of actions, each of which is infinitely small, and

---

[93] See above page 76, especially note 19; Leibniz apparently read Newton's *Principia* for the first time in Rome in the autumn of 1689 (Aiton, *Leibniz*, 160).
[94] "(...) mihi visum est hanc esse portam, per quam transire e re sit ad Metaphysicam veram (...)" (letter from Leibniz to De Volder, undated (Gerh. *Phil. Schr.* II, 195)). Loemker dates it in 1699 (Loemker, [Commentary], 521).
[95] "ex infinitis vis mortuae impressionibus continuatis nata" (Leibniz, "Specimen dynamicum," 238).
[96] Westfall, *Force in Newton's Physics,* 301.

therefore only has the drive to motion, without this being coupled with actual motion. But if there are no hindrances, these actions will change into a living force that is linked to a *finite*, and thus actual, velocity.

It is evident that Leibniz pictured the integration in this way from the analogies with which he began his explanation of the integration of dead force. These analogies relate to *kinematic* quantities, which are only 'mathematical entities', but are nevertheless good means for making accurate calculations.[97]

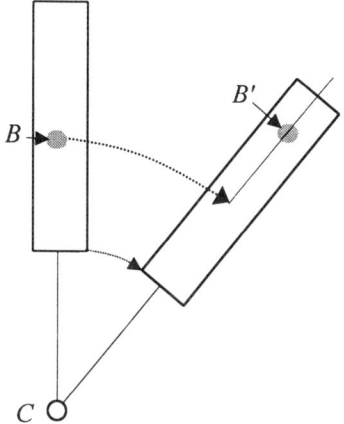

Figure 3.2 Impetus arises from infinitely many infinitely small conatuses

The first analogy is of the integration of impetus (mv) over time. His description of the process is that the result of the motion "is born by multiplying the impetuses, equal or unequal, that exist in the moving [body] during a [particular] time, with the time that they successively remain in them."[98] This is referring unmistakably to the integral $\int mv dt$, which Leibniz calls the 'quantity of motion'.

Leibniz's second analogy is the way impetus itself arises from the endless repetition of certain elements, which are only vaguely indicated. He illustrates this from the arrangement of a tube turning in a horizontal plane around a center C (see Figure 3.2). There is a ball, B, which is at first held in place in a tube while the tube turns. Suddenly it is freed to move freely in the tube. The "conatus" from the center (i.e., the radial velocity) immediately after the release is infinitely small in comparison to the velocity of rotation, or "impetus" in the words of Leibniz. In the same way, he says, the drive towards movement has two forms: one is an infinitely small drive

---

[97] Leibniz, "Specimen dynamicum," 238.
[98] "ex aggregato impetuum durante tempore in mobili existentium (aequalium inaequaliumve) in tempus ordinatim ductorum nascatur" (*ibidem*, 237).

("sollicitatio"), the second is the impetus itself, that is formed from the continuation or repeat of the elementary drives.[99]

In the second analogy it is striking that while the elements and the results of the integration differ infinitely in *magnitude*, they do not differ in *sort*. In fact both are a drive to motion. The integration is a cumulation or aggregation of infinitely many infinitely small elements. There is no question of integration over time or distance: this is integration over the variable 'sollicitatio' itself.

It appears natural to understand the integration of dead force as analogous to the origination of a finite velocity from infinitesimal velocities in the case of the rotating tube. After all, dead force is just like the first conatus in the bodies themselves, and so is similar to the product of mass and velocity. Living force, just like impetus, results from the aggregation of the successive conatuses.

Leibniz explicitly adds that living force and impetus are not proportional.[100] This remark appears to contradict what was said above, because the proportionality of dead force with the first conatus appears to imply that living force and impetus are also proportional. In fact this observation does not contradict my interpretation, but rather represents a contradiction in Leibniz's system itself. He recognizes that there is a proportionality between dead force and conatus in the mechanics of the five simple machines, but adds that this proportionality is due to the special condition that, when the movement is beginning, the distance traveled is proportional to the attained velocity.[101] Once a certain velocity has been built up, the proportionality no longer applies, because the obtained velocities are only proportional to the elements of the distance traveled, and not to the total distance traveled. However, living force is proportional to the total distance traveled, and thus differs from the velocity attained.

The contradiction in Leibniz's reasoning is visible here. On the one hand the transition from dead to living force is analogous to that from the first drive to impetus. On the other hand the proportionality of dead force to the first conatus is not reflected in the proportionality of living force to impetus.

Thus in my interpretation of Leibniz's linking of dead and living forces, I conclude that there is inconsistency in his thinking. Naturally this is a drawback, compared to Westfall's interpretation. But the question is, whether an inconsistency in this case really presents a difficulty. The concept of the infinitely small was understood in an intuitive rather than rational or systematic way until well into the eighteenth century. We can see this in the way Newton made the transition from impact forces to continuously acting forces, as described in Chapter 2. If Newton could overlook the incommensurability of these forces, although according to modern opinions he should not have been able to avoid facing it in his mathematical

---

[99] "formatum continuatione seu repetitione Nisuum elementarium" (*ibidem*, 238):
[100] *Ibidem*, 238-239.
[101] The term 'the five simple machines' refers to the five principles of the transmission of force known from antiquity, the lever, windlass, pulley, wedge and screw.

formulation, would it be reasonable to expect Leibniz to recognize the similar incommensurability between dead and living forces, in a treatment that was formulated only *qualitatively* and with general analogies?

However, a contradiction requires a basis. To identify it in this case, I refer back to my argument in Chapter 2, and to Leibniz's concept of force as explained in this and the previous section. Dead and living force are both derivative active forces. These reflect the momentary state of a substance and are related to the law of substances as the elements of a series are related to its formula. They consist of a drive to motion and thus are also the direct cause of their own development. The only difference is that one force is associated with actual motion, and the other with infinitesimal motion. They are thus the same in quality, but infinitely different in quantity. One can thus originate from the other through continual repetition, just as Newton's continuous acting force arose through the continual repetition of infinitely small forces of impact.

We see here that, although Leibniz distinguishes between dead and living force in a way that is reminiscent of the modern distinction between force and energy, the original unity in a qualitative sense nevertheless retains its validity and, what is more, its self-evidence. Leibniz's approach to forces thus forms a *transition* in the development of the concept of force. In a *quantitative* sense he uses the modern distinction between external force and kinetic energy, while in a *qualitative* sense the original unity of the concept of force is retained.

One important advantage of this interpretation of Leibniz's integration of the dead force is that it preserves the relationship between the concept of the living force and the scholastic idea of impetus, and so highlights the unity in Leibniz's concept of force, while also showing how the modern distinction can grow from that made by Leibniz.

3.5 CONCLUSION: SIMILARITIES BETWEEN NEWTON'S AND LEIBNIZ'S CONCEPTS OF MECHANICAL FORCE

Looking back at this and the previous chapter—for both Leibniz and Newton have, in their own way, reinstated the concept of force in mechanics—it is clear how much the mathematization of the concept of force in the eighteenth century took the substantial concept of force as its starting point.

Both Leibniz and Newton picture force as simultaneously the cause of external change and the cause of motion of the particle itself. The internal cause of motion is qualitatively equal to the external cause of change in motion. Integral calculus is used in such a way that the integrand and integral are qualitatively alike. The integrand is not integrated over time, distance, or anything else: the integral is no more than the aggregate of qualitatively equal, infinitely small parts. We can see this in Newton's mathematization of continuously acting forces, which treats these as an approximation to the limit of infinitesimal impact forces. For Leibniz, the dead force is living force with infinitesimal motion.

We could say that there is a primary and a secondary force: the secondary force is a derivative of the primary force and is mathematized accordingly. In Newton's case the primary force is the force of impact, in Leibniz it is the living force. Although the force of impact is in the first place an external force, while the living force is in the first place an internal force, both forces are similar to their correlate. The force of impact is analogous to internal forces, and the living force is analogous to external forces. Newton's innate force is both momentum and resistance to change in motion, depending on the observer's point of view. The momentum that an external force imposes on a particle is thus itself a force. Leibniz's living force is the cause of motion, and is at the same time the force acting in an impact and through which particles act upon one another.

Thus in mathematizing force, Leibniz and Newton assume the qualitative identity of internal and external force. As we have seen in Chapter 2, this had its historical origin in the scholastic identification of the force exerted by a pressing weight and by a body in motion. As Chapters 4 and 5 will demonstrate, the mathematization of force in the further development of mechanics in the eighteenth century will appear to abolish this identity, but at the beginning of the century it could still be upheld without objections. The manner in which integration was presented mathematically, as an aggregate of elements that are all similar to the result, made it possible to treat integration in accordance with this metaphysical principle.

One important component of this metaphysical principle is that forces can be attributed to substances. Both Leibniz and Newton combine the passive force of resistance to change in motion with the active force of perseverance in motion, inertial force and living force respectively. This reflects the scholastic idea of impetus: a body can only be in motion because an inner orce is continually causing this motion. Newton does not explicate this connection further, but Leibniz does, by presenting force as the essence of substance. In both cases the link means that the force that is lost in one substance can, and for Leibniz *must*, be transferred to another substance by the *action* of force.

The parallel between Newton and Leibniz can be depicted in a figure (see Figure 3.3). The idea underlying this figure can be simply presented using the water metaphor, as I showed in Chapter 2 for Newton's concept of force. Although Leibniz explicitly opposed the idea that substances affect one another, one function of his pre-established harmony was to explain how interactions can *appear* to be real. Therefore the task facing Leibniz differs from that of Newton in metaphysical, but not in mathematical terms. For both of them, the idea underlying the mathematization of force is that the cause of a *change* in motion in other particles must be traced back to some property of the particle itself, namely its *perseverance* in motion. The external force is that which *transfers* this internal force while at the same time it is itself the result of that internal force. Thus the internal force is like the water filling a barrel. When it is transferred to another barrel it is transformed into the internal force of that other barrel. However, the rate at which it flows to the second barrel depends on the internal force in the first barrel (see Figure 3.4).

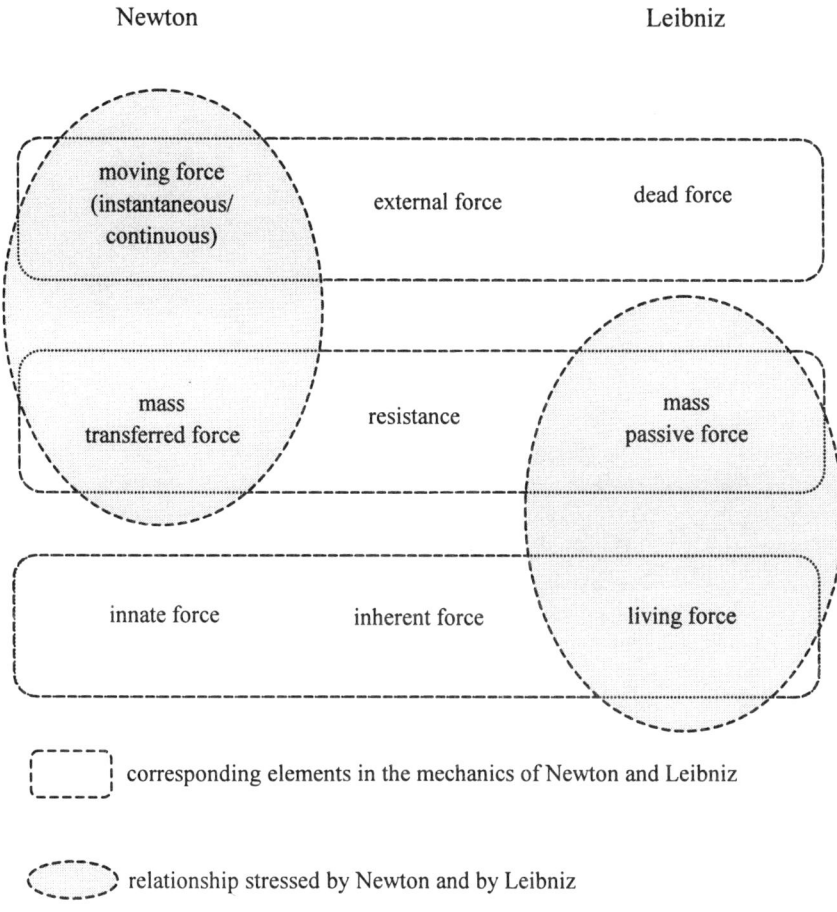

Figure 3.3 Parallels between Newton's and Leibniz's concept of force

The concept of substance takes many forms, and this applies equally to the connection between force and substance, which can be conceived in various ways. Leibniz's method is certainly not the only possible approach. Rather, his metaphysical approach to mechanics should be seen as a superb attempt at a coherent explication of this link, which in Newton's work remained implicit and so could later be buried, as it were, under a mathematical apparatus.

For Leibniz, substance is the individual thing, complete in itself, that incorporates all of its characteristics. This means that the connection between substance and force as Leibniz conceives it, must from the beginning support a well-nigh unbearable weight. He introduced the concept of force in his dynamics as the foundation of the material, mechanical world so as to explain the interactions of particles and their regularity. But force, when understood as the essence of substance, proves not to be capable of producing such effects.

The distinction between the fundamental and derived levels of reality provides a partial solution for this paradox, but also introduces a new problem into the promised foundation of the conservation of living force: the general conservation of living force must be based on conservation *per substance*.

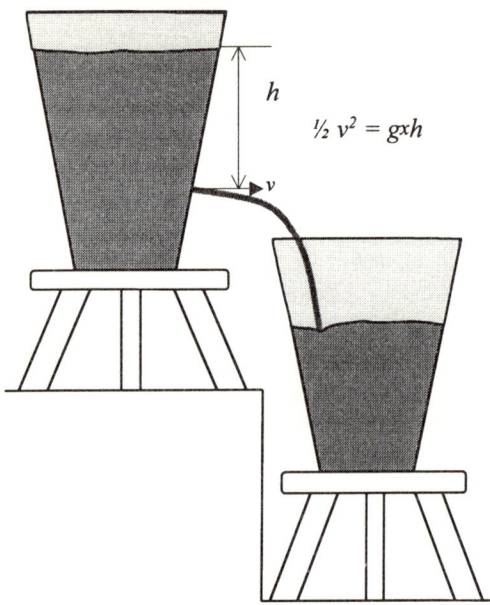

Figure 3.4 Metaphor of the action of force with the pouring of water from one barrel into another

The historical background to the controversies about the concept of force in the eighteenth century reveals what was at stake in metaphysical terms—which is not to say that this was equally clear for those involved. It was a question of the implicit unity of the concept of force, as both cause and transference, and along with this, sometimes implicitly and sometimes explicitly linked, the question of the foundation of this unity in substance. In the century between Leibniz's "Brevis demonstratio" and Newton's *Principia mathematica*, on the one hand, and Lagrange's *Méchanique analitique* on the other hand, both premises—as regards mechanics—melted like snow in the springtime. This came to the fore in the controversy concerning the true measure of force, leading to the distinction between 'force', 'momentum' and 'energy' that is now generally accepted. Let us then see what happened to the concept of force in this controversy, in a metaphysical sense.

B

TOWARDS A NEW METAPHYSICS

CHAPTER 4

FROM CAUSE TO PHENOMENON
D'Alembert's Foundation of Mechanics

> *The first causes are absolutely unknown to us, but they are subject to simple laws and constants, to be discovered by observation, and whose study is the aim of Natural Philosophy.*
> Jean-Baptiste Joseph Fourier[1]

4.1   INTRODUCTION

In the previous chapters I have shown that in Newton's view, force was a substance, and in Leibniz's view analogous to it. This means that the underlying idea about the action of force was that force is the cause of changes in motion and is simultaneously the change itself, which is, as it were, transferred. It would, however, be wrong to consider this double meaning as vagueness or ambiguity, since these are two aspects of one and the same thing. It is more appropriate to call the concept of force a unity-in-duality. To clarify this, I have used the metaphor of water pouring from one barrel to another. This metaphor justifies my calling this a remnant of the substantial concept of force in scholastics. After all, when scholasticism turned to the concept of impetus for a solution to the problem of motion, the cause of the *perseverance* of motion was equated to the cause of the *change* in motion.

This unity-in-duality in the concept of force provides us with a common denominator for some of the idiosyncrasies of Newton's mathematization of continuously acting external forces, and of Leibniz's attempt to mathematize living forces. For example, Leibniz's distinction between dead force and living force referred only to a quantitative difference—he had no intention of denying their qualitative similarity. Nonetheless, in retrospect one could say that this distinction had the potential to deconstruct the unity of force. After all, later thinking on the role of space and time in the integration of dead force to living force showed that a qualitative distinction

---

[1]   "Les causes primordiales ne nous sont point connues, mais elles sont assujetties à des lois simples et constantes, que l'on peut découvrir par l'observation, et dont l'étude est l'objet de la Philosophie naturelle" (Fourier, *Théorie analytique de la chaleur* (1822), v (opening sentence of the "Discours préliminaire")).

must logically follow the quantitative. In modern physics, this qualitative distinction is expressed through a dimensional distinction.

Looking back on the history of the concept of force, it is interesting to ask how this unity-in-duality was broken down, and how this could occur so thoroughly that the original unity can now be recognized only with great difficulty. An analysis of d'Alembert's significance for the development of the controversy of the true measure of living force, and especially his conception of force, offers a good starting point for answering this question.

4.2 BETWEEN LEIBNIZ AND D'ALEMBERT: A CHANGE IN THE DEFINITION OF THE PROBLEM

4.2.1 Atomism and Conservation Theories

The collision process served as a touchstone in the search for the true measure of force and its conservation. The change in the magnitude and direction of the velocity of the colliding particles necessarily implied that the force of the particles also changed. If one knew the laws governing impacts, it should be possible to discover a conserved quantity, which might serve as the true measure of force. Conversely, if one knew what the true measure of force was, one would also know what conditions the laws of impact had to meet. Therefore the search for the correct laws of impact was linked to the question of the true measure of force.

This link became more complicated when an attempt was made to bring the laws of impact into accordance with the process of collision. This resulted in serious difficulties regarding the concept of hard particles. In 1970, Wilson Scott described this development in the first chapters of *The Conflict between Atomism and Conservation Theory (1644–1860)*. Atomism played an important role in his interpretation, especially in the shape it was given by Newton.

In seventeenth century mechanical philosophy, there were four fundamental polar characteristics of material particles: soft-hard, elastic-inelastic, penetrable-impenetrable, and divisible-indivisible. However, the way these polar characteristics were linked was not immutable. According to some, a hard particle was elastic, according to others, it was inelastic; in the same way impenetrability could be paired with softness or hardness, or with elasticity or inelasticity.

Although Descartes described his particles as hard and impenetrable, he also said that they were brittle, i.e. that they were infinitely divisible. Hard inelastic particles could therefore rebound if they collided with one another. Conservation of mv, regarded as a scalar quantity, was thus not in conflict with his assumptions concerning the fundamental properties of material particles. For Leibniz, however, it was. Leibniz likewise viewed matter as infinitely divisible, but in his view this was paired with elasticity as a basic property. Consequently, mv^2 is conserved in impacts and he could even say that during *apparently* inelastic impacts, living force was transformed into the motion of the smaller internal particles of the colliding bodies. The 'directive' force was transformed into a 'relative' force.[2]

The discussion was greatly influenced by Newton's 1706 definition of hardness. In Query 23 of the Latin edition of his *Opticks*, he postulated his atom theory in which he characterized both hardness and softness as the incapacity to rebound:

> For Bodies which are either absolutely hard, or so soft as to be void of Elasticity, will not rebound from one another.[3]

Regardless of what Newton's exact intentions might have been with his Queries, it is certain that this characterization had a great influence on the discussion concerning the measure and conservation of force. This was because hardness, defined in this way, became a great obstacle for the theory of conservation of mv^2. However, it was not until after the publication of the sensational Leibniz-Clarke correspondence in 1717 that Newtonians became interested in the problem, although it is unclear whether this *post hoc* also entails a *propter hoc*.[4] Although Newton's characterization of hard bodies appeared to imply clear-cut support for Descartes's position, Newtonians were actually divided between the two positions, i.e. the Cartesian standpoint (conservation of mv), and the Leibnizian standpoint (conservation of mv^2).[5] The controversy now began in earnest.

---

[2] See also Westfall, *Force in Newton's Physics: The Science of Dynamics in the Seventeenth Century* (1971), 295.

[3] Newton, *Opticks: or, A Treatise of the Reflections, Refractions, Inflections & Colours of Light* (1704), 398. In the first English edition of 1704 there were only 16 Queries; Queries 17-23 were added to the Latin translation of 1706, were subsequently not included in the second English edition of 1717, but were published again in the third edition in 1721 and the following editions. In this and later editions, Query 23 is given the number 31. Here I have cited from the fourth English edition published in 1730. Papineau says that Newton's definition of hardness was already present in Wallis's prize essay about the laws of impact from 1669 (Papineau, "The *Vis Viva* Controversy" (1977), 121). However, although Wallis did make a remark of this type, it did not have much influence on the generally accepted view of hardness. For example, in his controversy with Leibniz between 1692-1694, Huygens maintained his view that hard bodies may rebound (Scott, *The Conflict between Atomism and Conservation Theory (1644-1860)* (1970), 13-15).

[4] A causal connection is made by Pulte, who refers mistakenly to Scott, *The Conflict*, who certainly did not interpret the relationship as causal (Pulte, *Das Prinzip der kleinsten Wirkung und die Kraftkonzeptionen der rationalen Mechanik* (1989), 66). Compare Papineau, "*Vis Viva* Controversy," 132-136.

[5] The Newtonian MacLaurin defended the conservation of scalar mv, the Marquise du Châtelet and Maupertuis, after initially being unclear about their positions, became advocates of the conservation of mv^2 around 1738. Beginning in 1722, 's Gravesande also supported the view that mv^2 remains conserved, to the great disappointment of many Newtonians who had found much support in his *Physices elementa mathematica experimentes confirmata. Sive introductio ad philosophiam newtonianam* (1720-1721). Hankins's idea that the "Newtonians joined the Cartesians in combating *vis viva*" following 1717 is therefore, generally speaking, incorrect (Hankins, "Eighteenth-Century Attempts to Resolve the *Vis Viva* Controversy" (1965), 282). However, there is something to be said for a shift from the Cartesian-Leibnizian opposition to the Newtonian-Leibnizian opposition, as well as for a geographical shift, as appears from a sketch made by Thomas Reid for his "An Essay on Quantity" from 1748; in this sketch he proposed that the controversy about the true measure could be solved by henceforth defining force as mv^2 on the continent and as mv in the English speaking regions (cited by Laudan, "The *Vis Viva* Controversy; A Post-Mortem" (1968), 139).

The Paris Academy of Sciences held two essay competitions for the years 1724 and 1726, concerning the laws of impact for hard and elastic bodies, respectively.[6] In their submissions to the competition, the Newtonian Colin MacLaurin and the Leibnizian Johann I Bernoulli based the laws of conservation on an analysis of the concept of hardness. MacLaurin used Newton's definition: hard bodies are "those whose parts do not yield at all during a collision"; he thus viewed hardness as a limit of non-elastic softness, and concluded from this that the *scalar mv* remains conserved only during collisions in the *same* direction.[7] However, like many other Newtonian *mv* supporters, MacLaurin no longer defended the *conservation* of force, but the proposition that *mv* is the *true* measure of force. Johann I Bernoulli based his analysis on Leibniz's continuity principle, which states that instantaneous changes in velocity are impossible. Bernoulli therefore did not interpret hardness according to Newton's characterization but rather as a limiting case of elasticity, concluding from this that the conservation of mv^2 is universally valid.[8]

Scott interprets this difference in answers as a conflict between atomism, at least in its Newtonian form, and conservation theories.[9] However, this resulted in a shift in the original problem. The question of whether there is a quantity that is *conserved*—and if so, which quantity—acquired an intrinsic interest, apart from the question of which quantity expresses the *true measure* of force. While some authors were interested primarily in the problem of how the qualitative concept of force should be mathematized, others were occupied mainly with the problem of which quantities—not necessarily linked with the concept of force—remained constant over time.

4.2.2 Experimental Modes of Response

Up to this point, we have only discussed the metaphysical and mathematical sides of the controversy: the question concerning the perfection of God, the definition of the concept of force, the nature of matter, and problems surrounding integral calculus. Around 1720, the discussion was unexpectedly given a new impulse: an attempt was made to find an answer through experimentation. Although this did not alter the basic principles of the discussion, it was greatly clarified by being made more tangible.[10]

---

[6] I derived the information about these competitions from Scott, *The Conflict*, Chapter 2.
[7] "ceux dont les parties ne cèdent point du tout dans le choc" (MacLaurin, *Démonstration des loix du choc des corps* (1724) (from Scott, *The Conflict*, 43n.16)).
[8] Johann I Bernoulli, *Discours sur les lois de la communication du mouvement* (1724–1726) (paraphrased in Scott, *The Conflict*, 43n3.).
[9] Scott, *The Conflict*, Chapter 2.
[10] According to Hankins, experimentation meant the "most decisive progress" in the controversy. It was very significant for d'Alembert and Desaguliers, for instance (Hankins, *Science and the Enlightenment* (1985), 31).

I will mention only the experiments conducted by Willem Jacob 's Gravesande during the 1720s.[11] Soon after the publication of his explanation of Newton's *Principia mathematica*, which was an important contribution to the acceptance of Newton's gravitational theory on the European continent, 's Gravesande turned to an investigation of the issue of living force. With an ingenious apparatus he determined the size of the hollows made by copper balls falling onto clay. The volume of the hollows proved to be proportional to the quantity mv^2 of the balls. In this way 's Gravesande became convinced that 'living force' was not just a word, but something real. He later wrote that, although the term 'force' often has many meanings, "what had been a mere argument about words at first, became an argument about the thing itself" in relation to collision processes.[12]

His conclusions caused a tumult among the Newtonians. The already waning dividing line between the competing camps of Newton and Leibniz supporters was under an even greater threat! In the following years 's Gravesande's success led other Newtonians to free themselves from Newton's metaphysical views.[13] The theological principles on which Newton based his rejection of the conservation of living force lost their absolute value. When 's Gravesande responded to Clarke's attacks, he emphasized that he followed the *method* and not the *person*.[14]

However, the 'method' also demonstrated the opposite. If two inelastic balls collide with one another, they come to rest if the quantity of motion in both balls is exactly equal. In 1722, 's Gravesande also carried out this experiment. Perhaps to force nature to pronounce judgment, he then developed an experiment in which *both* effects, the change in velocity and indentation, were achieved simultaneously. He allowed two equally large hollow copper balls with different masses, but with the same quantity of motion, to collide with a motionless hanging clay ball. His conclusion was that a moving body is associated with two effects: indentation and change in velocity, and accordingly with two causes, force and inertia, respectively.[15]

---

[11] 's Gravesande published his findings in "Essai d'une nouvelle théorie sur le choc des corps" (1722), "Sup[p]lement à l'essai sur le choc des corps" (1722), in the editions of his *Introductio* after 1723, and in "Remarques sur la force des corps en mouvement & sur le choc" (1729). He later developed yet another experiment which he published in "Nouvelles experiences sur la force des corps en mouvement" (1733). Concerning the apparatus used by 's Gravesande for his experiments see: De Clercq, "The 's Gravesande Collection in the Museum Boerhaave, Leiden" (1988). Even earlier than 's Gravesande, the Italian Giovanni Poleni performed similar experiments (Poleni, *De castellis per quae derivantur Fluviorum aquae (...)* Padua: Comini 1718). It is interesting that for him the context of the problem was not theoretical but technological: the management of river flows. In 1733 's Gravesande published a summary of Poleni's findings in his *Journal historique* (Poleni, "Epistolae mathematicae, second extrait," *Journal historique de la république des letters* 1733 (2, 2), 220-229). For the relationship between the experiments of Poleni and 's Gravesande see: Maffioli, "Italian Hydraulics and Experimental Physics in Eighteenth-Century Holland. From Poleni to Volta" (1989), 263-266 and De Clercq, *At the Sign of the Oriental Lamp* (1997), 169-171.
[12] "ce qui n'étoit d'abord qu'une Dispute de Mots, devient une Dispute sur la Chose même" ('s Gravesande, "Remarques sur la force," 197).
[13] See also above, note 5.
[14] 's Gravesande, "Remarques sur la force," 195.
[15] 's Gravesande, "Nouvelle experiences."

This shows that 's Gravesande clearly did not doubt the usefulness of the search for the true measure of living force. While his experiments did clarify what could be considered as an *effect*, they did not address the *concepts* of force and cause. His experimental discoveries and determinations of various effects did not lead him to conclude that the controversy was pointless; he simply divided force up, without accounting for the consequences.

For him the problem was not that he had to make a choice between mv and mv^2 as measures of force, but that *both* quantities must be given a place in mechanics. Thus we cannot say that he regarded the conflict as an argument about words, but rather that he wanted to prevent it from *becoming* an argument about words. Rather than attempting to solve the debate, he tried to clarify it.

4.2.3   The Importance of d'Alembert in the Vis Viva Controversy

At the time of the publication of d'Alembert's *Traité de dynamique* (1743, second edition 1758), the *vis viva* discussion was still vital and had in no sense been concluded. This is evident in the work on this topic by Boščovič from 1745 and 1758. Boščovič was well aware of the relationship between dead force F, the quantity of motion mv, and living force mv^2. Nevertheless, on metaphysical grounds he wanted to acknowledge only dead forces (gravitation and the like) as actually existing, in addition to motion itself. Although he realized that both mv and mv^2 are conserved, he interpreted mv as the true measure of force and mv^2 as only a mathematical quantity.[16] The publications of Immanuel Kant, Leonhard Euler, the Marquise du Châtelet, and Abbé Jean-Antoine Nollet, all during the 1740s, also illustrate the great importance of the problem of living force as an issue in both natural philosophy and mechanics.[17]

The answer provided by d'Alembert, that the controversy was merely an "argument about words,"[18] has been considered by many, up to the present day, as a triumph of natural science over the traditional link between metaphysics and natural science. Before 1743, the year of publication of the first edition of d'Alembert's *Traité de dynamique*, so the argument runs, no one had seriously thought of calling the controversy about the true measure of force a futile dispute about words.[19] Even

---

[16]   Rogerius Josephus Boščovič, *De viribus vivis dissertatio* (Rome: Monaldini 1745) and *Theoria philosophiae naturalis* (Vienna: Bernhard 1758, Venice ^{2}1763). Concerning Boščovič see Hankins, "Eighteenth-century attempts," 291-297; Pierre Costabel, "Le *De viribus vivis* de R. Boscovich" (1961); Jammer, *Concepts of Force: A Study in the Foundation of Dynamics* (1957), 170-178.

[17]   Immanuel Kant, *Gedanken von der wahren Schätzung der lebendigen Kraft* (1747); the Marquise du Châtelet, *Institutions physiques* (1740); Thomas Reid, "An Essay on Quantity" (1748); Leonhard Euler, "De la force de percussion et de sa véritable mesure" (1745); abbé Jean-Antoine Nollet, *Leçons de physique experimentale* (^{2}1749). See also Laudan, "Vis Viva Controversy," 132-135.

[18]   "dispute de mots" (d'Alembert, *Traité de dynamique*, xxii).

[19]   An idea from the rather grandiose-minded Szabó deserves to be briefly discussed here, especially because a much better historical craftsman like David Speiser copied the relevant passage in a shamefully uncritical fashion in his introduction to the collected work of Daniel Bernoulli. In 1726 Daniel Bernoulli presented an essay on the principles of mechanics at the Petersburg Academy, the

continued on next page

's Gravesande, who acknowledged that, given a meaning for the term 'force', an experimental-mathematical expression can be found for it, did not make any statements about the question of whether something like a *true* measure existed. It was d'Alembert who finally showed that metaphysical considerations are useless for natural science.

Viewed historically, the actual influence of d'Alembert's analysis can certainly be discounted, as was done at the end of the 1960s.[20] After all, the controversy did not stop after 1743, but continued for quite a long time.[21] Also, although the discussion among the more important physicists did cease at the end of the 1740s, it would for several reasons be excessive to attribute this to d'Alembert.

First of all, d'Alembert added an important argument only in the second edition of 1758.[22] Secondly, other thinkers such as Reid, Desaguliers, and l'Abelye—simultaneously and independently from d'Alembert—pushed the problem aside as a terminological misunderstanding.[23] Finally, none of the combatants referred to d'Alembert. Only MacLaurin, *mv* supporter and one of the most important adepts of Newton, discussed d'Alembert's view in 1748, only to reject it.[24] Laudan therefore concluded correctly that it is much more probable that the insight that "very little hung on the outcome of the debate," caused the quiet silencing of the discussion.[25]

---

continuation of former page
only essay he would publish on this topic. In this paper he reasoned that the proportionality of the living force to both the velocity and the square of the velocity follows from the hypothesis $pdt = dv$, where 'p' indicates the 'pressio' or the external force (per unit of mass). He showed that both proportionalities can occur, depending on what is defined as the effect (Daniel Bernoulli, "Examen principiorum mechanicae"). But it is impossible to conclude from this analysis what Szabó concluded from it: that Bernoulli judged the controversy to be a "Logomachie" (Szabó, *Geschichte der mechanischen Prinzipien* (1977), 71-72; Speiser, [Introduction to Daniel Bernoulli], 21-23). This is confirmed by the fact that Bernoulli shows no trace of any such view in two later articles about living force (Daniel Bernoulli, "Commentationes de immutatione et extensione principii conservationis virium vivarum (...)" (1737) and "Remarques sur le principe de la conservation des forces vives pris dans un sens general" (1748), respectively).

[20] Hankins, "Eighteenth-Century Attempts" (1965); Laudan, "*Vis Viva* Controversy" (1968); Iltis, "D'Alembert and the *Vis Viva* Controversy" (1970) and "Leibniz and the *Vis Viva* Controversy" (1971).
[21] See especially Laudan, "*Vis Viva* Controversy."
[22] Iltis focused attention on this fact ("D'Alembert," 138). However, I doubt whether this later argument would have made d'Alembert's points from 1743 more convincing. Iltis's argument that this is the case appears to be based on an overly simple interpretation of d'Alembert's concept of equilibrium (see note 100 below).
[23] Laudan, "*Vis Viva* Controversy," 139-140.
[24] MacLaurin, *An Account of Sir Isaac Newton's Philosophical Discoveries* (London: Miller & Nourse 1748), cited in *ibidem*, 134.
[25] Laudan, "*Vis Viva* Controversy," 140. See also Hankins, who says that fatigue and the awareness of a "disagreement over basic suppositions about the nature of force and matter," were to blame (Hankins, "Eighteenth-Century Attempts," 281). However, Hankins does not refer to the existence of the disagreement in itself, but to the fact that it concerned 'only' basic suppositions, and therefore allegedly did not concern mechanics itself. Incidentally, Thomas Reid had already referred to the unsatisfactory resolution in 1748: "It was dropt rather than ended to the no small discredit of mathematics, which hath always boasted of a degree of evidence inconsistent with

continued on next page

D'Alembert therefore did not have the great historical importance in the termination of the controversy that has been attributed to him since the nineteenth century. Perhaps we can, with Scott, claim that after d'Alembert we find an 'open season' during which the use of one measure or the other did not lead to great disunity.[26]

By placing him thus in perspective, we can now take a fresh look at d'Alembert. No longer does the self-image of analytical mechanics need to be reflected in d'Alembert's solution. Consequently, we can again ask about the actual content of his solution. In a book such as this, which is concerned with the changing relationship between metaphysics and mechanics, the question is obviously of the greatest importance. The point is how d'Alembert related physical concepts and laws to mathematical reality on the one hand and physical reality on the other, and, following on from this, what the meaning is of his radical attempt to abandon certain metaphysical presumptions while bringing others to the fore.

4.3   CAUSALITY IN MECHANICS

4.3.1   D'Alembert's Purification of Metaphysics

D'Alembert is famous not only in the areas of mechanics and mathematics; he is at least as well known as a man of letters. It is especially his work as the scientific editor of that bulwark of Enlightenment thinking, the *Encyclopédie*, that brought him great fame.[27] The way in which he viewed his role as editor—he wrote, for example, the famous "Discours préliminaire"—resulted in his becoming a member of the *Académie Française* in 1754 and led to his friendship with King Frederick II of Prussia.

However, this friendship did not prevent him from emphasizing the necessity of organizational independence for science in his *Essai sur la société des gens de lettres et des grands* (1753). In this essay he states that philosophy, which includes the sciences in his view, is at odds with aristocracy. After all, aristocracy derives its right of existence from traditional social relations incompatible with philosophy. According to d'Alembert, philosophers must always remain aware of the three concepts of freedom, truth and poverty.[28]

---

    continuation of former page
    debates that can be brought to no issue" (Reid, "Essay on Quantity," cited by Scott, *The Conflict*, 54-55). See also above, note 5.
[26] Scott, *The Conflict*, 55.
[27] He was a scientific editor only up to the seventh volume in 1757, when there was a quarrel concerning a disproportionately long article he had written about the city of Geneva. This quarrel led to his resignation as editor, leaving Diderot the difficult task of managing the job alone. However, he continued to write a number of articles which were important for mechanics, such as "Impulsion," "Matière," "Méchanique," "Percussion" and "Uniforme." The troublesome progress of the *Encyclopédie* between Volumes 7 and 8 is reflected in the "Avertissement" of Volume 8, which was not published until 1765, eight years after Volume 7.
[28] "liberté, vérité, et pauvreté" (D'Alembert, *Essai sur la société des gens de lettres et des grands* (1753), in: d'Alembert, *Œuvres complètes* IV (Paris: Berlin 1821–1822), 337-373, there 367-368).

The ideal expressed in these three concepts was one of the reasons that he felt obliged to decline Frederick II's flattering offer, in 1763, to become the president of his academy in Berlin. Regardless how close their friendship might be in other respects, the subordination of free philosophy to the hierarchical values of monarchy was unacceptable for d'Alembert. But this did not prevent him from providing the king with extensive and detailed advice, for example regarding the appointment of new members to the Berlin Academy.[29] In this, after all, he could remain free and independent.

D'Alembert's rejection of the organizational subordination of philosophy to higher powers can certainly not be separated from his rejection of a theological and metaphysical foundation in philosophy itself and its disciplines.[30] We must realize, however, that by 'theology' and 'metaphysics', d'Alembert is referring to these disciplines as they had historically taken shape in his own time. Although positivists may not like to acknowledge this, in expelling traditional metaphysics he simultaneously brought another form of metaphysics in through the back door. Moreover, his rejection did not apply to metaphysics and theology as a whole, but only to the *foundation* of mechanics in metaphysical and theological concepts and principles of which we have no clear conception ("idée nette"). In d'Alembert eyes, the use of such concepts was the most important cause of the sterility of dogmatic metaphysics and theology and of their inability to explain reality properly.

Thus d'Alembert strove for the purification of metaphysics, rather than its total rejection. Although, following Locke and Condillac, he advocates a sensualist epistemology, this does not prevent him from simultaneously taking a rationalist view of science: it is indeed possible to acquire necessary and certain knowledge founded on one or more fundamental principles—the 'elements' of the relevant science.[31] Although the fundamental concepts are based on sensory experience, they are "claire" and "simple," just as with Descartes. As a result, the way he founds the mechanical principles does not differ in this regard from the rationalist foundations from the previous century. To compare and distinguish it from the rationalism of the seventeenth century, we can characterize his philosophy of mechanics as *sensualist-rationalist*. Hence, d'Alembert's critique could also have been formulated as a positive assertion: a foundation can only be based on concepts of which we have a clear conception.

D'Alembert in fact strives to provide this, and consequently advocates a different type of metaphysics as a basis for mechanics. For example, he refers to a 'metaphysics of propositions' ("métaphysique des propositions"), which is nothing other than the "clear and precise explanation of the general and philosophical truths on

---

[29] The most important example of this is d'Alembert's later advice to the king to appoint Lagrange as director of the Berlin Academy. See Harnack, *Geschichte der Königlich Preussischen Akademie der Wissenschaften zu Berlin* (1900) I.1, 354-361.
[30] Terrall, *Maupertuis and Eighteenth-Century Scientific Culture* (1987), 125-133.
[31] See especially d'Alembert, "Élémens des sciences" (1755).

which the principles of science are based."[32] Simultaneously, the metaphysics of every science may not consist of anything other than the "general consequences resulting from observation."[33] It is apparently the task of metaphysics to complete the transition from observation-based propositions to science.

In a mechanics founded on this new metaphysics, there is no longer any place for the concept of force—at least in a formal sense. In a less formal sense the term 'force' does in fact appear in d'Alembert's mechanics. It is however no longer the same 'force' as the one he resisted.

D'Alembert is not a systematic thinker, and it is therefore necessary to bring his views together from various sources. However, the articles that he wrote for the *Encyclopédie*, the *Traité de dynamique*, and his prefaces for both of these, when taken together, make it possible to reconstruct a coherent picture of his thought in natural philosophy and metaphysics. I will begin with d'Alembert's articles about causes in the *Encyclopédie*, since the concept of cause provides an excellent illustration of the problematic relationship between metaphysics and science.

4.3.2   Final Causes

The article "Cause" from the *Encyclopédie* consists of one general article and six sub-articles, of which d'Alembert wrote only the first two: "Causes finales" and "Cause, *en méchanique & en physique*."[34] The general article is written by a certain Abbé Yvon and actually deals only with the mind-body problem, i.e. the question of what type of causality must exist between mind and body if the mind is to be free and active and matter determined and passive. The sub-article "Causes finales" can be viewed as a transition from the general article to the second sub-article. D'Alembert defines the concept of final cause only implicitly, by including it in a methodological principle: "The principle of final causation consists of seeking the *causes* of effects found in nature in the *end* that the Creator must have had in mind when producing these effects."[35] It is immediately striking that d'Alembert has limited the concept by reducing the possible purposes to external purposes. The objectives that nature might intrinsically strive for, which is what is meant by the Aristotelian-scholastic concept of final cause, fall outside his definition.[36] This limitation avoids an ontological contradiction between the concept of final cause and the mechanistic view of matter as passive and determined.

---

[32] "exposition claire & précise des vérités générales & philosophiques sur lesquelles les principes de la science sont fondés" *(ibidem*, 492).

[33] "conséquences générales qui résultent de l'observation" *(ibidem)*.

[34] The other four sub-articles concern the medical, judicial, and ecclesiastical use of the term.

[35] "Le principe des *causes finales* consiste à chercher les *causes* des effets de la nature par la *fin* que son auteur a dû se proposer en produisant ces effets" (d'Alembert, "Causes finales" (1755), 789).

[36] In the *Traité de dynamique*, xxx, there is a somewhat different formulation: "the aims that the Author of nature must have had in mind when instituting its laws" ("les vûes que l'Auteur de la nature a dû se proposer en établissant ces loix"). By using the term 'vûes' instead of 'fin', d'Alembert expresses his intention even more clearly.

D'Alembert's objection to the method of final cause is therefore not ontological but epistemological in nature. In the following line he says that the principle of final cause is equivalent to the derivation of phenomenal laws from metaphysical principles. And it is precisely there, in relation to the possibility of knowing and applying metaphysical principles, that we face insurmountable problems. To illustrate this, d'Alembert criticizes two such metaphysical principles. The first is that of Nature's 'abhorrence of a void' ('horror vacui'), which he calls sterile and absurd. The second principle is that of the simplest way or the shortest time, which will be discussed in the next chapter. According to d'Alembert this principle is misleading and even dangerous, because the relevant quantities in nature quite frequently show not a minimum, but a maximum. For d'Alembert the conclusion is clear: metaphysical, teleological principles are sterile, absurd, and dangerous for science.[37] No reliable knowledge about nature can be found with this approach.

D'Alembert's argument then takes a surprising turn in that he mentions the physico-theological views of Maupertuis in a distinctly positive manner.[38] Although he distorts what Maupertuis says, his remarks are interesting because they show an unexpected aspect of his own thinking. Surprisingly, the discovery of the *concord* of a final cause with phenomenal laws can be very useful, provided that the phenomenal laws themselves are founded, as they should be, on "clear and incontestable principles of mechanics."[39] Maupertuis's conviction that he had actually demonstrated this concord on the basis of a metaphysical principle, is omitted in d'Alembert's summary. For d'Alembert, the concord points not to a connection from metaphysics to mechanics and physics, but rather the other way around: it allows us to prove the existence of God. Thus his endorsement of Maupertuis's principle of least action is, paradoxically, an affirmation of the importance of a foundation based on mechanical principles. In d'Alembert's interpretation, the principle of least action makes it possible to base natural theology on mechanical principles.

4.3.3   Mechanical Causes

In the second sub-article, "Cause, *en méchanique & en physique*," d'Alembert discusses how the phenomenal laws can be derived. He defines a 'mechanical cause', or 'physical cause', as a cause of change in motion:

---

[37] "stérile," "absurde," and "dangereux" (d'Alembert, "Causes finales," 789). In this article, at any rate, d'Alembert does not make it clear to what extent he differentiates final causes from metaphysical principles.

[38] Maupertuis's principle of least action, along with Leibniz's principle ("Nature (...) always acts along the simplest and shortest routes" ("La nature (...) agit toûjours par les voies les plus simples & les plus courtes (...)")), which was rejected by d'Alembert, are dealt with in the following chapter. They are mentioned here because d'Alembert's commentary is important for understanding his own solution to the *vis viva* controversy. To avoid repetition, I will not consider Maupertuis's principle further at this point.

[39] "principes de méchanique clairs et incontestables" (*ibidem*).

everything that produces a change in the state of a body, i.e., that sets it in motion or that brings it to a stop, or that alters its motion.[40]

There are two forms of cause: one arises from the mutual action of particles, based on their impenetrability, i.e. momentum and causes derived from it. These are the causes that manifest themselves simultaneously with the effect.[41] This includes not only impact phenomena, but also the traction force of a rope and the pressing force of a rod. The impenetrability of matter is therefore one of the primary causes ("causes principales") of natural phenomena. As an advocate of clear basic principles, d'Alembert would have preferred to have written '*the only* primary cause', but he realized there were many phenomena for which it was very doubtful whether they could ever be derived from that single principle.[42] In such phenomena, the cause does not manifest itself; all we can know is the effect.[43]

It therefore is necessary to acknowledge a class of effects, and consequently a class of *causes*, in which momentum is not active, or is not manifest.[44]

This second, negatively defined, type of effects and causes includes gravitation and other actions-at-a-distance. It is uncertain whether the causes of such actions can ever be reduced to the action of momentum, and as long as this has not been done, this second form must be included in any complete analysis as a *possibility*.

D'Alembert now links the distinction between known and unknown mechanical causes to necessary and contingent truths in the laws of nature: the laws of impact are necessary because they are derived from the principle of impenetrability; the law of freely falling bodies is contingent because it can only be discovered experimentally. I will discuss this in greater detail below.[45]

However, in the last part of the sub-article, d'Alembert makes a surprising move that calls his antithesis between known and unknown causes into question again. The shift takes place in the midst of his argument against what was then the most important principle connecting metaphysical causes to mechanical phenomena, to wit, the principle that "effects are proportional to their causes."[46] Leibniz gave this principle a central role in his argument in favor of the conservation of living force,

---

[40] "tout ce qui produit du changement dans l'état d'un corps, c'est-à-dire, qui le met en mouvement ou qui l'arrête, ou qui altere son mouvement" (d'Alembert, "Cause, *en méchanique & en physique*" (1755), 789).

[41] D'Alembert, *Traité de dynamique*, x.

[42] See d'Alembert, "Élémens des sciences," 491, in which he outlines this as an ideal form for every individual science.

[43] D'Alembert, *Traité de dynamique*, xi.

[44] "Il est donc nécessaire de reconnoître une classe d'effets, & par conséquent de causes dans lesquelles l'impulsion ou n'agit point, ou ne se manifeste pas" (d'Alembert, "Cause, *en méchanique & en physique*," 790). Elsewhere he adds the agnostic argument that momentum is "something very obscure" ("quelque chose de fort obscur"), so that it may not be seen *a priori* as the only possible cause (d'Alembert, "Impulsion" (1765), 635).

[45] See pages 117-8.

[46] "les effets sont proportionnels à leurs causes" (d'Alembert, "Cause, *en méchanique & en physique*," 790).

and it is clear that d'Alembert criticizes the principle especially with an eye to this argument.[47]

In his objections, d'Alembert makes use of his antithesis between known and unknown causes, but gradually enters more dangerous waters. First of all he argues that the principle is *superfluous* for effects with *unknown* causes. "Because if one does not know the effect, one does not know anything; and if one knows the effect, one does not need the principle."[48] Consequently d'Alembert, like most eighteenth century natural philosophers, takes the standpoint that the *aim* of science is to know the *effects*. The contrast with Aristotelian-scholastic science and the rationalism of the seventeenth century now becomes very clear! Before the seventeenth century, after all, the knowledge of causes was thought to be the highest aim of every science. Likewise, in mechanistic ideology, every theory was subjected to the criterion of mechanical causality.

We could say that with this criticism, d'Alembert radicalized the turn that Newton in his *Principia mathematica* had given to explanation in natural sciences. Newton's method could be visualized as placing an intermediate mathematical level between material phenomena and their mechanistic foundation.[49] Thereby he split mechanics into two parts: on one side rational mechanics as we know it from the *Principia*, on the other side the speculative natural philosophy of the "Queries." D'Alembert now radicalizes this split, at any rate where phenomena with unknown causes are concerned, by breaking the link between the intermediate mathematical level and the mechanistic foundation. In fact, he was promoting the intermediate mathematical level to the status of a foundation.

But even if the cause is known, as is the case with colliding particles, the principle of proportionality of cause and effect is superfluous. Such causes consist of either a body in motion or a body that pushes another body. However, d'Alembert believes that the accompanying laws of motion and impact have been found to be independent of this principle.

Moreover, the principle is misleading. It neglects the problem of how cause and effect can be measured. For example, the fact that the cause of the force of impact of a moving body is found in both its mass and its velocity does not tell us in what proportions this is so: the relevant function could be mv, mv^2, or m^2v, for instance.

Moreover, even if we know that cause and effect are proportional to mv, for example, the principle does not say how the cause (i.e. the momentum of two particles before impact) results in a distribution of the effects (i.e. the momentum of the

---

[47] See Chapter 3 above. That d'Alembert has the *vis viva* controversy in mind, becomes especially clear at the end of the sub-article, where he says: "if one had never thought of saying that effects are proportional to their *causes*, one would never have disputed about the living forces" ("si on ne s'étoit jamais avisé de dire que les effets sont proportionnels à leurs causes, on n'eût jamais disputé sur les forces vives") *(ibidem)*.

[48] "Car si on ne connoît pas l'effet, on ne connoîtra rien du tout; & si on connoît l'effet, on n'a plus besoin du principe" *(ibidem,* 790).

[49] See I.B. Cohen, *The Newtonian Revolution. With Illustrations of the Transformation of Scientific Ideas* (1980).

particles after impact). The relation 'causa aequat effectum' (the cause is equal to its effects) indicates that the *sum* of the effects is equal to the *sum* of the causes indeed, but it is clearly incapable of saying something about their *distribution*.

4.3.4 Necessity and Contingency

The argumentation above, against the possibility of deriving effects from their causes, is somewhat at odds with d'Alembert's distinction between necessary and contingent truths. After all, he had made this distinction because some effects, those whose mechanical causes are known, can indeed be derived from their causes, and even necessarily so! It seems that d'Alembert himself suddenly became conscious of this tension, because he rushes to come up with a new distinction between true and apparent causes. He states that the necessity of the laws of motion and impact is not thrown into doubt in any way, because mechanical causes only *seem* to be causes, but are in reality "effects from which other effects result."[50] So the motion of a colliding body is apparently a cause, but is actually an effect. D'Alembert's argument, however, is directed against the derivation of effects from true causes (i.e., metaphysical causes), and not against deriving effects from other effects: "the *metaphysical cause*, the *true cause* is unknown to us."[51]

This turn in d'Alembert's thought about the fundamental concepts and principles of mechanics is crucial. Strictly speaking, the concept of cause no longer has any meaning in his mechanics. The term is retained, but it is given a different content. Where previously one could refer to cause and effect, now there are only effects. The mechanical 'cause' is itself also an effect. Actually, this is an unsatisfactory formulation. Cause and effect are a contrasting pair, like 'father' and 'son', so the concept of 'effect' has also become meaningless, along with the concept of 'cause'. If everything is an effect, nothing is an effect. With d'Alembert there are only phenomena.

But if the laws of impact and motion cannot be derived from metaphysical true causes, from where can they be derived? After all they are necessary! First of all it is interesting to understand d'Alembert's definition of necessity and contingency. The essay competition set up by the Berlin Academy on 3 June 1756 for the year 1758 tempted d'Alembert to formulate his thoughts on this subject. We see the result of this in the preface to the second edition of his *Traité de dynamique*.[52] D'Alembert

---

[50] "des effets desquels il résulte d'autres effets" (d'Alembert, "Cause, *en méchanique & en physique*," 790).

[51] "[l]a *cause métaphysique*, la *vraie cause* nous est inconnue" *(ibidem,* 790). The italics are d'Alembert's.

[52] D'Alembert, *Traité de dynamique*, xxiv-xxx. In his *Souvenirs d'un citoyen* (1789) Formey claims that d'Alembert himself also submitted an essay but that the prize was not awarded to him due to personal friction between d'Alembert and Euler (Formey, *Souvenirs* II, 239). However, no trace of this essay can be found in the archives of the Berlin Academy. It is therefore not unthinkable that Formey's memory has deceived him, and that he has confused this competition with that for the year 1752 concerning the resistance encountered by bodies in fluids, on which d'Alembert

continued on next page

clearly found this problem important—in fact he called it the "grand Problême Métaphysique."[53]

In a very elegant fashion, d'Alembert is able to link the necessity of the laws of motion with God's omnipotence and absolute freedom of will. Necessary truth is not that which cannot be untrue, but that which can be derived from foundations.[54] Consequently, the concept of necessity loses its absoluteness: the foundation in its turn is not necessary. Hence, that which d'Alembert considers to be necessary is, strictly speaking, also contingent. As a result, d'Alembert's concept of necessity does not limit the omnipotence of God because God can add other truths that do not follow from the foundations. D'Alembert refers to such truths as 'contingent' in order to retain the traditional antithesis with respect to the term 'necessary'. Thus the antithesis between 'necessary' and 'contingent' leaves room for God's freedom of will by linking metaphysical knowledge of the foundations to the experience of the results of God's decisions.

However, although d'Alembert describes concept of necessity only in terms of way its validity should be established, it is not an epistemological concept but an ontological one. A necessary truth follows from foundations, while a contingent truth is added to them from outside.[55]

4.4 D'ALEMBERT'S FOUNDATION OF MECHANICS

4.4.1 D'Alembert's Method

D'Alembert's foundation of the mechanical laws of motion begins with the method that he outlined in the introduction to his *Traité de dynamique*.[56] In this work, d'Alembert kept quite consistently to his own method, which is directed towards obtaining the clearest possible principles, as few as possible in number, and towards expanding the application of these principles as far as possible. With this method d'Alembert believed he could achieve his general aim, summarized succinctly as: "expanding the principles by reducing them."[57]

The method outlined has two elements. Firstly, knowledge of the more abstract and simple sciences, in this case geometry and algebra, must be applied. Secondly, the subject in question must be analyzed in the most abstract and simple manner possible. Thus we could say that the first element of d'Alembert's method states *that*

---

continuation of former page
published his *Essai d'une nouvelle théorie de la résistance des fluides* in 1752. (Compare Harnack, *Geschichte der Akademie* I.1, 339n.2).
[53] D'Alembert, *Traité de dynamique*, xxiv.
[54] *Ibidem*, xxv.
[55] For this reason the interpretation of d'Alembert's concept of necessary as 'hypothetically necessary' is incorrect.
[56] *Ibidem*, iii.
[57] "étendre les principes en les réduisant" *(ibidem, v)*.

the relevant science must be mathematized, the second element states *how* this mathematization must take place.

The subject of mechanics is 'matter in motion'.[58] In d'Alembert's view, we have an understanding of matter in motion insofar as it is left to itself.[59] In his opinion, the mere existence of matter implies principles of motion that we can determine with great precision. There are three such principles: the principle of inertia, the principle of composite motion and the principle of equilibrium. However, we cannot determine *a priori* whether these principles are in accordance with the actual ones because, as has been said above, we do not know God's will. It is only through experience that we know that some of them are indeed in accordance.

So these laws, for which we can derive the formula from our insight into matter itself, and whose validity can be proved from our experience, are necessary. The other laws of motion, for which both formula and validity must be demonstrated from experience (such as Newton's universal law of gravitation), are contingent.

D'Alembert's method raises the question, how it is possible that we *cannot* have knowledge of God and of many metaphysical causes and principles, but that we *can* have knowledge of matter? Whence this differentiation in our capacity to know? Is it a baseless differentiation that can be blamed on a century of mechanistic indoctrination? Or is there perhaps a categorical distinction in the nature of these things?

For d'Alembert, it is the latter. The fact that the concept of matter is not a metaphysical freak but is in accordance with something real, means at the same time that it has a nature which differs from that of metaphysical freaks. This can be clarified by contrasting d'Alembert's distinction and the Aristotelian-scholastic distinction between essential and accidental attributes. At first glance, the similarity is striking. In the same way that the true principles are inexorably linked with moving matter, the essential attributes are inexorably linked with their carrier: substance. In the same way that the contingently true principles are applied externally to matter, the accidental attributes are externally related to their carrier. However, there is an important difference in the nature of the concepts thus linked; whereas scholasticism was concerned with substances and their attributes, d'Alembert had other concerns.

To clarify this, we must go more deeply into d'Alembert's foundation of mechanics. In the course of this, we will also see how this foundation relates to his solution of the *vis viva* controversy.

---

[58] In the introduction to the first edition from 1743, d'Alembert refers only to motion as the subject of mechanics; whereas a passage concerning the Berlin essay competition for 1756 that was inserted in the edition from 1758 mentioned only matter ("matter left to itself" ("la matiere abandonnée à elle-même")) (*ibidem*, ix and xxv, respectively). In fact, however, both are part of the subject of mechanics, 'matter in motion'. Incidentally, it is strange that moving forces do appear in d'Alembert's definition of mechanics in the article of the same name in the *Encyclopédie* (d'Alembert, "Méchanique," 220).

[59] See note 58.

4.4.2 *The Subject of Mechanics: Moving Matter*

What is this moving matter on which the principles of motion are founded? Traditionally, a definition of the essence of matter would be required, from which the necessary properties could be derived. D'Alembert has a different starting point. For him, after all, it is impossible to determine the essence of matter. D'Alembert's view of matter lies between those of Descartes and Newton.

For Descartes, extension is the essence of matter. D'Alembert believes that not only the term 'essence' must be replaced by the term 'property', but also that extension as such is inadequate: what is really at stake is *impenetrable* extension.[60] As previously stated, he considers impenetrability to be a primary cause of the laws of motion.[61]

Later on in the *Traité de dynamique,* he refers to the Newtonian concept of attraction, but d'Alembert considers that it does not have the same level of clarity as 'impenetrability', and therefore cannot serve as a foundation for the principles of motion. Neither can it be derived from other causes, as shown by all the futile attempts to do so. He therefore prefers to consider attraction as a phenomenon expressed in the laws of gravitation. Nevertheless, he appreciates the relative nature of this distinction, as shown by his qualification concerning the clarity of the concept of momentum.[62]

I will show later that d'Alembert was indeed unable to link the impenetrability of matter to the laws of motion, at any rate not in a way that was satisfactory to his own standards.

In the *Traité de dynamique*, d'Alembert defines 'motion' as the continuous transition from one point in space to another, and 'rest' as remaining in the same place.[63] 'Motion' is a 'complex idea' which upon analysis appears to be comprised of the 'simple ideas' of space and time.[64] The term 'simple idea' ("idée simple") warrants some explanation. The simplicity of ideas does not rest in our perception or conception of them: considered in this way every idea is simple. Rather, simplicity lies in the thing itself, but not in the form of indivisibility, as is the case with a 'simple being' ("être simple"). The perfect equality of all parts, and the impossibility to further analyze the idea without destroying it, is what makes it simple.[65] The concepts of space and time are examples of simple ideas. Conversely, 'motion' and 'rest' are complex ideas.

---

[60] D'Alembert *Traité de dynamique*, 1. See also his article "Matiere" (1765), 189, in which he calls on the Newtonian Clarke for support.
[61] See above, page 115.
[62] "It is therefore a mistake to believe that the idea of *momentum* involves no obscurity at all, and to try to treat this force alone, excluding every other principle, as the one that produces all the effects of nature" ("C'est donc une erreur de croire que l'idée de l'*impulsion* ne renferme aucune obscurité, & de vouloir, à l'exclusion de tout autre principe, regarder cette force comme la seule qui produise tous les effets de la nature") (d'Alembert, "Impulsion," 635).
[63] D'Alembert, *Traité de dynamique*, 2 (Définition II).
[64] D'Alembert, "Élémens des sciences," 494.
[65] *Ibidem*, 493-494.

Moreover, it is remarkable that d'Alembert, in imitation of Newton, defines 'motion' and 'rest' in relation to *absolute* space, which he calls 'undefined space' ("espace indéfini").[66] This means that he cannot use the principle of the relativity of motion in his proofs. We will see below how difficult he made things for himself with this definition.

4.4.3 The Principles of Statics and Dynamics

Thus the subject of mechanics could be defined, according to d'Alembert, by nothing other than properties that can be expressed in space and time: extension, impenetrability, continuous displacement, etc. Forces, whether internal or external, are not part of his definition because they do not present themselves 'of their own accord'.[67]

During the actual argumentation, this rather flimsy starting point is supplemented with several principles, the most important being the principle of sufficient reason. In the case of d'Alembert, incidentally, one would do the principle more justice by calling it 'the principle of *in*sufficient reason'. D'Alembert is therefore exaggerating when he claims that the three independent principles—of inertia, of composite motion and of equilibrium—are derived solely from the analysis of "motion, considered in the simplest and clearest way," or of a "body left to itself."[68]

I will discuss d'Alembert's derivation of the first of the three principles in detail; for the second and third principles the broad lines will suffice. The three principles are structured by analogy to the model of Newton's three laws, but only in the case of the first (the principle of inertia) there is conformity in terms of content as well.

4.4.3.1 The Principle of Inertia

"I call (...) *force of inertia*, the property bodies have, to stay in the state in which they are."[69] The first principle confuses the reader right from the start: after all, had d'Alembert not clearly stated that he would disregard forces? What, then, makes him call the property of inertia a force? I must ask the reader to be patient as to the reasons for this and similar terminological confusions. At this point it suffices to say

---

[66] Nevertheless, d'Alembert reserves an escape route concerning the objective reality of absolute space; in his introduction he suggests that the concept of space has primarily an instrumental meaning: "We can always conceive an undefined space as the locale for the bodies, whether real or imagined (...)" ("Il nous sera donc toujours permis de concevoir un espace indéfini comme le lieu des corps, soit réel, soit supposé (...)")) (d'Alembert, *Traité de dynamique*, vi-vii). Contrast this with Briggs, "D'Alembert: Philosophy and Mechanics in the 18th Century" (1964), 39.

[67] D'Alembert says, "I have turned my back on *moving causes*" ("[J]'aie détourné la vûe de dessus les *causes motrices*"), and "I have entirely expelled inherent forces from the moving body" ("j'aie entièrement proscrit les forces inhérentes au Corps en Mouvement") (d'Alembert, *Traité de dynamique*, xvi-xvii).

[68] "Mouvement, envisagé de la maniere la plus simple & la plus claire," and "Corps abandonné à lui-même" *(ibidem,* xvi (fragment from 1743), and xxvii (fragment from 1756) respectively).

[69] "J'appelle (...) *force d'inertie*, la propriéte qu'ont les Corps de rester dans l'état où ils sont" *(ibidem,* 3). See also d'Alembert, "Force," 110.

that d'Alembert was certainly not concerned with a cause, but with a *phenomenon*, and with the foundation of this phenomenon in the idea of motion: "a body left to itself must remain eternally in its state of rest or of uniform motion."[70]

In his proof, d'Alembert splits the principle into two laws; the first formulated for the case of rest, the second for that of motion: "a body in rest will persevere in that state, as far as no external cause makes it change this state," and "a body once set in motion by some cause, must always continue it uniformly and in a right line."[71]

The proof of the first law is surprisingly simple and is based on the principle of sufficient reason. The proof is rooted in the idea of the uniformity of space: that the 'undefined space' is the same at every location and therefore has no preferred direction. However, d'Alembert does not specifically mention this fundamental property.[72] The proof reads: "for a body cannot not set itself into motion, because there is no reason for it to move in one direction rather than in another."[73] The passivity of matter at rest apparently derives from the uniformity of space.[74]

However, the extension of the first law to include the case of motion, an extension d'Alembert presents as self-evident with the words "from this it follows" ("delà il s'ensuit"), is anything but self-evident, since rest and motion have been defined in relation to absolute space.[75] Once a body moves, a difference between 'in front of' and 'behind' exists. Given this difference, the uniformity of space does not stand in the way of a striving for rest or a striving for increased velocity. The argument of the conformity of space can be used only to defend the impossibility of a *change* of direction, which is indeed what d'Alembert does.

To prove the principle of inertia for the case of motion, d'Alembert lapses into a circle. He inserts a corollary to the first law, which applies to the case of motion: "if a body receives a motion by any cause whatsoever, it will itself not be able to accelerate nor to slow down this motion."[76] This is logically incorrect and misleading: on the one hand it does not follow from the first law, as d'Alembert claims, and on the other hand it is made to serve as an important premise in his proof of the

---

[70] "un Corps abandonné à lui-même doit persister éternellement dans son état de repos ou de mouvement uniforme" *(ibidem,* xxvii).

[71] "[u]n Corps en repos y persistera, à moins qu'une cause étrangere ne l'en tire" (article 3) and "[u]n Corps mis une fois en mouvement par une cause quelconque, doit y persister toujours uniformément & en ligne droite" (article 6) *(ibidem,* 3-4).

[72] See also d'Alembert, "Élémens des sciences," 494: "every part of time and of space is absolutely identical" ("toutes les parties du tems & de l'espace sont absolument semblables").

[73] "[c]ar un corps ne peut se déterminer de lui-même au mouvement, puisqu'il n'y a pas de raison pour qu'il se meuve d'un côté plutôt que d'un autre" *(ibidem,* 4 (article 3)).

[74] In his proof of the principle of inertia in the *Encylopédie*, which rests on the passivity of matter, he makes the only appeal to theology that I know of—but which theology is it? He says, "religion assures us that matter can not think" ("nous sommes certains par la religion, que la matiere ne peut penser") (d'Alembert, "Force," 112).

[75] D'Alembert, *Traité de dynamique,* 4 (article 4).

[76] "si un Corps reçoit du mouvement par quelque cause que ce puisse être, il ne pourra de lui-même accélérer ni retarder ce mouvement" *(ibidem,* 4 (article 4)).

second law. Once the initial cause of some given motion has fallen away,[77] the velocity will remain steady if no other cause is acting, as follows from the preceding corollary to the first law, he writes.[78]

His proof is circular, but it should be noted that it is again based on the uniformity of space. This also appears to be the case with the extra argument that d'Alembert provides: once a body is moving uniformly and in a straight line, it will continue to do so. The core of this argument amounts to the following: if a body moves from A to B with a uniform velocity in t seconds, at every intermediate point it is in the same state as at point A. From every point we can therefore project a distance forward, which is equally long as the distance AB, and which also will be covered by the body in t seconds. The body will therefore *continue* to move uniformly and in a straight line.[79] This extraordinary argument is nonetheless important, because it once again states the identity of every location in space. Due to this property of space, says d'Alembert, once a motion is uniform and rectilinear, it will remain uniform and rectilinear.[80]

This particular interpretation of what d'Alembert is doing here, which is central to the entire argument of the present book, does require some qualification. For it is true that, in the *Traité de dynamique*, d'Alembert made a number of investigations into the possibility that inertial motion depends on the continuous action of a cause of motion. He takes this possibility explicitly into account not only in the proof of the principle of inertia, but also in the proof of the principle of composite motion. D'Alembert wants to show, with this open attitude, that mechanics is independent of assumptions about the way metaphysical causes act. The real (i.e. metaphysical) forces are simply not relevant because the same principles apply regardless of those forces.

A second qualification is that in the introduction to the *Traité de dynamique*, d'Alembert somewhat weakens his own basic assumptions by introducing two 'forbidden' principles: that a body cannot give itself speed and that the simplest motion must apply. The two laws of the principle of inertia allegedly follow from these two principles. In the case of the first principle there is no great problem: in the *Traité de dynamique* it is proved from the uniformity of space. The second principle, however, is a metaphysical principle of the worst kind, of the type he emphatically warned against in his article "Cause"!

---

[77] D'Alembert also distinguishes the possibility that the body, in order to move, requires the continuous action of a cause of motion. This possibility, however, cannot be brought into proper accordance with his further insights. He therefore treats it fleetingly and poorly: in that case there is no reason for the cause to increase or decrease and therefore the velocity remains uniform.

[78] *Ibidem*, 5 (article 6).

[79] *Ibidem*. This extra argument is not a proof of the corollary to the first law that was treated above, nor was this d'Alembert's intention. The difference between the two propositions is minimal, yet essential. The corollary is concerned with the problem of whether an imposed motion will *be* uniform, while the argument given here concerns the problem of whether a uniform, rectilinear motion will *remain* uniform and rectilinear.

[80] D'Alembert, *Traité de dynamique*, 5 (article 6).

Clearly d'Alembert's use of such principles was an incidental mistake rather than a change in viewpoint. He basically aspires to derive the principle of inertia from the uniformity properties of space and time without having to appeal either to knowledge of the nature of matter or to knowledge of forces.

4.4.3.2 The Principle of Composite Motion

The principle of inertia was a reformulation of Newton's first law, albeit one in which the term 'force' had a completely different content. D'Alembert could not do the same with Newton's second law. After all, this defines the relationship that exists between an external force and the resulting change in motion. D'Alembert's reinterpretation of the concept of external force led him to read Newton's second law as a mere definition, as I will discuss below. Its humble place is at the end of the chapter about the principle of inertia, as one of six notes.

Instead of simply allowing the place left vacant by Newton's second law to remain empty, d'Alembert uses the corollary to Newton's first and second laws as a starting point, and promotes it to the principle of the composition of motion (see figure 4.1).

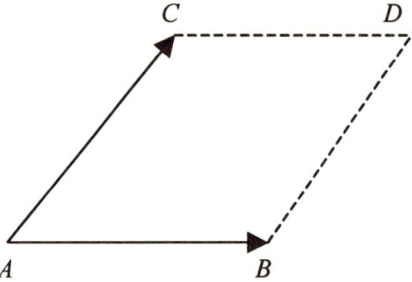

Figure 4.1 The principle of composite motion

> If two arbitrary forces act instantaneously on a body or point A to move it, one uniformly from A to B in a certain time, the other uniformly from A to C in the same time, and if one completes the parallelogram $ABCD$, I say that body A will pass the diagonal AD uniformly, and in the same time in which it would have completed AB or AC.[81]

Once again it is remarkable that d'Alembert refers to forces ("puissances"). However, he leaves these causes completely undefined except for the velocities which, taken separately, they would give to the body under consideration. Phrased in

---

[81] "Si deux puissances quelconques agissent à la fois sur un corps ou point A pour le mouvoir, l'une de A en B uniformément pendant un certain tems, l'1autre de A en C uniformément pendant le même tems, & qu'on acheve le parallélogramme ABDC; je dis que le corps A parcourra la diagonale AD uniformément, dans le même tems qu'il eût parcouru AB ou AC" *(ibidem,* 35 (article 28)).

modern language, a body that is instantaneously given two velocities will move with a velocity equal to the vector sum of the two individual velocities; in other words, velocities can be superposed upon one another.

This principle is not at all self-evident. Another possibility, for example, would be that the larger velocity would have disproportionately more influence in determining the resulting direction. In the Aristotelian-scholastic conception of mechanics, natural motions are distinguished from violent motions. The natural motion of a stone is directed towards the earth; when a stone is thrown upward, the motion is violent. To explain the resulting motion, however, it is assumed that the violent motion is initially the strongest, and prevents the natural motion to the earth. Later, when the violent motion is sufficiently weakened, the natural motion becomes the strongest; from that moment the body falls downward.

D'Alembert is therefore correct when he says that that the principle of composite motion requires proof. However, he does not realize that this requirement rests on a *distinction* between mechanics and mathematics, while the core of his proof actually consists of the *identification* of both. In this case, the identification is contained in the implicit application of the principle of relativity of motion.

Thus d'Alembert's proof can be seen as an indication of how natural the 'mathematization of physical space'[82] had become: the distinction still exists formally, but is no longer experienced as such. The transition from mathematical to mechanical space has *a priori* been made, under the tacit assumption that it can be made in the first place.

4.4.3.3 The Principle of Equilibrium

D'Alembert's third principle is really the reformulation of Newton's third law in kinematic form. Although the concepts of space, time, and velocity were sufficient for d'Alembert's first two principles, he cannot avoid adding a new concept to the third. After all, this concerns the interaction between different bodies. Contrary to what he has promised all the time, this new concept is not 'impenetrability'. Without wasting a word, he adds another quantity: 'mass'. Although this concept cannot possibly be derived from the ideas of space, time, velocity, or impenetrability, he applies it to characterize bodies and thus makes it a basis for the principle of their interaction:

> If two bodies whose velocities are inversely proportional to their masses have opposite directions, in such a way that the one cannot move itself without moving the other, there will be equilibrium between them.[83]

---

[82] According to Koyré (cited by Dijksterhuis, *The Mechanization of the World Picture*, section IV.147). See also above, section 1.2.1.

[83] "Si deux corps dont les vitesses sont en raison inverse de leurs masses, ont des directions opposées, de telle maniere que l'un ne puisse se mouvoir sans déplacer l'autre, il y aura équilibre entre ces deux corps" (d'Alembert, *Traité de dynamique*, 50-51). The definition of 'equilibrium between two
continued on next page

The principle is formulated for non-elastic bodies.[84] The way he proves this principle cannot be dealt with here, except to observe that the principle of sufficient reason again plays an important role.

4.4.4 Mechanics as a Science of Structure

The principle of equilibrium is the most important basis for the famous methodical principle that is named after d'Alembert, which allows us to reduce dynamical problems to static ones. I will not discuss this methodical principle here because its primary function is to solve dynamic problems with the help of the three principles discussed above.

D'Alembert's derivation of his three principles of mechanics shows how much he wanted to base mechanics on the structure of space and time. Only when this foundation proved inadequate, as with the principle of equilibrium (where he tacitly inserted the concept of mass), did d'Alembert seek it elsewhere. The foundation as a whole was entirely in accordance with the hierarchical ranking of algebra, geometry, and mechanics, as outlined previously.

Following this analysis of d'Alembert's foundation of the principles of mechanics, we can return to the question of the relationship between matter-in- motion and the principles of motion. We have seen that in d'Alembert's view moving matter is not a composite whole of substances which react to one another in a specific way according to their properties, thus yielding laws. For him, on the contrary, moving matter is made up of a collection of elements that are part of a time-space structure; its behavior must chiefly be derived from this structure.

It goes without saying that d'Alembert did not formulate all this in these terms. However, my thesis is not that d'Alembert viewed mechanics as a structural science, but that it is useful and clarifying thus to interpret his actual foundation of mechanics. But the most important argument for this interpretation is that it puts the remarkable position of the concept of force in d'Alembert's mechanics in a surprisingly clear light.

4.5 D'ALEMBERT'S CONCEPTION OF FORCE

4.5.1 Property, Function and Cause

It is paradoxical that, although d'Alembert had expressly stated, in the preface to the *Traité de dynamique*, that he would disregard both internal and external forces, he repeatedly fails to do so in the actual argument of his *Traité*. What is really going on here?

---

continuation of former page
bodies' can be derived from the previous article: 'equilibrium' is the mutual obstruction of motion (*ibidem*, 50 (article 45)).
[84] *Ibidem*, 45-46 (article 36).

First of all we must note that the terms 'force', 'power', and 'cause' ("force," "puissance," and "cause") are certainly not synonyms in d'Alembert's view, although he sometimes used them interchangeably. He makes an explicit distinction between 'force' on the one hand and 'power' and 'cause' on the other, a distinction which is very important for a correct understanding of his conception of force and of his solution to the problem of living force.

The first meaning of 'force' is expressed in the term 'force of inertia'. His explanation shows what he has in mind.[85] D'Alembert defines 'inertial force' not as a cause, but as a *property*, which only becomes manifest during a change of state. Why is he using the term 'property' and not something like 'power'? Simply because 'property' refers only to the observable effect, while 'power' would appear to indicate a "metaphysical and vague being, which resides in the body."[86] Moreover, states d'Alembert, such a definition would not allow any *proof* of the principle of inertia.[87]

D'Alembert therefore uses the term '*force* of inertia' in an identical fashion to *property* of inertia'. However, he may well have retained the term 'force' to maintain the link with Newton's generally accepted terminology. Just as with Newton, two properties, resistance and momentum, are merged into a single concept.[88]

A second meaning of 'force' is used in d'Alembert's mechanics much more frequently: one that is related to *change* in motion. I am purposely expressing myself somewhat vaguely here. It is in fact with respect to the *nature of the relationship* between 'force' and 'change in motion' that d'Alembert breaks with tradition. This relationship was traditionally causal: 'force' is the *cause* of the change in motion. With d'Alembert, however, this relationship becomes a *functional* one. He explains this in the first of the series of notes at the end of the chapter about the principle of inertia.[89] The formula $\varphi dt = du$, in which dt represents a time element and du represents a velocity element, is not a relationship between cause and effect but merely a mathematical relationship between velocity and time. The variable φ is therefore not a *cause*, but the "simple expression of the relationship between du and dt," or as we would say today: a *function* of time and velocity.[90] The concept of accelerating force

---

[85] D'Alembert, "Force."
[86] "être métaphysique & vague, qui réside dans le corps" *(ibidem,* 110). Hankins attributes this conception of force to Malebranche (Hankins, "The Influence of Malebranche on the Science of Mechanics during the Eighteenth Century" (1967), 208).
[87] D'Alembert, "Force," 110. His argument, however, is unclear and originates from an entirely different sphere of thought: a motion is new at every instant and isolated from the motion at every other instant; the body therefore continually requires the same cause. "One might think that it would have a continual tendency to return to rest, if the same cause which had set it in motion did not continue to pull it in one way or another" ("'[O]n pourroit croire qu'il tendroit sans cesse à retomber dans le repos, si la même cause qui l'en a tiré d'abord, ne continuoit en quelque sorte à l'en tirer toujours'") (d'Alembert, *Traité de dynamique*, 7).
[88] D'Alembert, "Force," 110: "[The inertial force] is called *resistance* or *action*, according to the point of view from which it is considered" ("(...) on lui donne alors le nom de *résistance* ou d'*action*, suivant l'aspect sous lequel on la considere").
[89] D'Alembert, *Traité de dynamique*, 22-27 (article 22).
[90] "simple expression du rapport de du à dt" *(ibidem*, 25 (article 22)).

("force accélératrice"), which is indicated by the variable φ, therefore undergoes *de facto* a transformation from a cause into a mathematical-mechanical function. It is the "quantity to which the increase of velocity is proportional."[91]

'Power' and 'cause' mean something quite different. Although d'Alembert uses these terms more frequently than the term 'force', they have much less significance for his mechanics, at any rate in a narrow sense. In the formulation of his principle of inertia, the term 'cause' is used primarily in a negative way, then meaning the *absence* of external influence.[92] When discussing the principle of the composition of motion, it was not so much the powers themselves as the resulting velocities that were important.[93] This applies equally to all other references to power or cause. In the equations of motion, the terms 'power' and 'cause' do not appear at all.

Although the terms could be dropped at the formal mathematical level, their frequent use indicates a meaning at a wholly different level. The distinction that we have previously seen d'Alembert make between metaphysical (true) causes and mechanical (apparent) causes returns here. For d'Alembert, the term 'power' initially comprises the actual, metaphysical cause of motion and changes in motion. In article 5, immediately following his proof of the principle of inertia, d'Alembert defines the 'power or cause of motion' as "anything that forces a body to move."[94] By this, he appears primarily to mean *changes* in motion, both continuous and instantaneous, but sometimes *perseverance* of motion as well—as in the proof of the principle of inertia. It is impossible to determine the relationship of these causes to the effects, but for d'Alembert it is equally impossible to allow the ultimate, true causes to disappear entirely and he therefore feels obliged to confirm their existence repeatedly, or in any case to leave this question open.

As a result, d'Alembert's concept of force appears to fall into two parts. Firstly, it can have the meaning of property or function. In the latter case, it is a mathematical quantity which serves only to express space-time phenomena more easily, but which can itself take no step beyond the horizon of these phenomena. Secondly, it can indicate an indefinable cause, in which case it is a metaphysical concept, about which we can say little more than that such a cause must exist.

At first glance this distinction appears to ignore the frequently condescending attitude that d'Alembert takes towards metaphysical causes. For example, in the article "Force," he writes, while discussing a collision process, that the term 'force' can be used to indicate either a perception, or a metaphysical being, or the effect itself, but that the "only reasonable meaning" ("seul sens raisonnable") is the latter.[95] If we were to conclude from this that he also categorically rejects the metaphysical meaning, then we must also, strictly speaking, make the same judgment about the

---

[91] "quantité à laquelle l'accroissement de la vitesse est proportionnel" *(ibidem)*.
[92] See above, section 4.4.3.1.
[93] See above, section 4.4.3.2.
[94] "tout ce qui oblige un Corps à se mouvoir" *(ibidem,* 4 (article 5)).
[95] D'Alembert, "Force," 111.

first, sensualist meaning. We would then have to say that d'Alembert disavows his own partially sensualist epistemology!

Another consideration is that if there was no metaphysical dimension in d'Alembert's concept of force, his principle of inertia would be a tautology.[96] If the concept of external cause is viewed as a mathematical function which by definition indicates change in motion, then indeed the principle says simply that motion does not change when it does not change. But if we view the concept as a metaphysical cause—and then in a broader sense, not only as a 'being'—the principle expresses something that is the *state* of matter. This turns the principle of inertia into perhaps the most important pillar of modern mechanics; it is certainly not something that is self-evident.

4.5.2 The True Measure of Living Force

This antithesis of metaphysical cause and mathematical function is also the basis for d'Alembert's solution to the problem of the true measure of living force. In the previous chapter we saw that the key to an understanding of the *vis viva* controversy was that force was seen as a substance—as divisible and as joinable as if it was water. Acknowledging this, we can see that the problem lay in a contradiction: forces are equal according to one criterion and unequal according to another. In d'Alembert's view, however, the concept of the force of a moving body can be meaningful only if we view it as "the property that moving bodies have to overcome obstacles they encounter, or to resist them."[97]

D'Alembert's definition allows him to determine force using the known mechanical laws and the experiments conducted by 's Gravesande and others. At the same time he separates the question of what quantity is the true measure of living force from the question of what quantity is conserved, thereby completing the process that was initiated by the mathematical approach to the collision process in the 1720s. That is why he discusses these two questions separately, the former in the preface to the *Traité de dynamique* and the latter in Chapter 4 of Volume 2.

Let us now look at the way d'Alembert thinks he can determine the magnitude of the force using 'obstacles'.[98] There are as many ways to do this as there are types of obstacles. D'Alembert arrives at a total of three, of which one turns out to be unusable. This is the type of obstacle that instantaneously destroys every motion regardless of its magnitude. For example, when an inelastic particle collides with an impenetrable wall, all velocity perpendicular to the wall is lost immediately.

An obstacle of the first type would be another moving body capable of instantaneously destroying the motion of the body in question, on the condition that it is in

---

[96] Hankins, who does not recognize this metaphysical aspect, indeed says that d'Alembert's principle of inertia is a tautology (Hankins [Introduction to d'Alembert's *Traité de dynamique*] (1968), xxvi).

[97] "la propriété qu'ont les Corps qui se meuvent, de vaincre les obstacles qu'ils rencontrent, ou de leur résister" (d'Alembert, *Traité de dynamique*, xviii).

[98] *Ibidem*, xix-xxi.

equilibrium with it.[99] According to the principle of equilibrium, this is the case if the quantities of motion, i.e. the products of mass and velocity, are equal. Therefore the quantity of motion is both the magnitude of the obstacle to be overcome and the magnitude of the force.[100] An obstacle of the second type would be, for example, a spring that slows a movement over time. We can first look at the *number* of springs that a moving body can simultaneously or sequentially compress; this number proves to be proportional to mv^2.

Consequently, d'Alembert concludes that different situations result in different measures of force. However, he is not satisfied with this and also wants to show that different measures can, with equal justification, be defined in the *same* situation. He therefore proposes to consider the 'total resistance'. This is certainly an obscure concept, which d'Alembert might well have scorned if it were applied by a metaphysicist such as Leibniz.[101] Moreover, he employs it in an ambiguous fashion. 'Resistance' is a Newtonian force, which, if multiplied by a time element, results in the change in the quantity of motion during that time. However, 'total resistance' is the sum of the products of the 'resistance' with the time elements. That is, it is the integral of the Newtonian force over time, which is also equal to the change in the quantity of motion. This third method of measurement, applied to obstacles of the second type, therefore results in the same measure of the force of moving bodies as that using obstacles of the first type.

The three methods then result in two measures of force, mv and mv^2. And although the quantity of motion wins as a measure because it is a common measure for static equilibrium and motion, the choice is otherwise a matter of preference and convenience. D'Alembert certainly does not want to claim that the quantity of motion is a better measure,[102] but he can only refute the thesis that living force is something real by, as it were, making the circle begin and end with the quantity of motion. These formulas do not express *causes*, but *effects*. Stated more precisely, the concept of living force is nothing more than an "abbreviated way to express a fact" ("maniere abrégée d'exprimer un fait"), where the 'fact' is naturally a resistance or an obstacle overcome. Consequently, it all belongs to the sphere of phenomena.

---

[99] 'Equilibrium' is defined by d'Alembert as the mutual obstruction of motion (see above, section 4.4.6). However, in d'Alembert's actual definition of this type of obstacle a more dynamic concept of equilibrium is applied. He literally speaks of obstacles which "have exactly the resistance that is needed to destroy the motion of the body" ("n'ayent précisément que la résistance nécessaire pour anéantir le Mouvement du Corps") (*ibidem*, xix). For the sake of simplicity I have used a concrete example here. A 'translation' to d'Alembert's own words would go too far at this point because d'Alembert makes a fluid transition from static equilibrium to kinematic equilibrium.

[100] Iltis's objection that d'Alembert is here incorrectly equating dead force with quantity of movement does not get us anywhere (—, "D'Alembert," 137). She apparently interprets d'Alembert's concept of 'equilibrium' as a merely static concept, which in the context of forces of bodies in motion is wrong! See above, footnote 100.

[101] See Chapter 5 below, in which Maupertuis rejects Leibniz's principle of least resistance as 'unclear'.

[102] As Hankins thinks ("Eighteenth-Century Attempts," 283).

4.5.3 D'Alembert's Transformation of the Concept of Force

Nevertheless, however strongly d'Alembert has to emphasize the functional character of living force, his argument would not be complete without adding something about the cause of the various effects. After all, "the body viewed in itself has nothing more in one case than in another."[103] One and the same cause can have a variety of effects. And this answer, the unity of metaphysical causes opposed to the diversity of mechanical effects, is what distinguishes him from 's Gravesande, who, despite his recognition of a diversity of effects, still clung to an unequivocal cause-effect relationship. At the same time, his answer shows a surprising similarity to Leibniz's distinction between a metaphysical and a phenomenal sphere of reality. The difference is that Leibniz claimed to be able to derive phenomenal forces from the metaphysical, whereas d'Alembert held this to be impossible.

D'Alembert's solution is therefore a negative one: he declares the question to be superfluous and useless for mechanics. The only useful talk is of effects, and of how to systematize the relationships between them using functions. The ultimate objective of science is a deductive system based on one or more fundamental principles and resulting in functional relationships between the effects. Causal relationships are not part of this.

D'Alembert's solution thus has two aspects, one related to his new definition of mechanical force (especially living force), the other related to the terms 'power' and 'cause'. Each aspect implies a radical change in the traditional relationship of metaphysics to mechanics. Only by taking account of *both* aspects can it become clear what d'Alembert's solution implied for the tradition: to what extent was it a *continuation*; to what extent does it present a *break*.

The first aspect implies that d'Alembert accepted the traditional distinction between a metaphysical-causal and a mechanical-effective sphere of material reality, but that—in contrast to others such as Leibniz—he denied *a priori* any possibility of bridging the gap. What for Leibniz is a distinction, is a divide for d'Alembert.

The second aspect entails seeking another way of founding the laws of motion and mechanical forces, namely, by means of the time-space structure of mechanical reality. This alternative foundation simultaneously implies another kind of metaphysics, although d'Alembert does not recognize it as such.

The two aspects, taken together, produce a revolution in the metaphysical foundation of mechanics. As a result, it is no longer possible to provide a foundation for causal relationships, only for principles of motion. Whereas the relations of substances to their properties and actions provided the initial model for the foundation of the laws of motion, the basis is now the relationship of the time-space structure to its principles.

---

[103] "le Corps considéré en lui-même n'a rien de plus dans un cas que dans un autre" (d'Alembert, *Traité de dynamique*, xxii). D'Alembert only added this consideration in the second edition of 1758. See above, footnote 22.

In this interpretation it is actually quite odd to state that d'Alembert 'solved' the *vis viva* controversy. It would have been much more in accordance with his own ideas about foundation if he had simply not allowed the term 'force' to arise in his theory. After all, it is too reminiscent of the indefinable cause-effect relationships. The term 'force', however, has become too commonplace simply to avoid using it. D'Alembert therefore adopts it by redefining it. An unintended consequence of this redefinition is that the concept is simultaneously transformed from a cause to a function. In this way d'Alembert solves a quite different problem, not that of the true measure of *causes*, but that of the true measure of *phenomena*.

This transformation of the concept of force, and d'Alembert's search for a new foundation of mechanics, entailed turning his back on substance metaphysics, at least as far as mechanics is concerned. Thus we saw that, where d'Alembert describes force as a 'property', we must not think of a property of a substance. 'Property' is used in the sense of the property of the time-space structure of moving matter. We could therefore summarize the transformation of force as the *transformation of a substantial concept into a concept of time-space structure*.

After d'Alembert, the theorists of mechanics continued to talk about 'force', without noticing that it had become a different concept. Because the concept of force had lost its earlier substantial character, the previous controversies had become incomprehensible. Likewise, the later principle of the conservation of energy differs fundamentally from attempts to define the true measure of force, because their roots lie in differing systems of metaphysics: the former in a metaphysics of structure, the latter in a metaphysics of substance.

4.6 Conclusion: Structuralization and Instrumentalization

In this chapter I have sought to clarify where d'Alembert, in his approach to the concept of living force, goes further than authors like 's Gravesande. Although 's Gravesande had also realized that the force of a moving body can be expressed in different ways depending on the type of resistance it has to overcome, he did not in fact draw any conclusions from this regarding the nature of the concept of force as such. In contrast, d'Alembert transforms the concept of force from a cause into a phenomenon, from a substantial concept into a structural one.

In the literature about d'Alembert the emphasis is frequently placed on his instrumentalist position. However, it is important to realize that the structuralization d'Alembert accomplishes, does not *necessarily* imply an instrumentalization as well. The present book is not so much concerned with the strictly individual solution that d'Alembert proposed for the problem of living force as with his contribution to the development of the relationship between metaphysics and mechanics.

Seen in this framework, the core of d'Alembert's argument is the insight that mechanical concepts must not be viewed as referring to individual substances, but to the time-space structures in which these substances are incorporated. To the extent that a metaphysical foundation is still possible, it must be of a structural nature. D'Alembert himself unintentionally provided an example of this with his foundation of the principle of inertia. Modern examples are the principle of symmetry or Einstein's relativity principle.

During the eighteenth century we see this new mode of metaphysical foundation return in the controversy over the principle of least action. This, then, provides the theme for our next chapter.

Chapter 5

From Efficient to Final Causes: The Origin of the Principle of Least Action

5.1   Introduction

5.1.1   The Greatest Possible Scandal at Court

Voltaire, the glorifier of reason and the world, must certainly have looked back on his time in Berlin with ambivalent feelings. After the death of his sweetheart the Marquise du Châtelet he spent some time at Sanssouci, Frederick the Great's country residence in Potsdam. He soon proved to be more imaginative and more entertaining than the then President of the Royal Academy, Maupertuis, so that they quickly became enemies.[1] The fact that Voltaire had been bypassed for the presidency six years earlier, in favor of Maupertuis, may also have played a role. Voltaire was bent on revenge, which can indeed be sweet.

The opportunity soon arose, in the form of a dispute about who could claim to be the originator of the mechanical principle of least action. Maupertuis, supported by Euler, defended his claim to discovery against the mathematician Samuel König, who claimed that the honor properly belonged to Leibniz. Voltaire then wrote a satire, the "Diatribe du docteur Akakia" (1752), which criticized Maupertuis for his pride: Maupertuis "had to be unbounced."[2] Although there were *two* attempts to burn the complete first printing of the booklet—at the command of Frederick the Great and with the 'co-operation' of Voltaire himself!—it was nevertheless possible to smuggle one copy out of Berlin.

Voltaire wrote a sequel in Dresden, in which König and Maupertuis signed a peace treaty. This contains the following passages, which also target Maupertuis's ally, the mathematician Euler:

> We no longer deride the Germans, and we confess that figures such as a Copernicus, a Kepler, a Leibniz, a Wolff, a Haller, a Mascau and a Gottsched

---

[1]   For the story of the relationship between Voltaire and Maupertuis, which developed from being fellow spirits to the greatest possible enmity, see Tuffet's 1967 introduction to Voltaire's *Histoire du docteur Akakia et du natif de St. Malo*, vii-cxxxi.
[2]   Compare the unbouncing of Tigger in Chapter 7 of A.A. Milne's *The House at Pooh Corner*.

knew a thing or two, and that we have studied under the Bernoullis, and are still studying. (...)

Moreover (...) our Lieutenant-General, Leonard Euler, empowers us to declare on his behalf (...) that he confesses in all innocence that he has never known anything about philosophy, and that he sincerely repents that he ever allowed us to persuade him of the possibility of knowing without having studied.[3]

This satire was written during the sequel to the discovery of the principle of least action. The scientific and philosophically discussion had already given way to the less exalted question of who should be regarded as the rightful discoverer of this principle which, it was claimed, linked metaphysics, theology and physics with one another and provided a common center for the whole of physics.[4] The booklet was the climax of Voltaire's involvement in this struggle for the honor of discovery.

His involvement began much more quietly in the autumn of 1752 with Voltaire's anonymously published response to the battle between Maupertuis and König and an answer, also anonymous, from Frederick the Great on Maupertuis's behalf—Maupertuis himself having fallen ill in the autumn.[5] Voltaire's satire is intended as an answer to Frederick the Great's open letter. Its secret publication not only led to a definitive break with Frederick the Great, it is also quite likely that, as Brunet writes, the controversy could have been simply ended had the participants calmed down, were it not for Voltaire's petulant intervention.[6] Voltaire's venom left its mark on the evaluation of Maupertuis and Euler at the time, and has indirectly influenced later evaluations. So there is much to say for Frederick's telling remark, in answer to the publication of the *Diatribe*: "If we consider your works, statues should be built in your honor, but if we considered your conduct, you would be thrown in chains."[7]

---

[3] "Nous ne rabaisserons plus tant les allemands, et nous avouerons que les Kopernick, les Kepler, les Leibnitz, les Wolf, les Haller, les Mascau, les Gotsched sont quelque chose, et que nous avons étudié sous les Bernoulli, et nous étudierons encore," and "De plus (...) notre Lieutenant-Général, Léonhard Euler déclare par notre bouche ce qui suit. I. Qu'il confesse ingénuement de n'avoir jamais apris la Philosophie, et qu'il se repent sincèrement de s'être laissé persuader par nous qu'on pouvoit la savoir sans l'avoir étudiée. II. (...)" (Voltaire, *Traité de paix conclu entre Mr le président de Maupertuis et Mr le professeur Koenig* (1753), 26-27 and 39.

[4] See Du Bois-Reymond, "Maupertuis" (1892); Harnack, *Geschichte der Königlich Preussischen Akademie der Wissenschaften zu Berlin* (1900) I.1, 331-345; Fleckenstein, [Foreword to Euler's *Commentationes mechanicae; principia mechanica*] (1957); Tuffet, [Introduction to Voltaire's *Histoire du docteur Akakia*] (1967); Costabel, "L'affaire Maupertuis-König et les 'questions de fait'" (1979).

[5] [Voltaire], *Response d'un académicien de Berlin à un académicien de Paris* (18 September 1752); [Frederick the Great], *Lettre d'un académicien de Berlin à un académicien de Paris* (11 November 1752).

[6] Brunet, *Maupertuis I. Étude biographique* (1929), 157.

[7] "si vos ouvrages méritent qu'on vous érige des statues, votre conduite vous mériterait des chaînes" (cited in: Voltaire, *Mélanges*, 1439).

Intermezzo
The Mathematical Formulation
of the Principle of Least Action

To give the reader some sense of the substance of the principle of least action, without having to assume any great knowledge of its original or contemporary context, I will introduce the principle in the mathematical form given to it by Euler and Lagrange. They understand it, in formal terms, as the principle that of all the possible mechanical changes, the change that actually occurs is that for which a certain quantity, called the 'action', is a minimum.

What does this mean for a simple example such as the trajectory of a ball that is thrown into the air? Or in more general terms, what does it mean for a moving particle that is subject to external central forces such as gravity (see Figure 5.1)

Suppose that the particle begins to move with an initial speed $\underline{v}_0$ at point $\underline{x}_0$ and time t_0. At time t_1 it will arrive at point $\underline{x}_1$, moving at speed $\underline{v}_1$. Now, it is possible to calculate the path it has traveled, point by point, on the basis of the forces at work and the initial conditions. We then determine the continuous changes that result from these two factors, by regarding the force as a succession of infinitesimal actions, and calculating their net effect by integration. But

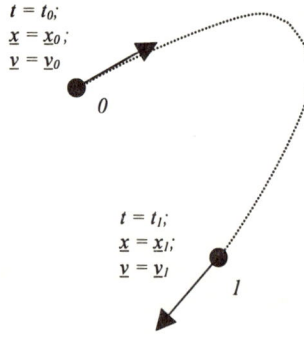

Figure 5.1 A particle in motion under the influence of external central forces

we can also calculate the path based on a principle of the *whole*. That is, for all possible paths, we determine the integral of a function of the characteristics of that path, known as the action. This integral is then minimal for the actual path taken.

In the case given, this integral has the form $_0\int^1 m\underline{v}d\underline{s}$, where 0 and 1 refer to the initial and final situations. However it will be clear that this principle entails some conditions regarding the possibility of variations. These are in part mathematical conditions, such as that the variations should be zero in the beginning and end situation. There are also physical conditions, such as that the conservation of mv^2 applies in all variations. Both conditions can already be found implicitly in Euler, and they were explicitly formulated by Lagrange (*Méchanique analitique*, 274-281).

For the sake of completeness it should be noted here that the conditions as found in Euler and Lagrange are not necessary. Nowadays the principle is presented in the more general formulation of Hamilton: $S=\int £(q, dq/dt, t)dt$ is a minimum. Here £ is the Lagrangian function, S the action, q a generalized coordinate and t is the time. For a mathematically-based introduction to the principle, see Whittaker, *A Treatise on the Analytical Dynamics of Particles and Rigid Bodies* (1904), 245-262; and for a more popular approach see d'Abro, *The Rise of the New Physics* I (1939), 256-268, or Feynman, *The Feynman Lectures on Physics* (1961–1963) I.26, 1-8 and II.19, 1-14.

5.1.2 The Significance of the Principle of Least Action for Mechanics

Voltaire's ridicule of Maupertuis's arrogant attitude may have been the basis of historians' often cynical attitude to Maupertuis's claim to a scientific reputation. But when we remove the remainders of the web that Voltaire wrapped around his prey, we can see that Maupertuis's principle of least action is not without interest, and that it has considerable significance for the science of mechanics. The historical significance of the principle lies in the first place in its being the first successful attempt to gain new insights in mechanics from an approach based on final causes. While such an approach was fiercely disputed in mechanical philosophy, it was nevertheless a common element in the natural sciences of the eighteenth century. However, teleological considerations were generally limited to reflections in natural theology. Except in optics, they were little used in the scientific theories proper. Maupertuis succeeded, by means of his principle, in making room for teleology in mechanics.

The principle was to suffer a remarkable fate after it was brought into the world by Maupertuis and Euler. Although the principle in all its variations and developments has borne and still bears the name that they chose to express its teleological character, this teleological character has been just as consistently denied, almost from the beginning. It has been recognized only as a mathematical principle. This is another reason why the principle of least action occupies a special place among eighteenth century teleological views.

Apart from the fact that it gave the physicists of the time an opportunity to vent their bile over metaphysics, the principle initially had no great significance for mechanics, not any more than for physics in general. Limiting conditions were formulated and the formulation itself was refined, but the principle was generally regarded as superfluous. As early as 1800 the principle seemed to have been forgotten. Hegel wrote nothing about this principle in his philosophy of nature, although he was well informed about the state of physics at that time, and the principle would have fitted consistently into his speculative dialectic philosophy, because of its unifying character.

Nevertheless the principle had a belated flowering, beginning around 1860, with Thomson and Tait in England and Helmholtz in Germany.[8] The principle seemed to develop into a universal overarching principle, from which all fundamental physical laws such as the laws of the conservation of energy and momentum and the equations of motion could be derived. But the revolutionary storms around the turn of the century prevented further growth. While it survived the theory of relativity with

---

[8] Thomson and Tait, *Treatise on Natural Philosophy* (1867). They predicted a great future for the principle, not only in dynamics, but also in "several branches of physical science now beginning to receive dynamical explanations" (*ibidem*, 337). In Germany, Hermann von Helmholtz extended the principle for domains beyond dynamics, such as electrodynamics, and pointed to the importance of the principle for all reversible natural phenomena (see Chapter 1, note 67).

flying colors,[9] in quantum mechanics its significance is reduced to providing the transition to classical mechanics.[10]

Although this later history does not concern us here, it is important to observe that, from this period on, the quantity of action became a purely mathematical magnitude, which could not really be visualized in the way that it was possible, up to a certain point, to visualize concepts such as mass, force and energy. In so far as action had any physical meaning, this could only be known through its mathematical expression.

As has been said, the situation at the beginning of the principle's development was different. In the years 1740–1750, when the principle was explicitly formulated by Maupertuis, Euler, König and some others, a physical and even metaphysical significance was attributed to it: the concept was closely linked to the action of forces. For example, the physical meaning of the concept of action was manifest in attempts to provide an ontological foundation for the concept and the associated minimal principle in particular situations. At that time the principle was part of a metaphysical debate—that to us counts as outmoded, but which must be seen as the historical basis of also the modern effort to find minimal principles in physics.

What concerns me here is mainly to clarify this physical and metaphysical issue, and to examine the significance of the principle for the development of mechanics. Events at the Berlin Academy between 1740 and 1751 were central to the physical and metaphysical issue. During that period a great deal of attention was given to the metaphysical and physical aspect of the principle, with Maupertuis and Euler playing key roles.

A new phase began in 1751, when König wrote an article in the famous Leipzig journal *Acta eruditorum*, in which he not only demonstrated that Maupertuis's

---

[9] In 1915 Max Planck even claimed that the principle "deserves the highest place among all physical laws" ("unter allen physikalischen Gesetzen die höchste Stelle einzunehmen geeignet ist"), because the principle, at least in its Hamiltonian form, has the characteristic of being Lorentz-invariant (Planck, "Das Prinzip der kleinsten Wirkung," 701).

[10] Modern handbooks generally do not deal with the principle explicitly. However, this need not mean that the principle cannot have any overarching significance. The principle is given a remarkably positive treatment in Landau and Lifshitz's series *Course in Theoretical Physics*. In the volume on *Mechanics* they regard the principle, in the form that Hamilton later gave it, as "[t]he most general formulation of the law governing the motion of mechanical systems" (—, *Mechanics* (^{2}1969), 2). The principle has the same status in their treatment of relativistic mechanics (—, *The Classical Theory of Fields* (1989=^{4}1975), 24). However, in other fields they do not mention the principle. It is only when they come to non-relativistic quantum mechanics that they are able to assign it any significance. This is in the approximation of the point of transition from quantum to quasi-classical mechanics, in which the quantity of action S is said to be proportional to the phase φ in the wave function $\psi = ae^{i\varphi}$, where Planck's constant, $\hbar = 1.054 \cdot 10^{-27}$ erg sec, is the constant of proportionality, so that $\psi = ae^{iS/\hbar}$ (—, *Quantum Mechanics; A Shorter Course* II, 24-26). However, Kemble considers that this 'action function' is not identical to that in the principle of least action, because the latter depends on the path whereas the former does not (Kemble, *The Fundamental Principles of Quantum Mechanics* (1937), 44). In recent years there has been renewed interest in the principle. It is accorded an important position in a modern textbook for relativistic quantum field theory: "it leads to the path integral formulation of quantum mechanics" (Wit and Smith, *Field Theory in Particle Physics* I (1986), 1).

proofs contained various mathematical errors, but also cited the last part of a 1707 letter from Leibniz in which he was said to have formulated the principle.[11] So began a vulgar dispute about priority, later turning into a court scandal, as described above. However, I will ignore this period because it gives us scarcely any idea of the principle or its background.

The historiography of the principle is a striking illustration of the dangers of a one-sided contextual approach, because the dominant emphasis on the dispute of the early 1750s about who originated the principle has seriously clouded our view of the substantial significance of the previous years. An incorrect picture of the relationship between Euler and Maupertuis has even been created. In a foreword to Euler's publications on the principle of least action, Fleckenstein says that the usual picture of Euler, that he incorrectly defended Maupertuis and in so doing betrayed König in a cowardly way, is at the most applicable only to the period after 1751.[12] For the 1740-1751 period, when there was still no question of a dispute about priority, and thus no question of displaying loyalty, this picture is not correct and Euler's eulogies of Maupertuis should be taken seriously. This implies that we must also take his *work* during that period seriously.

Although the principle of least action began as a teleological principle, its metaphysical meaning was not determined only by its teleological basis. Maupertuis and Euler sought an alternative to the Newtonian mechanics of forces, but in the light of later findings it must be said that the route they chose was not the smooth main road but a steep and winding path that finally came to a dead end. However, it would not be correct to suppose that they were only looking for the white hart that drew them onto this side path: a teleological explanation of apparently determined reality. To be fair to them, our interpretation of their alternative should do justice to the common features as well as the particularities of what is now the beaten path and of the path they cut for themselves. The history of the principle of least action shows that the ideas on the basis of the concept of force developed strongly during the eighteenth century. This development was possible because of a previous change in the concept of force, as described in Chapter 4, which I interpret as a shift in the metaphysical foundation of mechanics from substance to structure.

---

[11] König, "De universali principio aequilibrii et motus in vi viva reperto" (1751).
[12] Fleckenstein, [Foreword to Euler], XXV-XXVI. Fleckenstein says that Harnack (*Geschichte der Akademie* (1900) I.1, 331-345) and Kneser (*Das Prinzip der kleinsten Wirkung von Leibniz bis zur Gegenwart* (1928), 26-29) are responsible for this picture of Euler. In my opinion it would be interesting to evaluate what role Voltaire's manipulation of emotions may have played in creating this picture (cf. pages 134-135 above).

5.2 MAUPERTUIS'S NEWTONIAN BACKGROUND

5.2.1 From Soldier to Scholar

Before Maupertuis drew the derision of Voltaire, and with him of the whole world, through his fanatical efforts to claim the honor of first proposing the principle of least action, he was a famous and widely respected scholar. This respect was based mainly on the fame he had acquired in leading the army of the Newtonians in what could well be called the decisive battle against the Cartesians in France: the proof that the globe is flattened and not pointed at the poles.[13] The contrast between fame and derision arose not only from the capriciousness of the public, but also from a shift in Maupertuis's interests in natural philosophy. Although Maupertuis had always felt he was Newtonian, his research took him steadily further from the empirical and mathematical basis on which he had previously defended the Newtonian theory of attraction, and led him to a speculative metaphysical foundation that was generally rejected by Newtonians, and is in fact reminiscent of Leibniz's metaphysical approach to natural science.

This development implies that we cannot understand the meaning of the foundations Maupertuis's proposed later without first knowing why he rejected the old foundations. So the story must begin at the beginning, with an outline of Maupertuis's early years in physics.[14]

Maupertuis's youth was certainly not marked by any great interest in science. At the age of nineteen, after two disappointing years spent studying philosophy, he decided on a military career. It was only then that he was infected with enthusiasm for mathematics, under the inspiring influence of Guisnée, who was famous for his *Traité d'application de l'algèbre à la géometrie* (1705). Maupertuis became engrossed in a well-known discussion, in which Guisnée played a prominent role, concerning the fundamental problem of calculus: the existence of the infinitely small. It seems that this mathematical interest led him to turn his back on the military after three years of service, to try to win a place in the Paris *Académie Royale des Sciences*. Late in 1723 he was appointed as "adjoint mécanicien," and in the following years he presented a number of papers, including one biological and three mathematical studies.[15]

Up to this point, Maupertuis's career was hardly spectacular. However, his stay in London for six months in 1728 would radically change this, because he became an enthusiastic advocate of Newtonian mechanics and astronomy. In a famous passage in his *Lettres anglaises,* Voltaire gives pointed expression to the observation that France and England were actually two different worlds:

---

[13] For a detailed description of the associated 'journey to the pole' see Brunet, *Maupertuis I. Étude biographique*, 33-68.
[14] The biographical information for this outline is drawn from Brunet, *Maupertuis I. Étude biographique*, except where another source is indicated.
[15] For a bibliography of Maupertuis's work, see Brunet, *Maupertuis II. L'œuvre*, 461-465, and Tonelli, [Introduction to Maupertuis], XXIV*-XL*.

> A Frenchman who arrives in London will find things to have changed greatly, in philosophy as well as in other areas. He left the world full, and now finds it empty. In Paris the universe appears to be composed of vortexes of subtle matter, in London these are nowhere to be found. With us, the pressure of the moon causes the tides, but for the English it is the sea that is drawn towards the moon (...) for you Cartesians everything comes about through an impulse that one can scarcely understand, for Monsieur Newton it is through an attraction, whose cause is equally unknown; in Paris you imagine the earth shaped like a melon, in London it is flattened on both ends. For a Cartesian, light consists of air, for a Newtonian it travels from the sun in six and a half minutes (...) You can see how strident the contradictions are.[16]

Maupertuis's enthusiasm for Newton's theory of attraction played a very important role in the acceptance of Newtonianism in France. However, thorough preparation and a careful strategy were necessary to ensure that his ideas would be received in the French world. How was he to achieve it?

5.2.2 *Maupertuis's Defense of Newton's Theory of Attraction*

On returning from England, Maupertuis sought to win acceptance for Newton's theory of attraction in France. At first he gave no sign of this: he published a paper on differential calculus and carried out a field study of scorpions. At the same time, however, he was preparing himself to defend Newton's physics by studying the finer points of calculus for a year in Basel, under the great teacher Johann I Bernoulli, who had also been Euler's teacher. This resulted in several mathematical articles. In 1732, after several years in the lion's den, he finally dared to come out in public as an advocate of Newton's doctrine of attraction. He published the book *Discours sur les différentes figures des astres avec une exposition des systèmes de M. Descartes et M. Newton,* and a paper entitled "Sur les loix de l'attraction."[17]

In *Discours sur les différentes figures* Maupertuis defends the Newtonian theory of attraction versus the Cartesian vortex theory.[18] He precedes the mathematical and

---

[16] "Un Français qui arrive à Londres trouve les choses bien changées en philosophie comme dans tout le reste. Il a laissé le monde plein, il le trouve vide; à Paris on voit l'univers composé de tourbillons de matière subtile; à Londres on ne voit rien de cela; chez nous c'est la pression de la lune qui cause le flux de la mer, chez les Anglais c'est la mer qui gravite vers la lune (...) chez vos cartésiens tout se fait par une impulsion qu'on ne comprend guère, chez M. Newton c'est par une attraction dont on ne connaît pas mieux la cause; à Paris vous vous figurez la terre faite comme un melon, à Londres, elle est aplatie des deux côtés. La lumière pour un cartésien existe dans l'air, pour un newtonien, elle vient du soleil en six minutes et demie (...) Voilà de furieuses contrariétés" (Voltaire, *Lettres philosophiques* (1734): "Lettre XIV. Sur Descartes et Newton." In: Voltaire, *Mélanges*, 54).

[17] Later, in the third edition of his *Œuvres* in 1756, Maupertuis was to add this paper, with some changes, as the concluding section of the *Discours*, under the title "Conjectures sur l'attraction." I refer here to the 1768 edition, which is almost the same as the third edition of Maupertuis's *Œuvres* from 1756.

[18] For the history of explanations of universal gravitation see Van Lunteren, *Framing Hypotheses; Conceptions of Gravity in the 18th and 19th Centuries* (1991), especially Chapters 1 to 3.

experimental comparisons between the two theories with a section entitled: "A Metaphysical Discussion of Attraction" ("Discussion métaphysique sur l'attraction"), in which he seeks to demolish the idea that attraction is a "metaphysical monstrosity" and an "absurd principle." There is no indication in his argument of the arrogance with which he was to present his principle of least action in later years. On the contrary, he makes a most unassuming entry to the field where champions debate the question of whether gravitation should be explained on the basis of the principle of the force of impact and inertia, as in the Cartesian tradition, or is a principle inherent to matter, as some Newtonians say. He writes: "It is not my place to issue a judgment in a question that has divided the greatest philosophers, but I may well compare their ideas."[19] He says that he only wants to demonstrate that the concept of attraction is *metaphysically possible*, that it is *equivalent in terms of epistemology* to the force of impact, and especially that it is *mathematically more fruitful* than the Cartesian vortex. His main argument is that we do not have 'complete concepts' ("idées complettes") of things, we only know their characteristics on the basis of experience.[20]

From this beginning he develops various arguments. The first is that no other characteristics can be derived from known characteristics, except that characteristics that contradict those already known are not possible. While it does follow from this that we cannot attribute characteristics to bodies other than those experience has taught us, this does not imply that we may ignore particular characteristics in favor of others. Even where a characteristic seems bizarre and inconceivable, this can never be an argument for rejecting it: it is only habituation that makes some characteristics seem more evident and clearer than others; when it comes to the crunch, we can say of *all* characteristics that "they are perpetual mysteries for us."[21]

This counterargument of the mysterious foundation of all material characteristics is followed immediately by a proof of the untenability of the occasionalist objection to action-at-a-distance. For if one says that the force of impact is not inherent to matter itself, but that God continually effects the laws of motion through direct intervention in the motion of colliding bodies, it is also possible to say that God continually moves bodies towards one another.[22]

Maupertuis reinforces the empiricist approach in this epistemological argument with a classification of the characteristics that he then goes on to use against a metaphysical counterattack. The argument is that material characteristics can be subdivided into orders of various degrees of universality. The highest order is that of 'primordial', 'invariable', or 'primitive' properties, such as extension and impene-

---

[19] "[c]e n'est pas à moi à prononcer sur une question qui partage les plus grands Philosophes, mais il m'est permis de comparer leurs idées" (Maupertuis, *Discours sur les différentes figures (...)*, 90). Maupertuis consciously distinguishes these opinions from those of Newton, for whom the term "attraction" does not indicate the *cause* of weight but rather the fact itself, deliberately leaving open the possibility that attraction is an effect of a true impulsion (*ibidem*, 92).
[20] *Ibidem*, 94.
[21] "[c]e seront là toujours des mysteres pour nous" *(ibidem, 98)*.
[22] *Ibidem*, 99.

trability.[23] These are found in all bodies. The second order is that of situation-dependent characteristics, such as the characteristic of all moving bodies that they "move other bodies they meet," and the attributes of being at rest and being in motion.[24] The least universal order is that of individual characteristics such as having a particular form, color and odor.[25]

He then uses this classification to defend himself against the argument that, while we cannot demonstrate the necessary existence of forces of attraction, the forces of impact are said to be a necessary result of the impenetrability of matter. This argument only addresses the question of the *necessity*, and not the *possibility*, of attraction. However, he shows that the argument does not make attraction any less possible, by pointing out that the concept of necessity assumes that the various characteristics are ranked according to universality, and that a less universal characteristic is subordinate to one of higher order. Thus, motion in a collision is subordinate to the impenetrability of the colliding bodies.[26] Now, if the characteristic of attraction is not the demonstrable and necessary result of a characteristic of higher order, this does not contradict the possibility that the attraction itself may be a universal characteristic of the highest order, which is therefore not subordinate to any other.[27]

Finally, his mathematical argument goes without saying: it is based entirely on the evident success of Newtonian mechanics in comparison to Cartesian mechanics. However, what is of interest here is that something of the way in which he would later reconcile teleological causes with mechanics is already visible. It can be seen in "Sur les loix de l'attraction," in which Maupertuis mainly reiterates the defense that had been begun ten years earlier by agnostic Newtonians such as 's Gravesande.[28] However, he adds a new element to this line of argument. He asks whether there is a reason why gravitation has the precise form of inverse proportionality to the square of the distance.

He sees two possible explanations, one metaphysical and the other theological. Attraction may be derived either from the essence of material bodies or directly from the will of the Creator. The first possibility is not accessible for humans, but the latter offers us an important insight. In the actual form of the law of gravitation, the center of gravitation and the center of mass are the same, and the combined force of attraction of a system of particles has the same form as their individual forces of attraction.[29] The law is thus uniform for separate particles and for composite systems. Although this uniformity is not perfect,[30] this argument does indicate the es-

---

[23] "propriétés primordiales,", "invariable,", or "primitives" *(ibidem, 95)*.
[24] "de mouvoir les autres qu'ils rencontrent" *(ibidem, 96 and 101)*.
[25] *Ibidem,* 101.
[26] *Ibidem,* 102.
[27] *Ibidem.*
[28] See Van Lunteren, *Framing Hypotheses,* 68-73.
[29] Maupertuis, "Sur les loix de l'attraction," 167-168.
[30] Maupertuis recognizes that the uniformity does not appear to be perfect, because a body located within a spherical body is subject to a force that is proportional to the distance to the center of

continued on next page

sence of his later thought: the relationship between God and nature can be discovered through *universal regularities*.[31] Like Leibnizian mechanics, Maupertuis's teleological mechanics of the 1740s focused on universal regularities, and diverged in this respect from Newton's teleological ideas and even more widely from the usual physico-theological proofs of divine purposiveness from the specific characteristics of a species or even from individuals—the so-called 'argument from design'.[32]

However, it would not be correct to conclude from this that Maupertuis had no opinion concerning the actual reality of attraction. It is mainly a question of a textual strategy, intended to first help reluctant Cartesian readers to take a first step, and later to raise the question of real existence as a matter of fact.[33] In fact, attraction accords so closely to observations that he speaks of it in his conclusions as if it is a reality, although this is nowhere explicitly stated.[34]

5.2.3 The Expedition to the Pole

While I do not want to discuss the period between Maupertuis's first defense of Newtonianism and his observations about the principle of least action in any depth, this is precisely the period, roughly between 1732 and 1740, that ensured Maupertuis's fame as the person who introduced Newtonianism to France. We must bear in mind that the French discussion about Newtonian ideas was clouded by nationalism, as both d'Alembert and Maupertuis said. Maupertuis was the first person in France to dare to call himself a Newtonian. He became the support and refuge for the first popularisors of the Newtonian system in France, such as the Marquise du Châtelet and Voltaire.[35]

---

continuation of former page
mass. In fact this irregularity does not disprove uniformity, because there is no analogy between the attraction of spheres on bodies that are inside them and the attraction of the "ultimate particles of matter" ("dernieres parties de la matiere"), since the latter cannot contain particles and therefore can only affect particles outside themselves. Moreover, if attraction were proportional to distance, this would not be more uniform than the actual form of the relationship, and it would even be contrary to the universal order of nature, according to which effects decline as the distance from the cause increases (*ibidem*, 168-169). The argumentation may appear *ad hoc*, but it is not inconsistent with the general line of thought from which Maupertuis seeks his explanation.

[31] Pulte, *Das Prinzip der kleinsten Wirkung*, 43.
[32] For the agreement between Maupertuis's and Leibniz's physico-theology see Terrall, *Maupertuis*, 22-31. However, she deals only with the teleology in Maupertuis's later work (from 1744).
[33] He was, however, not successful, as he complains in *Lettres* (1752): "All this has been to no effect. Although the *Discours* has had some success in foreign lands, it has won me nothing but personal enemies in my own country" ("Tout cela fut inutile; & si ce Discours fit quelque fortune dans les pays étrangers, il me fit des ennemis personnels dans ma patrie") ("Lettre XII. Sur l'attraction," 285).
[34] See also Brunet, *Maupertuis II. L'œuvre*, 45.
[35] Brunet, *Maupertuis I. Étude biographique*, 23-28. Some years later Voltaire was to write a popularizing work, the *Elémens de Philosophie de Newton* (Amsterdam 1738). When he was staying for some time in the Netherlands for the purpose, he wrote to Maupertuis that Musschenbroek had told him that he regarded Maupertuis's *Discours sur les différentes figures* as "the greatest work that

continued on next page

His greatest contribution, however, was in the experimental sphere, in determining the shape of the globe. One of the most important criteria in the choice between Newtonian and Cartesian theory was the shape of the earth and of other planets.[36] According to Newton's theory of attraction, planets should be oblate spheroids, while the measurements of the Cartesian Jacques Cassini in a Paris laboratory seemed to demonstrate that it was an elongated sphere. Two expeditions were set up to determine the shape of the earth more accurately. The first, under the leadership of La Condamine, traveled to the equator in Peru in May 1735. The other, under the leadership of Maupertuis, went to the Arctic Circle in Lapland in May 1736.

The proof that these measurements provided of the superiority of Newton's theory of attraction also bestowed fame on Maupertuis. An ode from Voltaire provides a nice illustration of this fame:

> The globe we hardly knew, he knew to measure,
> Became the monument that made his name.
> His destiny's to demonstrate the earthly form,
> To please the world, and to enlighten it.[37]

5.3   THE BIRTH OF THE PRINCIPLE OF LEAST ACTION, IN MAUPERTUIS AND EULER

5.3.1   Minimal Principles in Statics

In the 1740s, Maupertuis entered a new phase in both his scientific position and his scientific work. In 1746 Frederick the Great appointed him as president of the renewed Royal Academy in Berlin. He was also leaning increasingly towards a metaphysical and speculative approach to natural science. He formulated the

---

continuation of former page
France has ever produced on the subject of physics" ("le meilleur ouvrage que la France eût produit en fait de physique") (letter from Voltaire to Maupertuis, January 1738, cited in Bunet, *ibidem,* 58). Which leaves the interesting question of how much of this is due to admiration for Maupertuis, and how much to contempt for Cartesianism.

[36] The other major problem, at any rate in France, related to the motion of the moon. The problems that this produced can be illustrated by the following anecdote: in 1747 d'Alembert, Clairaut and Euler calculated, independent from one another, that the precession of the apogee of the moon (that is: the shift in the point of the moon's orbit at which it is furthest from the earth) ought to be half of the observed value. In the end, all three proved to have made an error in the calculation! In the intervening years various proposals were made to adjust the gravitational equation, and the Cartesians seized on the problem to disqualify Newton's method (Heilbron, *Electricity in the 17th and 18th Centuries: A Study of Early Modern Physics* (1979), 60).

[37] "Le Globe mal connu qu'il a sçu mesurer,
Devient un Monument où sa gloire se fonde;
Son sort est de fixer la figure du Monde,
De lui plair, et de l'éclairer."
Written by Voltaire in 1741 as the caption for an engraving by Daullé, based on a portrait of Maupertuis by Tournière. The engraving serves as the frontispiece in *Œuvres* I. See also Tuffet, [Introduction to Voltaire], xxv and 5n.12 (49).

principle of least action in mechanics as a metaphysical principle from which all mechanical laws can be derived. In biology—or more accurately, in the study of life[38]—he formulated his epigenetic theory of the origin of species, which says that at the creation matter was given precisely the characteristics that make self-organization, reproduction and the emergence of varieties possible. Nevertheless Maupertuis kept his biological and physical work separated with such success that even now the connection between them is rarely made.[39]

His article "Loi du repos des corps," which Maupertuis presented at the *Académie Royale* in Paris on 20 February 1740 is an important step towards the first explicit formulation of both the term and the concept of the principle of least action. This article can be seen as the turning point in his scientific thinking. Maupertuis poses the question of whether a law similar to that in dynamics might apply in statics. In dynamics, the sum of the living forces (in formula: Σmv^2) is conserved. He distances himself expressly and radically from a rationalism that seeks to base itself only on evident principles. In fact he distinguishes between "clear and simple principles" and other principles that are neither simple nor generally proven, but that nevertheless, once discovered, can be of great utility. For him it is not *logical* certainty that is decisive in such cases, but a form of *inductive* certainty.[40] The principle of the lowest center of gravity in statics, and the principle of the conservation of living force in dynamics are examples of unproven but practically useful principles.[41] In fact, physics will never be able to provide an *a priori* proof.[42] That proof belongs to "some higher science" ("quelque science supérieure").[43]

---

[38] The science of biology only arose in its present form in the course of the eighteenth century. See above, section 1.2.2, and Hankins, *Science and the Enlightenment*, 113-157.

[39] Terrall, *Maupertuis*, 237-241. She explains this distinction as an attempt, possibly unconscious, by Maupertuis to avoid having to provide an explicit justification of the relationship between the two forms in which matter could exist, i.e., as passive matter—in his physical work—and as inherently active—in his biological work. If so, it might well be interesting to approach Maupertuis's principle of least action from the perspective of his biological works.

[40] "(...) nobody (...) who knows the power of induction, will doubt its truth" ("(...) jamais personne (...) qui connoîtra la force de l'induction, ne doutera de leur vérité") (Maupertuis, "Loi du repos," 268).

[41] As to what led to Maupertuis's interest in such principles, Brunet makes an interesting suggestion in *Maupertuis II. L'œuvre*, 237-244, which has been taken over by Pulte in *Das Prinzip der kleinsten Wirkung*, 67-68. The Marquise du Châtelet, whom Maupertuis served as a physics tutor, converted to Wolffianism in the late 1730s, partly as a result of the controversy concerning the true measure of living force. She supported the Leibniz-Wolffian party in this controversy. In a letter of 2 February 1738, she asked Maupertuis for his opinion on the question, referring explicitly to a treatise by Mairan ("Dissertation sur l'estimation et la mesure des forces motrices des corps" (1728)) and the competition essay by Johann I Bernoulli from 1727 ("Discours sur les lois the la communication du mouvement"), which has been mentioned above. Up to this time, Maupertuis had never expressed an opinion about living force, and the beginning of his reply, where he says that, so far as he was concerned, it was an 'argument about words', suggests that he was not interested. But he was referring here to the way in which Mairan had entered the lists, and the arguments Mairan used. The same letter shows that he had nevertheless adopted a position: that Bernoulli was right with his idea that mv^2 is conserved. Moreover, it is interesting that he considered the conservation of living force so fundamental that, on that basis, he denied the existence of perfectly hard particles. (On his previous recognition of hard particles see page 141).

The utility of these unproven principles is that they provide a shortcut for the intellect as compared to the arduous path that begins with the most simple principles.[44] This is particularly applicable in statics and dynamics, where

> (...) the complicated way that force is related to matter makes these refuges even more necessary, for spirits who have become exhausted or lost in their researches, than in the simple sciences. They can easily see whether their propositions have deceived them, and consider whether the [simple] principle is confirmed, or not.[45]

This remark on the utility of unproven principles is exceptionally important, because it shows what is at stake for Maupertuis or, to give him measure for measure, it shows what his personal Philosophers' stone could be.[46] In one motion he joins a fundamental goal to a practical one: we need a principle that provides understanding of the general rules governing the effects of forces and also enables us to avoid the often laborious calculations involved in using forces.

The difference and similarity with his mechanical views in the 1730s are clear. In the thirties he sought to extend mechanics using principles of the same ontological order, i.e., the actions and characteristics of material substances. Now he was looking for a further expansion with principles of a higher order, based on a 'higher science'.

What general and necessary law will apply for a static equilibrium? He presents a formula that states that the sum of the "forces of rest" ("forces du repos") is a maximum or minimum. We can only guess at the philosophical meaning of 'forces of rest' here. Maupertuis gives only a mathematical definition: the product of mass, m_i, external force, F_i, and distance, z_i, that is, $m_i \times F_i \times z_i$. It is assumed that there are central forces of the form fxz^n and that the dimensions of the masses are small in

---

continuation of former page

[42] The term *a priori* as used here and elsewhere by Maupertuis should not be confused with Kant's *a priori*. Kant focuses the concept on our capacity to know: a concept, idea or proposition is called *a priori* if it is prior to experience, i.e., if the experience only becomes possible given the concept. Before Kant it had a meaning in logic, referring to something that can be derived from principles. Maupertuis's and Euler's use of *a priori* must be read in this sense. In contrast, the complementary concept, *a posteriori*, has retained the same meaning: derived from experience or from effects.

[43] Maupertuis, "Loi du repos," 268.

[44] "Our spirit, being a thing of limited scope, often finds that the distance from the first principles to the point it aims at is too great. It tires or loses its way. These laws of which we have spoken allow it to dispense with part of the journey: it starts from them with all its strength, and often finds that it has but a little way to go to reach its goal" ("Notre esprit étant aussi peu étendu qu'il l'est, il y a souvent trop loin pour lui des premiers principes au point où il veut arriver, et il se lasse ou s'écarte de sa route. Ces loix dont nous parlons, le dispensent d'une partie du chemin: il part de-là avec toutes ses forces, et souvent n'a plus que quelques pas à faire pour arriver là où il desire") (*ibidem*).

[45] "(...) la complication qui s'y trouve de la force avec la matière, y rend plus nécessaires que dans les Sciences simples, ces asyles pour les esprits fatigués, ou égarés dans leurs recherches. Ils voyent facilement s'ils se sont trompés dans leurs propositions, en examinant si le principe s'y retrouve ou non" *(ibidem, 268-269)*.

[46] See his ridicule of the alchemical quest in "Lettre XX. Sur la pierre philosophale" in his *Lettres philosophiques*, 346-349.

comparison to the distances z.[47] A simple example of such a situation is the lever (see Figure 5.2).

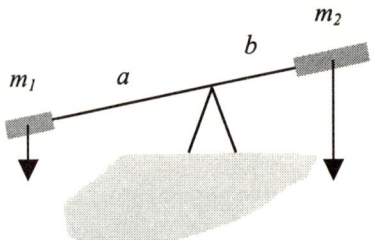

Figure 5.2 Lever subject to a central force working at an infinite distance (equilibrium for $m_1 \times a = m_2 \times b$)

Maupertuis provides a proof for two simple mechanical systems. The first is a system with one degree of freedom: a system consisting of a finite number of masses in a plane, which are rigidly linked to a fixed center, the resulting configuration rotating freely around one point. Each or these masses is subject to a central force, which is a function of the distance between the center of that force and its point of action, raised to the nth power (see Figure 5.3).

The second case is a system with two degrees of freedom. The arrangement is similar to the first case, except that the rigid connections are replaced with flexible connections in the form of non-elastic cords, and the center can move freely. Because in his proof Maupertuis uses the fact that, in the state of equilibrium, every cord lies on the line extending from the center of the system to the center of the central force that acts on the mass concerned, the fact that the masses can move does not increase the degrees of freedom of the system (see Figure 5.4).

---

[47] Maupertuis, "Loi du repos," 269: "Law of rest. Let there be a system of bodies that are either suspended or attracted from a center of forces, such that every force acts on every body in proportion to the nth power of its distance from the center. If all the bodies are to be at rest, the sum of the products of each mass multiplied by the intensity of its force and by the $(n+1)^{th}$ power of its distance from the center of this force (which one might call 'the sum of the forces of rest') must be either a maximum or a minimum" ("Loi du repos. Soit un systeme de corps qui pesent, ou qui sont tirés vers des centres par des Forces qui agissent chacune sur chacun, comme une puissance n de leurs distances aux centres; pour que tous ces corps demeurent en repos, il faut que la somme des produits de chaque Masse, par l'intensité de sa force, et par la puissance $n+1$ de sa distance au centre de sa force (qu'on peut appeler la *somme des Forces du repos*) fasse un *Maximum* ou un *Minimum*"). The forces that Maupertuis is referring to here are central *accelerating* forces of the form $F = f \times z^n$, with the effect of force being given by $F = dv/dt$, where f is the intensity of the force, z is the distance between the point of action and the center of force, and n is the power to which z is raised. Translated into modern quantities and units: the formula for the force of gravity on earth is $F_g = \gamma \times M_{earth} \times z^{-2}$, where γ is the gravitational constant (= $6.67 \cdot 10^{-11}$ N·m²·kg⁻²) and M_{earth} is the mass of the earth (= $5.98 \cdot 10^{24}$ kg). Thus $F_g = 3.99 \cdot 10^{14} x z^{-2}$ N·kg⁻¹, and the intensity of gravitation f is thus $3.99 \cdot 10^{14}$ N·m²·kg⁻¹.

I will not repeat Maupertuis's proof here. Suffice it so say that the foundation of his proof is the principle of virtual displacement, one of the fruits of his study under Johann I Bernoulli, who had formulated the principle in 1717.[48] The formulaic expression of the principle for this case is $\Sigma m_i x f_i x z_i{}^n dz = 0$.

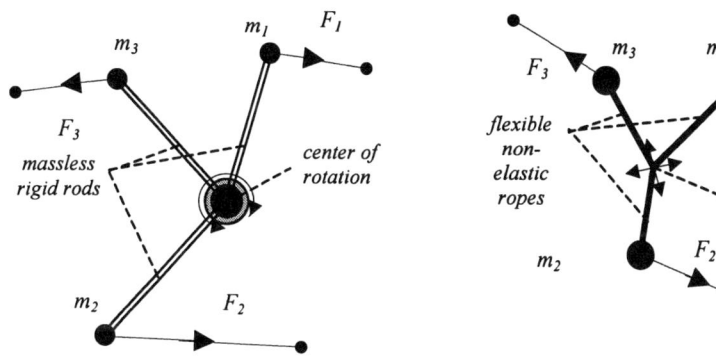

Figure 5.3 System with one degree of freedom

Figure 5.4 System with two degrees of freedom

Pulte quite rightly remarks that Maupertuis did not at this point seek to generalize the principle as a minimal principle in dynamics.[49] He is only seeking a necessary principle in statics that would be analogous to the principle of the conservation of living force in dynamics. The analogy between the static and the dynamic principles is not in the content, but in their universality and necessity. Moreover, Maupertuis was not yet as certain that such principles can be proved as he was a few years later. Both aspects were to change in the following years, as I will explain below.

Thus Maupertuis's "Loi du repos" makes it clear that his intention is to move from a mechanics in which each force is individually significant to one in which these forces are subordinate to higher principles. It was not yet very evident that this shift in the mechanical foundation of the laws of motion would at the same time entail a shift in the metaphysical foundations of mechanics. Maupertuis was soon to change that.

5.3.2 The Principle of Least Action in Optics

Even before he became acquainted with Euler, who by that time was already developing his ideas in the direction of an extremum principle,[50] Maupertuis published an

---

[48] Regarding this principle, see Hiebert, *Historical Roots of the Principle of Conservation of Energy* (1962), 7-57; and Lindt, "Das Prinzip der virtuellen Geschwindigkeiten" (1904).
[49] Pulte, *Das Prinzip der kleinsten Wirkung*, 49.
[50] See below, section 5.3.4.

article in which he goes one step further in the direction outlined above. The paper was entitled "Accord des différentes loix de la nature qui avoient jusqu'ici paru incompatibles" ("The harmony of various natural laws which have thus far appeared to be incompatible"). On 15 April 1744 he presented it to the Paris *Académie Royale*, which he had joined the year before. In this article he seeks a higher principle in the field of optics rather than in statics. As in "Loi du repos," he is looking for an alternative path for the derivation of physical regularities, using a higher science. The thought was that this approach might also be fruitful in optics, because of the optical laws of rectilinear propagation, reflection and refraction only the first two are analogous to mechanical laws. The law of rectilinear propagation is analogous to the principle of inertia, and the law of reflection is analogous to the law governing an elastic impact with a fixed wall. The law of refraction differs from the law that rules the corresponding mechanical phenomenon, i.e., a ball passing through different media. Therefore if he could find a reconciling principle, the power of his approach would be strongly demonstrated.

Attempts had already been made to explain the laws governing light using only metaphysical principles, which Maupertuis equates here with

> Those laws to which Nature herself appears to have been subjected by a higher intelligence which, in producing its effects, causes Nature always to act in the most simple way.[51]

Of these attempts, Maupertuis discusses only that of Fermat, with Leibniz's explanation in its wake.[52] Unfortunately, he says, Fermat's elaboration of the concept 'simple' was incorrect. Fermat supposed that the *shortest time* is the simplest way that nature can work. He could only do this on the basis of the 'erroneous' assumption that light travels more slowly in denser media, whereas 'in fact' it would travel

---

[51] "ces loix auxquelles la Nature elle-même paroît avoir été assujétie par une intelligence supérieure qui, dans la production de ses effets, la fait toûjours procéder de la manière la plus simple" (Maupertuis, "Loi du repos," 276). Maupertuis's description of metaphysical principles appears rather strange, because it provides only an example but is presented as a definition. This gives the impression that Maupertuis *equated* metaphysics with teleological explanations. But although Maupertuis's metaphysics is indeed teleological, his horizons are broader than that. This can be seen for example in his *Essay de cosmologie* (1750), in which metaphysics has a much broader meaning.

[52] While he does mention Leibniz it is only as a follower of Fermat. Just as he had in "Lois du mouvement" in 1746, he accuses Leibniz here of adjusting his physical hypotheses to his metaphysical principles. In "Lois du mouvement" the issue was whether hard particles existed. Here the question is whether light goes faster or more slowly in denser media. Moreover, Maupertuis here identifies Leibniz's concept with Fermat's concept, which is incorrect as he later discovered for himself. In the 1756 edition (*Œuvres* IV, 23) he says that in 1744 he only knew Leibniz's idea through Mairan's "Mémoire sur la réflexion des corps" (1723). Therefore he adds an appendix with part of Euler's article "Sur le principe de la moindre action" of 1753, in which he criticises Leibniz's article "Unicum opticae, catoptricae et dioptricae principium" (1682). The combination of caricatured criticism of Leibniz and strong admiration for him is typical of the eighteenth century attitude to Leibniz. We can clearly see how forced this was in Maupertuis's *Lettres* (1752).

faster.[53] Maupertuis concludes from this that neither time nor distance can be a universal criterion. Moreover, he says, time and space have the same value. What grounds could there be to support the minimality of one rather than the other?

No, the real expense ("dépense") that nature seeks to economize on in the motion of light, is the 'quantity of action'.[54] This solution appears to the modern reader to be plucked out of the air. However, Maupertuis was drawing on a well-known line of thought, so that a few words were enough for his audience. He explains the principle in a few sentences. Despite its brevity, this is the clearest account he gives of the meaning of the quantity of action. Therefore I will discuss his remarks in some detail.

In the first place it is important to know that the concept of action originates from mechanics. Evidently Maupertuis assumes that there is an analogy between light and moving particles. I will return to this point shortly. The concept of action is based on the idea that motion needs a continual cause:

---

[53] The quotation marks are intended to indicate that physics later agreed with Fermat. However, at that time there was no experimental proof for either idea. In 1772 Béguelin stated that he did not expect that it would ever be possible to measure changes in the velocity of light, because of the enormously high speed involved (Hakfoort, "Nicolas Béguelin and his Search for a Crucial Experiment on the Nature of Light (1772)" (1982), 302). Maupertuis's preference here for Descartes's thesis, that light goes faster in denser media, is not based on experiments, but on the fact that, according to Maupertuis, all plausible explanations of refraction phenomena are based on this thesis or are in accord with it (Maupertuis, "Accord des différentes loix," 278). In any case, there is a certain irony in the fact that, if Maupertuis had assumed Fermat's theory of light, action would always have been a *maximum*, and it seems likely that he would never have developed his own form of the principle, which in turn would mean that optics and mechanics would not have been linked, or at any rate that the link would have been created in a different way.

[54] Maupertuis, "Accord des différentes loix," 278. The term 'action' had been in use in mechanics for some time, but had no well-defined meaning. Newton used it in his *Principia mathematica* in several places, the clearest being in the third law where action and reaction are said to be equal in magnitude and opposite in direction (see Westfall, *Force in Newton's Physics*, 548-550). A contemporary of Maupertuis, the Cartesian d'Arcy, wrote a criticism of Maupertuis in which he attacked not only the possibility of a purposeful nature but also the nominal definition of the concept of action. Apparently the term was linked with something real in Cartesianism as well. The discussion between Maupertuis and d'Arcy took place in four articles: (1) Sir Patrick d'Arcy, "Réflexions sur le principle de la moindre action de Mr. de Maupertuis" (1753). (Note that d'Arcy had only presented this article in 1752, but it was given priority in the publication of the *Mémoires*, which indicates the great interest of the Paris academy in the problem); (2) Maupertuis, "Réponse à un Mémoire de M. d'Arcy sur la moindre action" (1754); (3) d'Arcy, "Réplique à un Mémoire de Mr. de Maupertuis sur le principle de la moindre action" (1756); (4) L. Bertrand, "Examen des Réflexions de M. le Chevalier d'Arcy sur le principle de la moindre action" (1755, presented 3 October 1754).

Finally Leibniz's concept of action must be mentioned here, in particular because Helmholtz and Kneser consider it to be the origin of Maupertuis's concept of action (Helmholtz, "Zur Geschichte des Princips der kleinsten Action" (1887), 250-255; Kneser, *Das Prinzip der kleinsten Wirkung* (1928), 1-26). However, Maupertuis's concept of action and its function in mechanics is essentially different to that of Leibniz (see Gueroult, *Leibniz: Dynamique et métaphysique* (or.1934), 110-154 and 215-235; and Pulte, *Das Prinzip der kleinsten Wirkung* (1989), 56-64).

> A certain action is required to transport a body from one point to another. This action depends on the speed of the body and the distance it travels (...)[55]

A certain action is necessary to move a body: is there not an obvious relationship to impetus, the force of inertia, and living force? Action appears to be a measure of what is required to move a body from A to B at a particular velocity and along a particular path. Maupertuis concludes from this that the action, for a single particle, is proportional "to the sum of the distances, each multiplied by the velocity at which the body passes through it," i.e., to the sum of the product of the velocity and distance traveled by a particle. In a formula this is $\Sigma v \Delta s$.[56]

However, there is a difference as compared to the concept of force, in that the concept of action does not express force itself but rather the activity of force,[57] i.e., its effort. The principle is no longer conceived in terms of the forces themselves, as the causes of their effects, but the effects are dealt with in another way. It is not a kinematic principle either. A teleological principle has replaced force, and this teleological principle is the intermediary between ourselves and our knowledge of the effects.

Not only the law of refraction, but also the laws of reflection and of the rectilinear propagation of light follow from this principle.[58] From a comparison between his own principle and that of Fermat, Maupertuis concludes that it is not the use of the final cause itself that leads to errors in natural science; errors arise from confusing the principle and the outcome.[59] This is reminiscent of d'Alembert's argument in the controversy concerning the true measure of living force, where he said that a quan-

---

[55] "Lorsqu'un corps est porté d'un point à un autre, il faut pour cela une certaine action, cette action dépend de la vîtesse qu'a le corps et de l'espace qu'il parcourt (...)" (Maupertuis, "Accord des différentes loix," 279).

[56] "à la somme des espaces multipliés chacun par la vîtesse avec laquelle le corps les parcourt" (ibidem).

[57] Jacobi has already emphasized that the principle was intended by Maupertuis to mean that "Nature arrives at its effect with the least possible effort" ("die Natur ihre Wirkungen mit dem kleinsten Kraftaufwand erreiche,") and not that "forces in nature necessarily always produce the least possible effects" ("in der Natur die Kräfte nothwendig immer die kleinste Wirkung hervorbringen müssten"). He attributes the latter idea incorrectly to Euler (Jacobi, "Das Princip der kleinsten Wirkung" (1842–1843), 43-44). Mayer and Helmholtz, independently, have expressed this contrast more clearly using the terms of activity and passivity (Mayer, Geschichte des Princips der kleinsten Action (1877), 12-13; Helmholtz, "Zur Geschichte des Princips der kleinsten Action" (1887), 252). Fleckenstein concluded from this that it is not correct to translate 'actio' with the German 'Wirkung': while 'actio' refers in principle to the "active deed of realizing, the effort" ("aktive Tätigkeit des Wirkens, der Aufwand"), the concept 'Wirkung' indicates the "passive result" ("passiv Erwirkte") (Fleckenstein, [Foreword to Euler] (1957), xxx).

[58] Some years later d'Arcy and König showed that the outcome is not always a minimum: sometimes the action is in fact a maximum. This had no influence on the development of the principle by Maupertuis, and scarcely any on that by Euler (see below, sections 5.3.4 and 5.4.2). The criticism did not become important until later, and further conditions for minimality were then formulated (see Jacobi, "Das Princip der kleinsten Wirkung").

[59] "(...) they have been led astray not by the principle but by their own haste, which has led them to identify something a principle, when it is only a consequence of the principle" ("(...) ce n'est pas le principe qui les a trompés, c'est la précipitation avec laquelle ils ont pris pour le principe ce qui n'en étoit de conséquences") (Maupertuis, "Accord des différentes loix," 280).

tity such as mv or mv^2 expresses an *effect* and not the *cause* itself.[60] Maupertuis makes the same distinction, with two important differences.

The first difference is that d'Alembert saw the true cause as transcending human knowledge and as something that in principle could not be expressed in a mathematical formula, whereas Maupertuis proposes a hierarchy of formulas in which the highest, and universal, formula is the direct expression of the true cause. The second difference is that Maupertuis uses a completely different sort of cause: teleological or final rather than efficient causes. The consequence of these two differences is that Maupertuis does not retrace d'Alembert's steps in developing his principle, but rather goes one step further. Where d'Alembert had shifted the question from *efficient cause* to *effect*, Maupertuis goes from *effect* to *final cause*.

But can final causes, unlike efficient causes, be known just like that? I should again emphasize that Maupertuis expressed himself in a much more nuanced way before the controversy with König than after that. Although the principle of least action is a 'necessary' principle, metaphysically speaking, he frankly concedes that it would be quite possible to make an error in identifying the physical quantity with which nature is most economic. The teleological method can only be verified by consistency with the usual methods. This is because the efficient and final cause must be in harmony with one another: "No-one can doubt that all things are regulated by a supreme Being who, when he implanted the forces that demonstrate his power in matter, also ordained that they should produce effects that demonstrate his wisdom."[61]

Thus, although he emphasizes the metaphysical necessity of the principle of least action, he does not claim that the concept of action can be deduced by human reason from the divine wisdom. The converse line of thought—that a better notion of that divine wisdom, and even an argument for the existence of God, can be derived from the concept of action—is completely in accordance with what has been said thus far, but Maupertuis was only to defend this several years later. I will return to this point below.

5.3.3 Extending the Principle of Least Action to Mechanics

As has been noted, Maupertuis's frequent comparisons between *optical* laws and phenomena, on the one hand, and *mechanical* laws and phenomena on the other hand, point to his interpretation of light as a mechanical phenomenon. His acceptance of a mechanistic theory of the propagation of light is another indication of this. Perhaps its clearest expression is the fact that he explains the concept of action by analogy with the motion of a particle. In the light of the later extension of the principle of least action to the laws of motion, it is interesting to examine the basis of this

---

[60] See Chapter 4, pages 117 and 129.
[61] "On ne peut pas douter que toutes choses ne soient réglées par un Estre suprême qui, pendant qu'il a imprimé à la matière des forces qui dénotent sa puissance, l'a destinée à exécuter des effets qui marquent sa sagesse" (Maupertuis, "Accord des différentes loix," 280).

comparison and whether Maupertuis was already anticipating the possible extension of the principle to mechanics here.

We have seen that in the "Loi du repos" there was still no indication of a possible extension. This has now changed. Maupertuis is continually looking for analogous phenomena in optics and mechanics that could be explained by common laws. The rectilinearity of light corresponds to the principle of inertia in mechanics. The law of reflection covers the impact of an elastic ball with an unmoving surface. While the third law of optics, the law of refraction, does not apply to the corresponding motion of a particle passing through media of differing resistances, this does not exclude the possibility that the principle on which this law of optics is based could also be valid in mechanics. The suggestion that he supposed this to be the case, is supported by his explanation of the concept of action in terms of particles..

However for Maupertuis, correspondences in *behavior* (following regular laws) are not based on any correspondence between the *nature* of the phenomena. This can be seen from the way he justifies his preference for a metaphysical explanation. In discussing the various explanations of the law of refraction that have been proposed, he begins by showing that the phenomena are quite different. Then he divides the explanations into three classes, the first two classes being based on principles that also apply to mechanical particles. Of these, the first covers all explanations that only admit the "most simple and commonplace mechanical principles."[62] Here he is referring mainly to Descartes and his followers. Although this approach has failed, he shows some appreciation of its value.[63] The second class comprises those explanations that assume, in addition to mechanical principles, that light "is striving towards bodies," either as a characteristic of matter, or as the effect of some cause.[64] Here he is referring to Newton and Clairaut respectively. In contrast to his treatment of Descartes, he recognizes that the latter have been successful in explaining phenomena of refraction, but says nothing further about them. Finally, the third class, which seeks to use only metaphysical principles, has thus far not produced any successes, but nevertheless inspires Maupertuis to follow this path further.

Maupertuis does not explain why he ignores an approach which has produced successful explanations to follow a route that has not yet produced any results, or why he chooses the metaphysical path in preference to mechanical principles. Only his agreement with Fermat, and the thoughts that he attributes to him, would lead one to suspect what the reason was:

> [Fermat] has apparently also despaired of deriving the phenomena of refraction from the behavior of a ball thrown against obstacles or moving through resistant media; but he has not turned instead to the atmosphere around the body or to attraction; although it is evident that the latter principle would not

---

[62] "principes les plus simples et les plus ordinaires de la Méchanique" *(ibidem, 275)*.
[63] "[Descartes] a toûjours le mérite d'avoir voulu ne les [i.e., the material phenomena, JCB] déduire que de la Méchanique la plus simple" *(ibidem, 276)*.
[64] *Ibidem*, 275.

have been unfamiliar or unacceptable to him; rather he has sought an explanation for these phenomena in a completely different and purely metaphysical principle.[65]

Where universal gravitation and the force of impact played a central role in his defense of Newton in the 1730s, at this point the principle of least action was playing the same role. In effect Maupertuis has replaced the direct approach, by means of forces, with an indirect approach by means of principles. He states that he does not want to attempt a theory of the *nature* of light, but only to examine its *metaphysical principles*. What Maupertuis means is, if the simple and direct path fails, because the individual forces are unknown or are not sufficiently known, we do not have to resort to complex characteristics—as attraction was generally assumed to be. We can also try an alternative route, based on metaphysical principles.

Although Maupertuis seems here to be making force antithetical to principle, this is only the case on the methodological level.[66] In ontological terms, the principle is not only a principle of actions, but also of the forces themselves: it is the law that the forces must obey. However, this assumes that force has a basis in the structure of material reality. We can see from this that Maupertuis's principle also constitutes a transition from a substantial to a structural foundation for force, as we have already seen in relation to d'Alembert's transformation of the concept of force. Therefore Maupertuis can extend the concept of action to mechanics not because light and matter have a similar substantial nature, but because they have a similar structure.[67]

5.3.4 Euler's Response to Maupertuis's "Law of Rest"

Now is the time to summon up Euler, because this moment marks the beginning of a degree of co-operation between Euler and Maupertuis. Although both had lived in Berlin since the beginning of the 1740s, they had evidently delayed their first meeting until etiquette required it. They met at about the time of the ceremonial opening of the Berlin Academy on 8 October 1745. Late in 1745, Euler made a courtesy call on Maupertuis. It was probably only during this visit that they learnt about one another's activities with regard to extremum principles.[68] Maupertuis showed Euler a copy of his "Loi du repos"—but remarkably, not of his "Accord des

---

[65] "[Fermat] avoit aussi désespéré apparemment de déduire les phénomènes de la réfraction de ceux d'une balle qui seroit poussée contre des obstacles ou dans des milieux résistans; mais il n'avoit eu recours ni à des athmosphères autour des corps, ni à l'attraction; quoiqu'on sçache que ce dernier principe ne lui étoit ni inconnu ni désagréable; il avoit cherché l'explication de ces phénomènes dans un principe tout différent & purement métaphysique" *(ibidem,* 276).

[66] In *Das Prinzip der kleinsten Wirkung,* 44, Pulte interprets Maupertuis's principle as a search for a *"force-free* description of nature" ("*kraftfreien* Naturbeschreibung").

[67] I believe this refutes Terrall's understanding, in which this expansion was attributed to Maupertuis's implicit concept of the particle character of light. See Terrall, *Maupertuis,* 37: "Maupertuis's defence of a Newtonian corpuscular theory of light ultimately led him to a decidedly non-Newtonian mechanics."

[68] See Costabel, [Introduction to *Correspondance*], 13-14.

différentes loix"!—or perhaps he only told him where to obtain it. In the previous year Euler had himself written on the occurrence of mathematical extrema in nature, in two annexes ("Additamenta") to his *Methodus Inveniendi* (1744). On 10 December 1745 Euler wrote a letter to Maupertuis in response to his "Loi du repos," but he interpreted the principle that Maupertuis formulated there as a universal principle, as Maupertuis himself had since done in "Accord des différentes loix."[69] Euler did not yet know the "Accord," but he projected his own opinions onto "Loi du repos." In short, in 1744-1745 Euler, looking for *mathematical extrema*, and Maupertuis, looking for *universal principles* seem to have found themselves on the same track.

In his letter Euler expresses his conviction that "all of nature behaves according to some principle of a maximum or a minimum," and that research in this direction would be much more valuable than studying separate problems.[70] Once again we see that metaphysics had not been made redundant, but that it needed to be reformed. Not only is metaphysical research useful to clarify our knowledge, it is also necessary to discover "the true principles of metaphysics."[71] Euler agrees with Maupertuis that the extremum principle in its general form (i.e., as 'the simplest path') is universal, but claims that the quantity concerned must be "dug up" separately for each individual case.[72] In the "Additamenta" Euler had written something similar, that the extremum quantities cannot simply be derived from metaphysical principles but can be found from the results of the usual methods. Despite this fundamental objection, Euler adds, one should at the same time seek to generalize. The principle in "Loi du repos" for example can be extended to systems of multiple particles that are subject to random forces,[73] and Euler thinks that also the extreme quantities that he himself used in the "Additamenta" can probably be derived from Maupertuis's principle.[74]

The "Additamenta" were intended as applications of Euler's *Methodus inveniendi lineas curvas maximi minimive proprietate gaudentes* (1744) in mechanics. This work contains a method developed by Euler to find curves that, under particular constraints, exhibit a particular characteristic to the greatest or smallest degree. This method, together with other methods, would later be given the name of variational calculus. Such a method became necessary at the end of the seventeenth century, when Johann I Bernoulli drew the attention of the scholarly world to the puzzle of the brachistochrone (the curve of fastest descent). The problem here was to find the planar curve on which a point mass subjected only to the force of gravity will slide

---

[69] Letter from Euler to Maupertuis, 10 December 1745 (*Leonhardi Euleri opera omnia* IVa.6, 56-57). *Leonhardi Euleri opera omnia* is abbreviated below to *EOO*.
[70] "par tout la nature agit selon quelque principe d'un maximum ou d'un minimum" *(ibidem, 56)*.
[71] "les veritables principes de la metaphysique" *(ibidem)*.
[72] *Ibidem*. Euler used the term 'deterrer' ('to dig up') here, which according to Costabel refers to a metaphor used by Viète in 1591, of the mathematician as a miner who excavates the earth with his pick to reveal hidden truths, which are as objective as the jewels that lie in the earth (*ibidem*, 56n.4). Euler possibly intended to emphasize that the mathematician had a greater role to play on this point than the metaphysician.
[73] *Ibidem*.
[74] *Ibidem*, 57.

(without friction) from point A to a lower point B in the least possible time. Many mathematicians, including Leibniz and Jakob Bernoulli, developed a solution, but no-one was able to generalize his solution so that it would also apply to other problems. In the decades that followed, solutions to similar problems were found, but they continued to be particular solutions only.[75] Euler's method was the first general solution for such problems.

The first of the two "Additamenta" relates to the curvature of elastic ribbons ("laminae"), the second to projectile motion in a friction-free medium.[76] However, Euler goes beyond this. In both "Additamenta" he provides a natural philosophical approach that discusses the occurrence of maxima and minima and how the relevant quantities can be determined.

Because the universe is ordered in the most perfect way, a maximum or minimum must underlie everything that happens. So it should be possible to calculate all actions from final causes, with the aid of the *Methodus inveniendi*, just as well as they can be calculated from efficient causes.[77] Like Maupertuis, but independent of him, Euler calls calculation based on efficient causes the 'direct' method, while calculation from final causes is the 'indirect' method.[78] Each has its own value, because sometimes the efficient causes are unknown, and sometimes the final causes.[79] In the first "Additamentum" Euler uses a quantity provided by Daniel Bernoulli, the potential force ("vis potentialis," $\int ds/r^2$). In the second "Additamentum" he outlines a method for finding such quantities. We must begin, he says, with the results that we have obtained using the direct method. By means of a thorough study of curves, and particularly of what can be regarded as effects, produced by moving forces in the form of the curves we may be able to discover the extremal quantity.[80] Now, since the effects consist of induced motions, he wishes in the first place to examine whether the minimalization of these motions, or rather the

---

[75] For the history of these see, e.g., Stäckel (ed.), *Abhandlungen über Variationsrechnung* I.*Theil*; Carathéodory, "Einführung in Euler's Arbeiten über Variationsrechnung" (1952); H. Goldstine, *A History of the Calculus of Variations from the 17th through the 19th Century*. Berlin: Springer 1980.
[76] The titles are "Additamentum I: De curvis elasticis" and "Additamentum II: De motu proiectorum in medio non resistente, per methodum maximorum ac minimorum determinando," respectively.
[77] "There is no possible doubt that all effects in the world can be determined equally well from final causes using the method of maxima and minima, or from the efficient causes themselves" ("Quando scilicet (...) dubium prorsus est nullum, quin omnes Mundi effectus ex causis finalibus ope Methodi maximorum et minimorum aeque feliciter determinari queant, atque ex ipsis causis efficientibus") (Euler, "Additamentum I. De curvis elasticis," 231).
[78] Cf. page 154 above.
[79] "In fact when the efficient causes are very much concealed [and] the goals are less difficult to grasp, it is usual to solve the question with the indirect method; in contrast, the direct method is used wherever it is possible to determine the effects from the efficient causes" ("Quando scilicet causae efficientes nimis sunt absconditae, finales autem nostram cognitionem minus effugiunt, per Methodum indirectam Questio solet resolvi; e contrario autem Methodus directa adhibetur, quoties ex causis efficientibus effectum definire licet") (Euler, "Additamentum I. De curvis elasticis," 231).
[80] "effectus a viribus sollicitantibus oriundus" (Euler, "Additamentum II. De motu proiectorum," 298).

"aggregate of all motions," yields the desired result.[81] Although this method does not provide a foundation of principles, the agreement between the results using the extremal quantity obtained in this way and results from the direct method does provide sufficient certainty.[82] The quantity concerned, $\int mv ds$, proves in fact to agree with Maupertuis's action, $\Sigma v \Delta s$.

So much for Euler's approach before he came in contact with Maupertuis's work. It is interesting to note that Euler speaks about "effectus," where Maupertuis talked about "actio." The difference between these two terms has been discussed in relation to the translation of 'actio' with the German 'Wirkung': 'effectus', it is said, refers to the passive result, the effect, and 'actio' to the effort.[83] However, once he had become acquainted with Maupertuis's work, Euler adopted his term as well.[84]

The method that Euler outlines shows how eagerly he is looking for a way in which such quantities can be determined *a priori,* and how much Maupertuis's "Loi du repos" appears to offer at least a potential answer. So he is pleased to show, in his letter to Maupertuis, that the quantity that is valid for the curvature of elastic ribbons, $\int ds/r^2$, can be derived from his principle for statics.

Mutual influence is demonstrable here. One notes on the one hand that for Euler, as for Maupertuis in the "Loi du repos," the minimality of the quantities looked for is not at stake. In his first "Additamentum," and in his letter to Maupertuis, he consistently refers to a "maximum or minimum." We should not be mislead by the fact that his second "Additamentum" deals only with the minimum of the integral $\int mv ds$, nor by the fact that, later in his letter of 10 December, Euler writes twice only "maximum."[85] It is clear that at this time he was indifferent toward the distinction between maximum and minimum clearly. However, some years later Euler comes to regard only the *minimum* as real—in his articles "Recherches sur les plus grands et plus petits" (1748) and "Réfléxions sur quelques loix générales" (1749); this refinement can presumably be attributed at least in part to the influence of Maupertuis's "Accord des différentes loix," with which he had in the meantime become acquainted.

On the other hand Euler's response implies a challenge to Maupertuis. Previously, Euler had spoken of 'digging up' the extremum quantities, and claimed that he found only the same solutions as those produced by the usual calculations. Nevertheless, he would definitely attach great value to a proof of the accuracy of the formulas used. This point, the question of whether extremum principles should be sought through metaphysical or mathematical argumentation, must have been an important stimulus for Maupertuis to develop his teleological method further.

---

[81] "aggregatum omnium motuum" *(ibidem).*
[82] *Ibidem.*
[83] See note 57.
[84] First evident in his letter of 24 May 1746 to Maupertuis (*EOO* IVa.6, 63).
[85] *EOO* IVa.6, 57.

5.4 THE FIRST STEPS OF THE PRINCIPLE OF LEAST ACTION

5.4.1 Maupertuis's "Lois du mouvement" (1746)

Maupertuis must have been elated over Euler's letter; Euler had in fact produced an extremum quantity that corresponded exactly to his own concept of action. This was also a stimulus to now generalize his principle to cover both optics and mechanics. At the same time, as noted above, he must have felt challenged to provide a metaphysical basis for the minimum quantities discovered. He attempts this in his last scientific article in this field, "Les lois du mouvement et du repos, déduites d'un principe de métaphysique." He presented this at the Berlin Academy on 6 October 1746. In addition to an attempt to elaborate on what he had begun in "Accord des différentes loix," that is, the search for a common principle for various domains and laws, this article also contained an attempt to use this principle to provide a proof of the existence of God.[86]

Although in the present context I am interested primarily in the first of these, I cannot ignore this secondary goal, which may well have been much more important to Maupertuis. It appears in fact that the secondary goal was an additional motive for Maupertuis's—deliberate or unconscious—focus on the possibility of deriving physical laws from metaphysical principles. Maupertuis based his proof of the existence of God on the universal laws of motion.[87] He attributes evidential value to these laws on the basis of mathematical evidence and the universality and generality of the relevant laws. If these first laws can be derived from the "all-knowing and all-powerful Being"[88]—Maupertuis's God—this is a powerful proof of the existence of God.

---

[86] It seems likely that two articles that have thus far gone largely unnoticed, by Euler and Daniel Bernoulli respectively, were the direct inspiration for this article. However, since this origin has little influence on my argument they need not be considered here. The articles concerned are: Daniel Bernoulli, "De variatione motuum a percussione excentrica" (presented in 1737, published in 1744), and Euler, "De communicatione motus in collisione corporum sese non direct percutientium" (presented in 1737, published in 1744). See David Speiser, [Introduction to Daniel Bernoulli], 57-61.

[87] "The supreme Being is everywhere, but He is not equally visible everywhere. We see him better in the simplest objects: we look for Him in the primary laws which He has imposed upon Nature, in these universal rules that govern the conservation, distribution [i.e., in impacts, JCB] or destruction of movement (...)" (Maupertuis, "Lois du mouvement," 290). In part I of "Lois du mouvement," 282-289, Maupertuis rejected the traditional Proofs of God's existence. The uniformity that Newton thought he had discovered in astronomy and biology is not *perfect*, and might therefore be the result of chance. The functionality of the biological organs (which Maupertuis calls agreement with the needs of the whole), however persuasively it may seem to speak of purposeful design, could be the result of the natural sifting of an immense number of individuals created by blind destiny. Apart from the lack of evidential value, one major shortcoming of such an argument is that it does not demonstrate the *wisdom* of God, only his *design*. He calls the quest of many contemporary physico-theologians for order and coherence in "the most minute details of Nature" frankly ridiculous and blasphemous (*ibidem*, 285-286).

[88] "Etre tout puissant et tout sage" *(ibidem,* 293).

The proof is reminiscent of the erroneous argument 'if A→B, and B is true, then A is true'. However, its value as a proof of the existence of God is not important here. What is important is that, in seeking a foundation from which the laws of motion can be derived, Maupertuis looks for a basis that does not subsist in the nature of bodies or in the empirical realm, but rather in the higher principle which he had been aiming at in his articles of 1740 and 1744. What he then called an alternative path has become more important: it is now the main road. The derivation that Maupertuis gives assumes a general principle:

> General Principle: Whenever any change occurs in nature, the quantity of action required for the change is as small as possible.[89]

The 'quantity of action' here is defined as the "product of the mass of the body, its velocity and the distance it travels," i.e., as $\Sigma mv\Delta s$.[90] Although the quantity of action is defined just as it had been two and a half years earlier in "Accord des différentes loix," except that the mass of the body has been added, its meaning is now quite different. The quantity of action is now equal to the *change* in the quantity $\Sigma mv\Delta s$ rather than to that quantity itself.

From his general principle, Maupertuis derives laws of impact for simple cases, for both elastic and hard, inelastic bodies. He also provides a new proof for the law of leverage in statics, in which the position of the point of equilibrium is determined by a minimum.[91]

In "Lois du mouvement" he no longer asks why 'action' is defined in this way, as he had in "Accord des différentes loix," or why it is always a minimum and never a maximum quantity. Moreover, Maupertuis does not seek to cover this omission by reference to the concept of action from his optics. While he gave no elaborated justification of this quantity there, at that point he did at least leave the question of whether he had derived the concept mathematically from the laws of optics or had drawn it from metaphysical insights open. In "Lois du mouvement" he claims to do the latter, but fails to actually do so. We might wonder whether the arrogant modesty that characterizes Maupertuis's writing style in this period is the result of his inability to accept that a proof of the existence of God demanded more than he could fulfill.[92]

---

[89] "Principe General: Lorsqu'il arrive quelque changement dans la Nature, la Quantité d'Action, nécessaire pour ce changement, est la plus petite qu'il soit possible" *(ibidem,* 298).

[90] "la produit de la Masse des Corps, par leur vîtesse et par l'espace qu'ils parcourent" *(ibidem).*

[91] It is not clear why he later decided to omit this part of the application of the principle in his *Œuvres* of 1756, especially since Euler in his "Harmonie entre les principles généraux de repos et du mouvement de M. de Maupertuis" (1752) had actually derived the validity of the principle in dynamics from its validity in statics! The proof was retained in the previous editions of Maupertuis's *Œuvres* (1752 and 1753 respectively) (Tonelli, [Introduction to Maupertuis], lxxx*).

[92] For example, after giving a short history of the principles of mechanics, he writes: "After all the great men who have worked on this matter, I hardly dare to claim that I have discovered the universal principle on which all laws are based" ("Après tant de grands hommes qui ont travaillé sur cette matière, je n'ose presque dire que j'ai découvert le principe universel, sur lequel toutes ces

continued on next page

What *is* clear is that calculations based on the concept of action are independent of considerations about force. This point, which is crucial in determining the meaning of the principle of least action, only became clear to Maupertuis as a result of Euler's commentary on the drafts of his article. In a letter of 24 May 1746, Euler responds to the manuscript that Maupertuis had evidently sent to him for his comments.[93] Part of his criticism focuses on Maupertuis's attempt to relate the principle of least action to the actions of forces. Euler writes: "the determination of the quantity of action bears no relation to any force whatever."[94] Therefore the unresolved question of how much force is required to give a body a certain velocity can simply be ignored.

Not only has Maupertuis removed the relevant passage from the final version, he even added a tirade against the concept of force—hich strongly resembles d'Alembert's criticism—when he later incorporated the article in his *Essay de Cosmology* (1750).[95] The concept of force is "a feeling in our soul" that is used for simplicity's sake to indicate the influence bodies have on one another, but which can only be measured through visible effects.[96] The "motive force, the capacity of a moving body to move other bodies," is only a word, coined as a means of summarizing our knowledge. It is a description of the phenomena, without any underlying reality.[97] A comparison with his work from the 1730s clearly shows how his increased emphasis on the pursuit of universal principles was accompanied by a fundamental change in his ideas about force.

A second point of criticism in Euler's letter is that, in the application of the principle, the distinction between velocity and distance disappears, because the formula for the quantity of action uses the term 'distance' to refer to the distance-traveled-in-a-unit-of-time, which by definition has the same value as the velocity. Euler points out that this would mean that the concept of living force could be used instead of the quantity of action.[98] Because Maupertuis always assumed discontinuous changes in velocity in his works, and ignored acceleration, he saw no reason to adjust the definition.[99] This left Euler with the question of what precise role the

---

continuation of former page
loix sont fondées") only to continue, without any trace of doubt, to formulate the principle and sing its praises (Maupertuis, "Lois du mouvement," 295).
[93] Letter from Euler to Maupertuis, 24 May 1746 (*EOO* IVa.6, 63-65).
[94] "la determination de la quantité d'action n'a egard à aucune force" *(ibidem,* 64).
[95] Maupertuis, *Œuvres* IV, 27r.11 - 31r.9. This tirade was present from the first edition of the *Essay* in 1752 (Tonelli, [Introduction to Maupertuis], lxi*).
[96] "un sentiment de notre ame" (Maupertuis, *Œuvres* IV, 30).
[97] *Ibidem,* 31.
[98] Euler's letter to Maupertuis, 24 May 1746 (*EOO* IVa.6, 63-65, this passage on pages 63-64). In fact in "De universali principio" (1751), König suggested, as an alternative to Maupertuis's principle, that what was actually at stake was a minimum *change* in the living force.
[99] In my opinion it is not correct to regard the definition that Maupertuis gave in "Accord des différentes loix" as better than that in "Lois du mouvement." Although the definition in "Accord" is more adequate where the velocity is not piecewise uniform but, for example, uniformly accelerated, the difference is not important here, because Maupertuis spoke only about situations in which motion is proportional to time except for interruptions, and the velocity is thus constant. It is

continued on next page

distance traveled played in the quantity of action. In later correspondence between Euler and Maupertuis in 1748, Euler showed clearly that this role is linked to the ranking of the formulas for the quantity of action in various situations. I will explain this below.

A final point that clarifies the meaning of 'universal principles' for Maupertuis is the way he deals with the question of the characteristics of matter. In "Loi du repos" he still considered that only elastic bodies exist in reality. In "Lois du mouvement" he argues that both elastic and hard bodies can exist.[100] As in his defense of attraction, he seeks in his argumentation mainly to remove experimental and metaphysical objections.[101] Awkwardly, he also *disavows* his earlier support for the universality of the conservation of living force. He accuses the supporters of a universal conservation of living force, especially Leibniz, of preferring to deny that hard bodies exist than to give up their principle.[102] Accordingly, for the new edition of "Loi du repos" he modified the text concerning the conservation of living force, to narrow its range to elastic bodies.[103]

It could be that this change in position arose out of a broad debate concerning Leibniz's monadology, the theme of the first philosophical prize contest held by the revitalized Berlin Academy for the year 1747.[104] Euler was among those who defended the impenetrability of primary particles. Maupertuis may have been influenced by this discussion. In 1738 he was still denying the existence of hard particles, but in 1746 he writes: "(...) simple bodies, primitive bodies, which constitute the

---

continuation of former page
    therefore not correct to suppose, on the basis of the fact that he did not see that the change in definition was a backwards step from a modern point of view, that he "did not have any precise idea in mind" ("n'avait pas en tête une idée précise") (Costabel, [Introduction to *Correspondance*], 65n.2).

[100] Compare his *recognition* of hard particles in 1732 (see above page 141), and his *denial* of them in 1738 (see above, page 145n.41).

[101] The impossibility of the existence of hard bodies was demonstrated with both experimental and metaphysical arguments. Experiment showed that all matter can be compressed. In metaphysics, the principle of continuity meant that change in motion must be continuous, which is impossible for hard bodies. Against the first argument, Maupertuis points out that compression could also be explained by the fact that the body is composed of many simple bodies, which are in themselves hard. He refutes the metaphysical argument by appealing to our ignorance about the origin and transfer of motion. In doing so, he undermines the principle of continuity. Nevertheless, the metaphysical argument for the existence of hard particles is a tautology: starting from the impenetrability of matter, he concludes that it is hard (Maupertuis, "Lois du mouvement," 293-295).

[102] "(...) the fact that the conservation of living force can only apply in collisions between elastic bodies confirms the opinion that there are no other kinds of bodies in nature" ("(...) comme la conservation de la Force vive n'avoit lieu que dans le choc des Corps élastiques, on s'est confirmé dans l'opinion qu'il n'y avoit point d'autres Corps que ceux-là dans la Nature") (*ibidem*, 295).

[103] Compare the text of the first version: "tout systeme de corps en mouvement" (*EOO* II.5, 269) with that from the *Œuvres*: "tout système de corps élastiques en mouvement" (*Œuvres* IV, 47). It seems likely that Maupertuis introduced this restriction as early as the first reprint in 1752. Tonelli, strangely enough, says nothing about this (Tonelli, [Introduction to Maupertuis], lxxxi*).

[104] For the significance of the philosophical prize contest of the Berlin Academy see section 6.2.

elements of all other bodies, must be hard, inflexible and unalterable."[105] However, a more important factor could be the fact that the principle of least action provides Maupertuis with a much more powerful instrument than he had had in 1738 and in 1740, when the only general principle he knew was the conservation of living force. Now that the principle of least action has been shown to apply to both hard and elastic particles, he sees the conservation of living force as less fundamental, because it does not apply to collisions between hard particles.

This change in position should not be regarded as mere opportunism: if in fact it can be shown that the principle of least action also relates to phenomena that are improbable but nevertheless possible, it has greater universality and should be accorded the higher status. In 1738 the universality of the principle of conservation of living force made him to deny the existence of hard particles on the basis of their improbability. By 1746 in contrast, upholding the logical possibility of the existence of hard particles provided him with an extra argument for subordinating the principle of the conservation of living force to the principle of least action.

The latter explanation of his about-face, that is, that his defense of the existence of hard bodies is based on the universality of the principle of least action, also implies that speculations about the nature of material substances are for him secondary to speculations about the principles that determine their structure. This primacy of structure over substance is in striking agreement with his approach to optics, as we have already seen, and is in fact an extension of it.[106] Just as the laws of light can be better explained using a metaphysical principle of structure than on the basis of forces that emanate from substances, we can see here that his recognition of the existence of hard bodies is based on placing greater reliance on structural principles than on speculations about substance.

However, the accusations Maupertuis levels against Leibniz are a sign that he cannot yet completely accept the primacy of structure. By accusing Leibniz of opportunism he hopes to exonerate himself from something that is in fact the basis of his approach!

5.4.2 Euler's Insights

Although Maupertuis did not reveal any new insights into the principle of least action after the publication of "Lois du mouvement," his role had not yet come completely to an end. His subsequent involvement was in part negative, in the dispute about priority with König, in which Maupertuis did himself no credit, as I have already mentioned in section 5.1. On the other hand, Maupertuis played a positive role through his commentary on a draft of Euler's article "Recherches sur les plus grands et plus petits qui se trouvent dans les actions des forces" (1748). There was a correspondence between them concerning this in 1748, of which only

---

[105] "(...) les Corps simples, les Corps primitifs, qui sont les élémens de tous les autres, doivent être durs, infléxibles, inaltérables" (Maupertuis, "Lois du mouvement," 294).
[106] See page 154.

Euler's letters have survived. Nevertheless these give a good notion of Maupertuis's influence.[107] While it may appear in the first letters as if Euler is only trying to explain to a slow-witted correspondent how the correspondent's own principle looks when put in mathematical form, in the sixth of the seven letters as a whole it suddenly becomes clear that something fundamental is afoot, in the midst of the mathematical problems.[108]

The fundamental issue is the question of how the various forms of the principle—the formulation in statics referring to point masses and fluids, the formula for the curvature of a rod, and the dynamic formulation—are related to one another. Previously, Euler had spoken of the various forms as if they were part of a common order, as if they all "arose from the same principles."[109] Maupertuis's answers showed him that there is a certain hierarchy among them. The formulas $\int dt \int V dv$ and $\int ds \int V dv$ are *applications* of the 'absolute quantity of action' $\int V dv$ in different static and dynamic situations in which space and time play a role. The true action is thus the integral of a single central force, V, over the distance v of which it is a function, with the integrand being positive if the force is directed centripetally. The mathematical form of the concept of the absolute quantity of action is identical to what was later to be called 'potential energy'. The subordinate forms are special instances of the absolute case, and are derived *a posteriori*. From this, it is clear why both Euler and Maupertuis could see Euler's work as an application of the theory of Maupertuis.

It is likely that Euler modified the draft of his "Recherches sur les plus grands" on the basis of this correspondence, but how is not known.[110] The topic in "Recherches" is a chord fastened at either end, subject at every point to random central forces.[111] If the chord is entirely flexible, the quantity of action is equal to $\int ds \int V dv$ (where the sign of the force V is positive if it is centripetal). This, Euler says, is in accordance with Maupertuis's formula. If on the other hand the chord is elastic, there is an additional factor, $\frac{1}{2} A \int ds/r^2$, which Euler calls the 'quantity of the action of elasticity' ("quantité d'action de l'élasticité") to distinguish it from the 'quantity of the action of forces' ("quantité d'action des forces").

After working out some problems involving the two formulas, Euler presents a striking discovery in the concluding section. He shows that, although the two formulas look very different, a common form can be uncovered by rewriting

---

[107] These are the letters written between 26 April and 14 June 1748 (*EOO* IVa.6, 101-119).
[108] Letter from Euler to Maupertuis, 8 June 1748 (*ibidem*, 115-118).
[109] "partir des memes principes" (letter from Euler to Maupertuis, 9 May 1748 (*ibidem*, 107-109, this passage on page 108)).
[110] In his letter of 8 June 1748 Euler says that he will only need to change 'a few words' ("quelques mots"). However, he emphasizes that he does not just speak empty praises, but actually adopts Maupertuis's ideas (*ibidem*, 117).
[111] Euler did not use the term 'central force' here, but rather describes what he means as: "any forces directed either towards a fixed point, or towards several, being proportional to any functions of these distances" ("forces quelconques, qui soient dirigées, ou vers un point fixe, ou vers plusieurs, étant proportionnelles à des fonctions quelconques de ces distances") (Euler, "Recherches sur les plus grands," 4).

$\frac{1}{2}A\int ds/r^2$ as $\int ds \int A/r dr^{-1}$. We can now regard A/r as an elastic *force*, and the differential dr^{-1} as the *path* along which the force of elasticity moves the chord. According to Euler, this reveals the analogy between the formula $\frac{1}{2}A\int ds/r^2$ and $\int ds \int V dv$.[112] In this way Euler shows that Maupertuis's formula is much more general than was at first thought, because it applies not only to the action of external central forces, but also to the action of the forces of elasticity in the chord itself.

This change is also visible in another article written around that time, "Réfléxions sur quelques loix générales de la nature."[113] In this article, Euler sets himself the task of finding the minimum quantities of mechanics using his *a posteriori* method. He first finds the solution of a particular problem using the *direct method* and then looks for the quantities for which the solution represents a maximum or minimum value. Then he derives from this the formula for the quantity of action that applies in that case, in accordance with Maupertuis's metaphysical thesis. The accuracy of the formula is thus verified by the parallel use of the *indirect method*. Finally he shows by generalization that the various formulas are analogous, and can be regarded as applications of the single principle provided by Maupertuis.[114]

However, in this article Euler is more concerned with forming a better understanding of the principle itself and the concept of action. Unlike Maupertuis, he does not try to do this by complementing the mathematical calculation with a metaphysical approach, but rather by using physical intuition. How can we see that the quantity of action of forces on their point of application really is equal to $\int V dv$, and that this action is also minimal in the state of equilibrium? The basis of his argument is a physical analogy, i.e., a static system of springs (see Figure 5.5). A number of identical springs, each with a variable length x_i and—for the sake of argument—a constant force of elasticity, all act on a point Z. The springs are divided into three groups, in such a way that each of the groups pulls on point Z in the same plane but in a different direction. It is now evident, says Euler, that the total length of the springs, Σx_i is a minimum in the state of equilibrium, and that thus the differential $\Sigma dx_i = 0$. In the same way, in the actual situation with variable forces, $\Sigma \int V_i dv_i$ is also minimal, and $\Sigma V_i dv_i = 0$.[115]

Euler's proof leaves many questions unanswered. In the first place it is circular, in that it assumes the very thing which is to be proved, and the consequences of the shift from a constant to a variable force are not sufficiently elaborated. More important, it does not actually provide an answer to the original questions, of why the action of a force follows this formula and not another, and why the action is a

---

[112] *Ibidem*, 36-37.
[113] Euler describes this article to Maupertuis in the same letter of 8 June (*ibidem*, 117). The article was presented in the Berlin Academy on 6 February 1749 (*Die Registres der Berliner Akademie der Wissenschaften*, 134-135).
[114] Euler, "Réfléxions sur quelques loix générales de la nature qui s'observent dans les effets des forces quelconques," 39-40.
[115] Euler, "Réfléxions sur quelques loix," 51-52.

minimum. The reason that Euler nevertheless considers that this argument clarifies matters must therefore be his preference for the direct method, in which problems are solved on the basis of the usual principles of mechanics. He must have felt that this approach allows us to understand why Maupertuis's principle has that precise form.

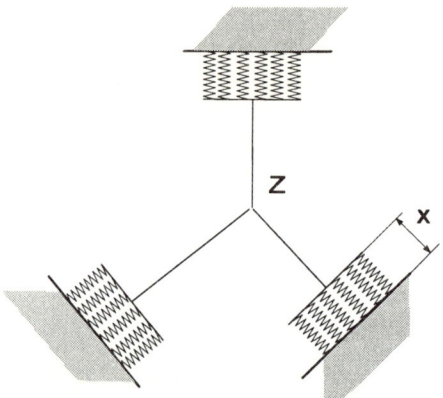

Figure 5.5 Euler's analogy with a static system of springs

What is more interesting is the continuation of his 'proof', in which he shows how the concept of action can be extended to three-dimensional bodies. The influence of his correspondence with Maupertuis can be seen very clearly here, because the proof not only deals with the expansion of the mathematical formula but also addresses the logical structure. Even more than in "Recherches," Euler clarifies the connection between the application of the principle of least action in situations of any type, on the one hand, and the concept of the action of a force on a single point mass on the other hand. All applications are based on the formula $\Sigma \int V dv$—the action of a force on a point mass—which can be applied to a three-dimensional body by integrating over space, or to a body in motion by integrating over time. Thus $\Sigma \int V dv$ is the *absolute* quantity of action, while $\int ds \Sigma \int V dv$ and $\int dt \Sigma \int V dv$ are *applications* of the formula, depending on the nature of the problem concerned. Later, in "Harmonie entre les principles généraux de repos et du mouvement de M. de Maupertuis" (1752), this relationship was further strengthened by deriving its validity in dynamics from its validity in statics.

There is at first sight a contradiction between the metaphysical *a priori* and the mathematical-physical *a posteriori*, since the first recognizes only minima, while the second yields both minima and maxima. However, this contradiction is only apparent and in fact demonstrates the surplus value of metaphysics: the concept of action is not just a mathematical quantity, it is a quantity that says something about the *essence and purpose* of reality.

The paradox arises because a mathematical truth has been coupled with a metaphysical truth. The mathematical principle is, that "all actions produced by natural forces always imply a *maximum*, or a *minimum*."[116] The metaphysical principle is, that "nature always works towards a certain goal, which it seeks to achieve using the most economical means," which is "the least quantity of action."[117] The metaphysical principle is superior to the mathematical one, it is as it were the plumb-line "to which all actions orient themselves."[118] Formulas found mathematically must be manipulated changing the sign so that they do produce a minimum for the actual situation, if they are to yield the quantity of action.

In "Réfléxions" Euler is so convinced that the action is always a minimum that he excludes even the mathematical possibility of a maximum.[119] Almost four years later, in "Harmonie entre les principles" (1752), he admitted that in the static case a maximum is also a mechanical, or at any rate a mathematical, possibility. But he explains such an equilibrium to be completely different in nature to the equilibrium corresponding to a minimum, just as the equilibrium of a cone is different if it is standing on its tip than when it is standing on its base. The latter is self-correcting, the former is not. Clearly Euler is referring here to what we now call *stable* and *unstable* equilibria, respectively.

5.4.3   Euler's Foundation of the Laws of Collisions

Of the articles about the principle of least action that Euler wrote after 1749, only one is important in relation to metaphysics. This is his "Recherches sur l'origine des forces" (1750). In this article he attempts to derive the *laws* of impact by providing a revised mechanical foundation for the *forces* of impact.[120] The problem of the foundation of the laws of impact has been discussed in detail in previous chapters. In Chapter 2 we have seen how Descartes considered that matter's characteristic of extension provided sufficient basis for the formulation of the laws of impact. Chapter 3 showed how Leibniz, who was aware that it was not possible to derive something like the actions of forces simply from extension, attributed an inner force to matter. In Chapter 4, d'Alembert rejected such a force and replaced it by adding impenetrability as a characteristic of matter. He claimed that he could then derive the laws of impact. However, in the actual derivation he appeared to use only the structural characteristics of space and time. The present chapter has shown that Maupertuis recognized that the impenetrability of matter makes the *existence* of laws

---

[116] "toutes les actions, qui sont produites par les forces de la nature, renferment constamment un *maximum*, ou un *minimum*" (Euler, "Recherches sur les plus grands," 1). The italics are Euler's.
[117] *Ibidem*.
[118] *Ibidem*.
[119] Euler, "Réfléxions sur quelques loix," 50-51.
[120] A later and more adroit version can be found in letters 69-79 of Euler's *Lettres à une princesse d'Allemagne* (1760), 149-173.

of impact and forces of impact necessary, but that he did not go so far as to also seek to derive their *form* from it.

Euler's attempt can now be seen as a continuation of this line of development.[121] He uses more or less the same reasoning as d'Alembert in his proof of the laws of motion. For example, Euler, like d'Alembert, proves the principle of inertia from the symmetry of time and space.[122] However, when he comes to the foundation of the laws of impact, he adds the concept of the force of impact.[123] In his opinion, impenetrability entails the concept of force,[124] at least if we define force as "any cause that is capable of changing the condition of a body."[125] However these forces are not permanent, they exist only if "(...) the bodies, if they were to maintain their state of motion, would have to penetrate one another."[126]

By subordinating force to impenetrability Euler can link force to its action without making them equivalent: a force exists only if it *must* act, and thus *acts*.[127] The force of impact is by nature no more than a consequence of the impenetrability of matter. In fact we can say nothing more about force than that it maintains impenetrability. Its magnitude and direction remain indeterminate.

Nevertheless Euler can proceed by adding two principles. The first of these enables us to ascertain the magnitude of the force of impact: "In the impact between bodies, their impenetrability produces only the minimum force required to protect them from penetration (...)"[128] This principle certainly resembles Maupertuis's principle of least action, which becomes even clearer from the remainder of the sentence: "(...) and no doubt it is this fact that is the basis of the general principle that all changes in the world are produced at the least expense possible, or with the smallest forces capable of producing the effect concerned."[129]

The second principle relates to changes in direction: the colliding bodies are always "deflected from penetration as directly as possible."[130] In other words, the

---

[121] For Euler's concept of matter see his anonymous essay *Gedancken von den Elementen der Cörper* (1746), as well as his *Lettres à une princesse d'Allemagne*, 149-153 (letter 69-70). Although Euler follows Descartes in his characterization of matter, in the sense that he rejects 'occult qualities' and psychic characteristics such as wanting, feeling and thinking as explanations of material characteristics, he adds mobility and impenetrability to the characteristic of extension, because these characteristics are independent from one another but are nevertheless necessary to explain material reality.

[122] Euler, "Recherches sur l'origine des forces," 110-111 (sections 3-5).

[123] *Ibidem*, 115 (section 18).

[124] *Ibidem*, 116 (section 20).

[125] "toute cause qui est capable de changer l'état des corps" *(ibidem,* 111, 115 en 117 (sections 8, 18 and 22 respectively)).

[126] "(...) les corps en continuant leur état se devroient pénétrer mutuellement" *(ibidem,* 116 (section 21)).

[127] *Ibidem*, 117 (section 23).

[128] "Ainsi dans le choc des corps leur impénétrabilité ne fournit toujours que la plus petite force, qui est capable de les garantir de la pénétration (...)" *(ibidem,* 118-119 (section 26)).

[129] "... et c'est sans doute sur cette circonstance, qu'est fondé ce principe si général, que tous les changemens au monde sont produits aux moindres dépens qu'il est possible, ou avec les plus petites forces, qui sont capables de cet effet" *(ibidem).*

[130] "le plus promptement détournés de la pénétration" *(ibidem,* 119 (section 27)).

force is perpendicular to the tangent plane at the point of contact. Euler then uses these additional specifications of the forces involved in an impact to derive the principle of equality of action and reaction, as well as the laws of impact for non-elastic and elastic bodies.[131] I will not describe these derivations here: the important point is to see how Euler deals with the principle of least action, and this has been made clear now.

Both of these principles that Euler adds to the characteristic of impenetrability derive from the principle of least action, and are at the same time the foundation of that principle. Both force and action are subject to the same minimum principle.[132] Euler's foundation of the force of impact shows even more strongly than Maupertuis's argument that the final cause and efficient cause have the same metaphysical basis. Where for Maupertuis this basis was the oneness of God, for Euler the reference to God is as it were reflected onto matter itself.

It is remarkable that Euler locates this material foundation in impenetrability, i.e., in a characteristic of material substance. This immediately raises the question of how this differs from the foundations of his mechanistic predecessors, who sought to derive the laws of impact from characteristics such as hardness, impenetrability, and elasticity. Have they simply not been sufficiently penetrating? On the contrary! It is striking that Euler seems to forget here that his derivation includes two extra principles, neither of which is derived from material substance.

Euler's presentation of his derivation puts his earlier distinction between the 'direct' and 'indirect' method in a deceptive light. On the one hand, he recognizes a fundamental difference in approaches to material reality; yet he also wants to derive the ontological foundation of both methods from material substance, via the action of force. It is as if Euler wants to use the light of his mechanistic foundation to unmask the indirect method as a shadow of the direct method, and so show that ultimately the laws of motion can only be derived from efficient causes.

Euler's pursuit of a mechanistic foundation should not be surprising however: he also seeks to base the Newtonian theory of attraction on material impenetrability. Euler thus combines a Newtonian scientific practice with a mechanistic explanation of natural phenomena.[133] However, there is also an important difference: while attraction had already proved its merits, the principle of least action was still at the beginning of its illustrious ascent. Euler's explicit derivation of the principle from a mechanistic foundation is therefore less important than the new contribution he made by elaborating its implications in mechanics.

What is concealed in his foundation but is nevertheless visible in his elaboration is that the minimality of the integral of action does not say anything about the separate material substances, but about their structure. In the end, the forces of

---

[131] *Ibidem*, 119-121 (sections 28-31) and 121-125 (sections 33-44) respectively.
[132] This is even clearer in his later *Lettres à une princesse d'Allemagne* (1760), because he identifies the term 'action' in Maupertuis's principle with the forces acting between colliding particles (Euler, *Lettres à une princesse d'Allemagne*, 170 (letter 78)).
[133] Van Lunteren, *Framing hypotheses*, 104-111.

impact are not derived from the characteristics of the material substances, but from a principle of material structures: that the interactive effects are as small and as direct as possible.

5.5 CONCLUSION: TELEOLOGY AND STRUCTURE METAPHYSICS

The principle of least action was only one of many mechanical principles proposed in the middle of the eighteenth century. At the time it caused some amazement, but was nevertheless taken just as seriously as any other principle. However, in later historiography it has become one of most controversial principles of mechanics. This is partly because of the scandal that arose out of the dispute about who had first proposed it, and is partly attributable to the inability of historians to locate the principle in any context other than this scandal. In this chapter, I have tried to explain the principle in its metaphysical context and to identify from this its significance in the development of mechanics. I have limited myself to the work of Maupertuis and Euler, without considering the criticisms of contemporaries such as d'Arcy, d'Alembert and König. However, Chapter 7 deals with the way in which Lagrange was to integrate the principle in his mechanics.

Maupertuis's introduction of teleology into mechanics should be regarded as a continuation of his defense of the Newtonian theory of attraction against the continental mechanical philosophy. Although he was very successful in this defense, the germ of his later radical change of course was already evident there. The traditional approach, in which the material substance is the starting point, proved unsuitable to provide a satisfactory foundation for the concept of force. The idea of universal regularity offered him an alternative way of providing explanations. Initially this took the form of a principle of extrema in statics. This principle did not say how large the individual forces would be, but did provide a constraint that the effects of force would always meet. Some years later he formulated a principle in teleological form, first in optics, and later in mechanics. This principle had two forms: 'the *sum* (or integral) $\Sigma mv\Delta s$ is minimal', and 'the *change* in $\Sigma mv\Delta s$ is minimal', respectively. Two years later he approached the argument from the other direction, and developed a proof of the existence of God from the principle of least action.

This sketch shows clearly that Maupertuis's later teleological approach is a particular form of a new, much more general approach, which focuses directly on structure. The three successive phases of Maupertuis's approach share common principles that do not base the effects of forces on material substances as such, but on their *structure*. This shows how the change from a mechanics of substances to a mechanics of structure was followed, with Maupertuis, by a change in the corresponding metaphysics from a metaphysics of substance to a metaphysics of structure. Compared to this development, it is not so important that he chose a variant that was only one of the possible variants, and was soon found to be inappropriate in natural sciences, perhaps because it was *only* possible.

Euler did not go through a radical development comparable to that of Maupertuis. In metaphysical terms he always presented himself as a mechanistic scholar.

Nevertheless, he was prepared to adopt and even develop approaches of other types, as his involvement in the principle of least action showed. His interest in this principle was certainly not only mathematical: he sincerely believed that nature was purposive and that God was involved in this. However, the most important difference between Euler and Maupertuis was the foundation of final causes: for Maupertuis, the goal was to be located in God's wisdom while for Euler it was ultimately to be located in material substance. In the end, Euler's goal was to base structure in substance, and to base the mechanics of structure in a metaphysics of substance.

However, the significance of the principle of least action lies not so much in the specific way in which Maupertuis and Euler tried to provide a basis for it, but in the general way in which they continued along the path that we saw d'Alembert begin in the previous chapter: the transition from a substantial to a structural approach to force and action.

C

BETWEEN METAPHYSICS AND MECHANICS

CHAPTER 6

THE CONCEPT OF FORCE IN THE 1779 BERLIN ESSAY COMPETITION

> *The whole problem of philosophy appears to consist in this, that we study the forces of nature on the basis of the phenomena of motion; [and] then derive the other phenomena on the basis of these forces.*
>
> Isaac Newton[1]

6.1   INTRODUCTION

At the end of the eighteenth century, natural science awaited the synthesis of its philosophical principles in Kant's transcendental philosophy. It would be difficult to over-estimate the extent to which this synthesis has influenced ideas about the foundations of natural science. For example, theories of knowledge such as empiricism and rationalism, that were respectable in the eighteenth century, came to be regarded as naive and dogmatic, respectively, after Kant. It almost appears that the Berlin Academy had a presentiment of this decline in eighteenth century philosophical thinking, and as a homage to it announced a competition question on the foundation of force for the year 1779, although they probably knew in advance that the competition question would be greeted with derision by many in the world. In any case, the competition gives us a good view of the way in which force was conceptualized in the 'République des Lettres', in both ontological and epistemological terms. This competition allows us to gauge how far the change in the experience and understanding of force in mechanics had penetrated to the average scientist and interested scholar. This will complement the picture drawn in Chapters 4 and 5, where the controversies about the principle of least action and the true measure of living force

---

[1]   "Omnis enim philosophiæ difficultas in eo versari videtur, ut a phænomenis motuum investigemus vires naturæ, deinde ab his viribus demonstremus phænomena reliqua" (Newton, *Principia mathematica*, "Auctoris praefatio 1686," 16).

were discussed, and interpreted as a change in the foundation of mechanics and as a first exploration of the structural nature of mechanics, respectively.

6.2 THE BERLIN ACADEMY'S PHILOSOPHICAL COMPETITIONS

6.2.1 An Innovation for the Academic World

Frederick the Great is notorious for his domineering politics and famous for his literary, philosophical and musical ambitions. At an early age, his feeling for the aesthetic drove him to resist the Spartan upbringing of his father, Frederick Willem I, the 'soldier king'. As a teenager he even made an unsuccessful attempt to run away. It was only when he was king, after the death of his father in May 1740, that he had the opportunity to achieve his ideals. His efforts included the reorganization of the former Scientific Society (*Sozietät der Wissenschaften*) and its fusion with the Society of Literature (*Société littéraire*) to form the Royal Academy of Sciences and Literature (*Académie royale des sciences et des belles-lettres*). This academy was unique among eighteenth century academies in having a department of speculative philosophy in addition to, and with equal status to, the mathematical, experimental physics and literary departments. Frederick adopted the idea of holding an annual competition from the Paris *Académie royale*. The person who provided the best answer on a theme selected by the Academy would be awarded a prize of 50 ducats.[2] Although the Paris academy had already initiated such a competition, the presence of a department of speculative philosophy resulted in the emergence of an entirely new phenomenon in the history of science: the philosophical competition. Like the existing competitions, one important goal of the Berlin competition was "to foster the progress of the sciences in the broadest sense, and to guide it along the correct lines."[3] Moreover, the Berlin philosophical competitions were promising because of the presence in the Academy of the two most important schools of philosophy: English empiricism and German rationalism. For the first time in history, the followers of these schools were given the chance to debate their ideas in public. It is therefore no wonder that the announcements of the questions for the competition were anticipated with great curiosity throughout Europe.[4]

However, the high status of the philosophical competitions is not in itself a reason for attributing any great importance to the essays that were submitted in response. Although there may be some interesting essays buried among the

---

[2] *Die Registres*, 99 ("Réglement de l'Académie," art. 19). The "Réglement de l'Académie" is printed in the *Histoire de Berlin* for 1746. The original has been preserved as part of the 'Régistres de l'Académie' (2 June 1746) in the *Akademiearchiv* (the former *Zentrales Archiv*), and has been reprinted in Harnack, *Geschichte der Akademie* I.1, 299-302, and in *Die Registres*, 96-99.

[3] "(...) Fortschritt der Wissenschaften im Grossen zu befördern und in richtigen Bahnen zu halten" (Harnack, *Geschichte der Akademie* I.1, 396).

[4] At least in Germany, France and Switzerland. There appear to have been no submissions from England or the Netherlands.

submissions for the philosophical competitions in general, their significance does not lie primarily in their possible influence on the development of science.[5] References to the essays for the philosophical competition, in contrast to those for the mathematical competitions, are rare and are found in authors of the second rank. The essays should therefore be considered mainly as a portrait of the era, revealing the ideas of the public i.e., of educated laymen, regarding what were then live issues.

The reason for this may well be that the choice of both the competition question and of the winner were very strongly determined by power politics, with the result that even the essayists themselves were sometimes influenced in the process of writing. Naturally the practice of science, like any other 'way of life', entails a certain variant of the struggle for power, in the forms of alliances and maintaining or redirecting the flows of funds etc., even if the goal is only to be able to prove the truth of one's conviction in an entirely legitimate way. But in the case of the philosophical competitions, at any rate at the time of the government of Frederick the Great, we can confidently assert that these were a political rather than a scientific activity, and politics of a rather low order at that.[6] This is also true of the competition question regarding the foundation of force that the Academy announced for 1779, on which this chapter will focus. A reconstruction of the condition of the philosophical competition in 1779 should make this clear.[7]

---

[5] Examples include Mendelssohn's *Abhandlung über die Evidenz in metaphysischen Wissenschaften* (1764) and Kant's *Welches sind die wirklichen Fortschritte, die die Metaphysik seit Leibnizens und Wolf's Zeiten in Deutschland gemacht hat?* (1796).

[6] Thus Winter's suggestion, that "the history of the prize contests held by the Berlin Academy of Sciences should be given more weight in the history of the European Enlightenment" (Winter, [Introduction to *Die Registres*] (1957), 33), cannot be sustained without some reservations.

[7] The sources for the reconstruction, in the *Akademiearchiv*, the Archive of the Berlin-Brandenburg Academy of Sciences and the Humanities (previously: Central Archive of the Academy of Sciences of the German Democratic Republic), are in the first place the competition essays themselves. These will be referred to below using only their archive code, consisting of the letters 'I–M' and a sequential number. The competition essays were catalogued in 1956 by Lore Ulbricht and R. Faber, in the 'Findbuch Preisschriften' (D1/14). Werner Krauss's plan to publish them has unfortunately remained confined to the prize contest for 1780: *Est-il utile the tromper le peuple?* (Krauss ed. 1966). Addition interesting information can be derived from the 'Régistres the l'Académie' (I-IV-32). Some of these, covering the period 1746-1766, are printed in *Die Registres der Berliner Akademie* (Eduard and Maria Winter eds. 1957). There are also a number of interesting letters in (I-VI-10), cited by Buschmann, "Die philosophischen Preisfragen und Preisschriften der Berliner Akademie der Wissenschaften im 18. Jahrhundert" (1989). Formey provides a retrospective eye-witness account in his *Souvenirs d'un citoyen* (1789). The scant secondary literature is limited to Harnack, *Geschichte der Akademie* (1900), I.1, 396-422, and II, 305-310 (the latter is an overview of all academic prize contests); Biermann, "Aus der Geschichte Berliner mathematischer Preisaufgaben" (1964); Müller, *Akademie und Wirtschaft im 18. Jahrhundert. Agrarökonomic Preisaufgaben und Preisschriften* (1975); Bongie, [Introduction to Condillac] (1980) and Buschmann, "Philosophischen Preisfragen" (1989).

6.2.2 The Status of the Philosophical Competition at about 1779

The article in the regulations that stipulated the obligation to hold competitions only indicated the outlines of the procedure. The details, that each of the four departments ('classes') of the Academy should hold a competition in turn; that the jury would usually consist of the members of the department concerned; that there should be two years between the announcement and the competition award; all these were to develop from practice. The way in which the particular question would be chosen and the possibility of an honorable mention—the so-called 'Accessit'—were also not stipulated.

In the first years each of the four departments presented a number of questions to the weekly general meeting, from which one was selected by vote.[8] It appears that the selection by vote quickly became a formality.[9] We know that this was the case for the question for 1779, because Lagrange, who was present at the meeting to which the question was presented,[10] had to obtain the formulation from d'Alembert.[11]

The proposals for the questions might well come from other departments. For example, Euler, who was then the director of the mathematical department, was closely involved in drawing up the question for the first philosophical competition, which asked for a critique of the theory of monads.[12] Moreover, the anonymous and strongly-worded publications emanating from Euler, and from Formey in response to Euler, give us grounds to suspect that they saw this first competition as a means of carrying on their dispute.[13] This was also apparent in the evaluation by the jury, which consisted of members from all four departments.[14] Thus Euler could also evaluate the essays that had been submitted.[15] Only his influence could explain the fact that the essay written by one Justi, which was later to receive very negative reviews but had an advantage over the other submissions in that it expressed Euler's opinions very well, was awarded the prize. This state of affairs could only bring

---

[8] The 'Régistres de l'Académie' for Tuesday 9 June 1746 report that Formey, the secretary of the Academy, presented three questions on behalf of the department of philosophy. The question concerning the theory of monads was then chosen (*Die Registres*, 99).

[9] In 'Régistres de l'Académie', 21 May 1750, it is stated that "The Permanent Secretary (...) has indicated the subject that the department of literature has proposed for the same year, 1752" (*Die Registres*, 150). Thus the Department of Humanities *selects* a topic: their proposal is simply presented and approved.

[10] 'Régistres de l'Académie', 3 June 1777 (*Akademiearchiv* I-IV-32, 262).

[11] Letter from Lagrange to d'Alembert, 27 January 1778 (*Œuvres de Lagrange* XIII, 334-336, this passage on 336).

[12] This first philosophical competition question was selected on 9 June 1746, and the prize was awarded on 1 June 1747 (*Die Registres*, 99 and 112 respectively). Euler's involvement can be seen from the protocols of the old *Société Littéraire* (Winter, [Introduction to *Die Registres*], 33).

[13] [Euler], *Gedancken von den Elementen der Cörper* (1746); [Formey], *Recherches sur les éléments de la matière* (1747).

[14] Harnack, *Geschichte der Akademie* I.1, 403.

[15] His remarks on this were preserved in manuscript and were published in 1862 under the title 'Différentes pièces sur les monades'.

philosophy into disrepute, and it took the award of ten Accessits to cool the heated tempers in the Wolff camp.[16]

The regulations said nothing about the composition of the jury. Presumably a jury was chosen by the meeting, with the members initially coming from all departments, as in 1747, but in later years it increasingly came to be assumed that they would originate only from the department concerned. Such, at least, can be deduced from a letter from Sulzer in 1754, in which he writes: "Heinius, Formey, Merian and myself actually comprise the whole class of philosophy in the Academy. The first two are avowed Leibnizians. Merian on his own can do nothing."[17] From the implicit equation of the members of the jury and the members of the philosophy department, we can see how natural this had become after just eight years.[18]

The citation from Sulzer also shows that a struggle was underway in the department of philosophy between supports and opponents of Leibniz's and Wolff's philosophy. But it would be too simple to say that the balance of power between Wolffians and anti-Wolffians[19] determined the outcome of the prize contest. This is evident from the same example, the prize contest of 1755: Formey changed his mind out of fear for Maupertuis, according to Sulzer, or by conviction, according to Premontval. Sad to say, the first possibility is the more probable.[20]

Naturally the formulation of the competition questions had prepared the way for the partisan spirit that played such a large role in the judging. It is striking that all the philosophical competitions during the reign of Frederick the Great that were regarded as controversial, related to a metaphysical position in the tradition of Leibniz and Wolff. The first three of these were critically formulated, requiring that a Leibnizian-Wolffian idea should be defended or attacked. The question for 1747 related to the foundation and physical meaning of the theory of monads; for 1755 to the significance and truth of the idea 'everything is good'; and for 1763 to the degree of certainty of metaphysical, as compared to mathematical, truths. Although the questions for 1768 and 1779 also related to Leibniz and Wolff, these had a quite different character. The question for 1768 was intended only as an accolade to Leibniz, while the question for 1779, as we will see, actually posed a problem from Leibnizian-Wolffian metaphysics rather than asking for a critique!

This change in the nature of the competition questions could be due to the changing composition of the department of philosophy, since Merian, being a Newtonian and a Hume translator, moved to the Department of literature in 1771.

---

[16] The pieces were all published in one Volume: Justi et al., *Dissertation qui a remporté le prix proposé (...) sur le système des monades avec les pieces qui ont concouru* (1747). See also Bongie, [Introduction to Condillac] (1980).
[17] Letter from Sulzer, 22 September 1754 (cited in Harnack, *Geschichte der Akademie* I.1, 405).
[18] Maupertuis was in fact present in the commission, but he abstained from voting. It is not known whether, as president of the Academy, he was the formal chairman of the jury but had no voting rights, or he abstained voluntarily.
[19] While Sulzer says that Heinius and Formey were 'more Leibnizian', Leibniz was seen mainly through the eyes of Wolff (see above, section 3.1).
[20] Harnack, *Geschichte der Akademie* I.1, 405.

However, it could also be explained by Euler's departure in 1766. In any case, there are grounds for supposing that changes in both the formulation of the questions and the judging reflected shifts in the balance of power between Wolffians and their opponents in the Academy, and especially in the department of philosophy. Harnack writes that during the time that Maupertuis was President the Newtonians had the upper hand, and after that the Wolffians, under the influence of Sulzer, prevailed. However, it must be remembered throughout that these schools of thought were followed in an eclectic way: "This eclectic attitude in philosophy, with a pointed rejection of the materialists combined with an acceptance in principle of empirical methods, albeit with dogmatic provisos, is characteristic of the last ten years of the Frederician Academy (...)"[21]

All this makes it plausible that the partisan spirit in the previous years, originating in a struggle for power, continued in later years as the prejudice of exhaustion. It is therefore hardly surprising that the jury of 1779, consisting of the members of the department of philosophy, wanted to see a Wolffian answer.[22] Moreover, it had become a public secret over the years that the jury itself supported a more or less Wolffian position. This can be seen in a letter from Hißmann, one of the participants, to Merian.[23] Hißmann apologizes for his Leibnizian essay, arguing that this was the only way of having a chance at winning the competition. He also accused the winner, Pap de Fagaras, of only saying what could be derived directly from Leibniz.

We can therefore conclude that the philosophical question for the year 1779 stemmed from the work of the members of the department of philosophy itself, and that they simply announced at the weekly general meeting that they had selected a

---

[21] "Diese eklektische Haltung in der Philosophie, mit scharfer Abweisung der materialistischen, mit principieller Zustimmung zur empirischen Methode, aber mit dogmatischen Vorbehalten, charakterisiert die letzten zehn Jahre der fridericianischen Akademie (...)" *(ibidem,* 445).
[22] The members who actually judged the essays were:
    Nicolas de Béguelin   (25.6.1714—3.2.1789)   member from 2.11.1747
    Jean Henri Samuel Formey (31.5.1711—8.3.1797)   member from 23.1.1744
    Louis de Beausobre   (22.8.1730—3.12.1783)   member from 27.2.1755.
Two other members took part in formulating and choosing the competition question, but had already died when the competition was judged:
    Johann Georg Sulzer   (16.10.1720—25.2.1779) member from 29.10.1750
                                                     Director from 8.8.1776
    Leonhard Cochius   (28.1.1718—28.4.1779)   member from 26.4.1770.
Finally there were two members of the Academy, Johann Heinrich Lambert and J.B. Merian, who were not part of the Department of Philosophy but had published in the field, and they may therefore have had some influence (biographical data derived from *ibidem,* 466-481).
[23] Letter from Hißmann to Merian, 29 September 1780 *(Akademiearchiv* I-VI-10, 80/81). Merian had been Director of the department of literature since 7 (or 8) February 1771. Before that he had been a member of the department of philosophy (see also note 22). Thus Hißmann was writing to an insider. Supposing that he knew this, and given that his letter shows him to be an opportunist, it is very unlikely that he is seeking to win Merian's favor by appealing to something that Merian would disagree with. Presumably, then, Merian shared his criticism.

question to be announced by the Academy. The jury that judged it also consisted of the members of the department of philosophy.

The time at which the question was set was exceptionally opportune in relation to the development of the concept of mechanical force. In the previous decades there had been a radical change in the metaphysical dimensions of the concept of mechanical force. However, the book that would weld mechanics into a clear and cogent unity on the basis of this change, and would simultaneously set out the relationship between mechanics and metaphysics in an apparently unambiguous way, had not yet been published. This book, by Lagrange, will be discussed in the following chapter. The competition for 1779 may be able to show us whether the change in thinking that was apparent among the advance guard in mechanics had penetrated to a wider public. Was the old thinking about the concept of mechanical force still being maintained? Were alternatives being developed? In short, how does the concept of mechanical force that had developed *within* mechanics relate to opinions about it on the part of people *outside* the field of mechanics?

6.3   THE SIGNIFICANCE OF THE COMPETITION ON THE FOUNDATION OF FORCE

6.3.1   Parisian Arrogance

Like many of its predecessors, the question for 1779 was highly controversial. D'Alembert wrote to Lagrange that the whole Paris academy burst into laughter when Condorcet, the secretary, read out the question.[24] Lagrange himself then apologized to d'Alembert for this very German and very Leibnizian question, although he had not had any part in it.[25] The king also rejected the question indignantly, perhaps under the influence of d'Alembert. Such abstract philosophical questions were no longer interesting for Frederick: his long reign had confronted him with moral problems, and he now expected his academicians to contribute to solving them. So not surprisingly the competition for 1779 was the last in which the Academy was granted such wide freedom. The king prescribed a substitute question, originating from d'Alembert, and the substitute was if possible even more controversial: "Can it be useful to deceive the people?" ("S'il peut être utile de tromper le peuple?") This was finally published as an *extra* question for the year 1780. It caused quite a stir, because a constitutional dilemma had been openly expressed, and moreover by a royal academy! Soon after, Frederick actually sent an official order to the Academy, stating that henceforth only "very interesting and useful questions" would be set.[26]

---

[24]   Letter from d'Alembert to Lagrange, 22 September 1777 (*Œuvres de Lagrange* XIII, 330-332, this passage at 332).

[25]   Letter from Lagrange to d'Alembert, 27 January 1778 (*Œuvres de Lagrange* XIII, 334-336, this passage on page 336). Presumably Lagrange is pretending here that he knew nothing, unless he was asleep during the meeting in which the question was announced. According to the attendance record, he was present then! ('Régistres de l'Académie', 5 June 1777 (*Akademiearchiv* I-IV-32, Bl. 216b)).

[26]   "questions très-intéressantes et très-utiles" (Harnack, *Geschichte der Akademie* I.1, 417).

The question regarding the foundation of forces ("Fundamentum virium") was announced in the public meeting of the Berlin Academy of 5 June 1777 by its secretary, Formey.[27] If we may judge by the widely scattered origin of the essays that were submitted, it was then published in many newspapers and journals. The deadline for submissions was 31 May 1779.[28] The winner and the Accessit were announced on 3 June 1779.[29] Despite the jeers of the Paris Academy, there were a considerable number of responses: sixteen essays.[30] If we compare this to the average of ten essays for the academic competitions between 1745 and 1788, we can see that the theme could still elicit reactions, and was certainly not a thing of the past, as was thought in Paris.[31]

6.3.2 Formulation and Analysis of the Question

How, precisely, was the question worded? I will cite the first part from the French in Harnack, *Geschichte der Akademie*, and where he stops I will continue from the German translation as given in Rehberg's essay (for the original text, see the annex at the end of the chapter):

> Effects can be observed throughout nature; thus there are forces.[32] But these forces must be determined if they are to act, which implies that there is something real and durable, susceptible to being determined and it is this real and durable something which we call the fundamental and substantial force. Therefore, the Academy asks:
>
> What is the distinct idea of this fundamental and substantial force which, when it is determined, produces an effect? In other words: what is the foundation of force? Now, to imagine how this force could be determined, it is necessary either to prove that one substance acts on another, or to demonstrate that fundamental forces determine themselves.[33]
>
> In the first case the Academy expects a clear idea of the original passive force, and the way it suffers from another force. In the latter case it should be clearly stated what determines these forces and how it does so. How is it that the same force sometimes produces a certain effect, which it cannot produce

---

[27] 'Régistres de l'Académie', 5 June 1777 (*Akademiearchiv* I-IV-32, Bl. 216b).
[28] This date is not drawn from a primary source, but was given in *Akademiearchiv* I-M724, 2 (giving as its source the *Hamburgischen neure Zeitung*), and in *Akademiearchiv* I-M733, 2.
[29] 'Régistres de l'Académie', 3 June 1779 (*Akademiearchiv* I-IV-32, 262).
[30] See section 6.4.1.
[31] Between 1745 and 1788 the Berlin Academy announced 44 prize contests, including those that were not satisfactorily answered (Harnack, *Geschichte der Akademie* II, 305-309). Of the essays that were submitted, 346 have survived ('Findbuch Preisschriften'). Adding two essays per competition—for the winning essay and the Accessit—gives an average of ten essays per competition.
[32] At this point, the German translation contains the addition "oder wirkende Ursachen" ("or efficient causes"). The equation of forces with efficient causes is certainly not derived from Wolff, who makes a clear distinction between these two concepts. See below, pages 183-184.
[33] Harnack, *Geschichte der Akademie* II, 308. Harnack does not indicate his source for the formulation of the question. The only French essay, I-M729, does not provide a verification, since it gives only a much shorter version.

at other times? How can it happen, for example, that one clearly understands something that another person is explaining, while one would not have understood it without this teaching? Why is it that a person may not be able to reproduce concepts which he has forgotten, even though he did produce them the first time, and it is surely a basic principle that the will and capacity joined together must necessarily result in the effect? Finally, if the original force produces everything out of itself, what is the real and true distinction between these two cases: although one clearly comprehends a very artistic piece of music composed by a great musician, or grasps the solution to a very difficult mathematical problem set by someone else, one nevertheless cannot compose the music or solve the problem himself, however hard one tries?[34]

How exactly did this question originate? Since all the members of the department of philosophy were more or less Wolffian, it is not likely to have been the result of an internal controversy.[35] We can assume however that it originated in a problem that one or more members were interested in themselves. Indeed Hißmann writes in 1783 that he had heard from Sulzer that Cochius had raised the question. However, the other members are said to have doubted whether it could be answered.[36] So its origin would be a dispute with a philosophical school from outside the Academy. It is very likely that this school is in fact Hume's epistemology which, for example, sharply criticizes the possibility of actually knowing causal relationships, however useful they may be in everyday life. Although Kant was inspired by Hume's epistemology to provide a new foundation for physics, consisting of the transcendental conditions of knowledge, the members of the Academy apparently saw it only as a danger: one could sink into epistemology and lose sight of the possibility of an ontological foundation.[37]

---

[34] Rehberg, *Abhandlung über das Wesen u[nd] die Einschränkungen der Kräften*, 3-4. In the manuscript (I–M731) sent to the academy, Rehberg omits the wording of the question. The printed text is the same, except for punctuation, as that given by Hißmann (I–M734, 4-5, and "Versuch über das Fundament der Kräfte," 9-11). Since the authors lived far apart (in Leipzig and Göttingen respectively), so that they presumably had different sources, we have to conclude that the translation was made not by the journal or newspaper, but by the Academy itself.
[35] The Department of Philosophy in or about June 1777 consisted of Cochius, Beguelin, Formey, L. de Beausobre, and the director, Sulzer (see note 22).
[36] Hißmann, *Versuch über das Fundament*, 8n.
[37] This is also Cornelia Buschmann's interpretation. She goes even further: the defence against Hume's epistemology actually lay behind the question. But she does not give any good arguments for the theory that the question was consciously formulated in two layers. Her only argument is a reference to a 1773 essay by Cochius: "Examen de la Question: si toute succession doit renfermer un commencement?" (Buschmann, "Philosophischen Preisfragen," 216). In this essay Cochius gives a negative answer to the question in the title, "Whether all successions contain a beginning." He carefully refutes a range of arguments with epistemological, logical and ontological objections. How then is this article an attempt to provide ontological foundations for causality? The argument contains not so much as a word on Hume's epistemology, let alone his criticism of causality. Moreover, the evidence does not support her suggestion that we can see in this article "that it was especially the demands posed for a philosophical foundation by the fact that history becomes ever more scientific that could not be met by epistemological considerations alone, since the latter do not yield an understanding of actual historical development" (*ibidem*).

continued on next page

> Nevertheless, the formulation of the question clearly indicates that it derives from Wolff's metaphysics.[38] I would like to look at it more closely, reformulating it in such a way that its central ideas stand out clearly.

> The question itself is preceded by two propositions. First, from the fact that *effects* are observed, it is concluded that *forces* exist. Then it is assumed that the *acting* of a force entails that something *real* and *permanent*, that is: a *substantial and fundamental force*, is *determined*.

These two introductory propositions contain much that is implicitly assumed. The meaning of the term 'to determine' ("déterminer," "bestimmen") is particularly unclear. We can only say that the concept of force entails both that there is something real and permanent, and that it should be 'determined'. A force must be 'determined' in order to act. Thus 'acting', at the level of the phenomena, corresponds to 'determine' at the level of the force. The substantial and fundamental aspect of force is seen here as that which must be 'determined'.

In this way, it is suggested that there are two forms of force, just as with Leibniz and Wolff: a primary and real form on the one hand, and, on the other hand, a secondary form that is only analogous to the primary form. Force in the primary sense is undetermined, while force in the secondary sense is determined. This can be illustrated for the forces in mechanics, such as gravitation and living force; these are forces in the secondary sense, because while it is true that they have an observed action, they are not physically real—i.e., they are transitory. These mechanical forces are determined forms of force in the primary sense, that is, the *substantial and fundamental force*.

> The Academy's question then focuses on this *substantial and fundamental force*. What is the 'distinct idea' of this force, that is, what is its true nature, what is the foundation of forces? The Academy outlines two incompatible approaches to answer this question, both relating to the way in which the fundamental force is determined: either, 'fundamental forces determine themselves' or 'one substance acts on another'.

However, this antithesis is not in itself clear. Is the Academy saying here that substance and force are identical? This was not the case for Wolff![39] Wolff defined

---

continuation of former page
However, the ideology of 'history becoming scientific' is not only wide of the mark in view of the actual historical person of Cochius, it also misses the actual historical intention of the piece: what Cochius claims in his article is nothing more or less than that epistemology provides a powerful argument against the dogmatic idea of a world with a beginning and an end.

[38] Hißmann also says this in his letter to Merian of 29 September 1780: "The whole competition was completely drawn from the Leibnizian system, at the insistence especially of the late Cochius's" (*Akademiearchiv* I-VI-10, 80b).

force in a general sense as the source of changes ("principium mutationum"), while the efficient cause ("causa efficiens") refers to the thing that "generates something," or "raises the potential to actuality through its act." That is, the efficient cause is the substance in which the force is located rather than the force itself.[40] It is plausible that something like this is intended here, i.e., that 'force' refers to the principle of change and 'substance' to that which contains the force. The two possibilities for the determination of the fundamental force are therefore that it is determined from the substance itself—thus by the fundamental force itself—or from another substance. The first possibility is in accordance with the Leibnizian-Wolffian idea of substance, the second appears rather to be derived from Newton's concept of force.

> Next, the two possible approaches are explained further. If substances can act on one another, it is necessary to provide a *distinct idea* of the original passive force and how it is affected by another force, i.e., how it is determined.

This can be explained as follows: if the original force is determined by another substance, there must be an original passive force. This can be illustrated by Leibniz's idea of inertia, although the Academy does not mention it. Leibniz supposed that inertia, a passive force, is a determination of an original passive force. The substance that is acted upon contains a fundamental *passive* force, which is in its own way *susceptible* to being acted on by the fundamental *active* force of another substance. The question of *what* this original passive force is and *how* it is determined follows logically from this. It is however remarkable that the Academy fails to ask what the relationship between the original passive and the original active force may be. After all, it is the latter which interests them most!

> As for the second approach, in which the fundamental force is determined by itself, no further general remarks are made. However, all the following questions and examples appear to be intended as elaborations on this second approach. These refer to the phenomenon that a force has a certain effect at one time, which it does not have at another time. At the same time, the examples indicate that the persons who set this question did not see this as a problem for material forces such as gravitation so much as for intellectual forces: concepts, memories, musicality and mathematical understanding. The terminology in the examples continually shifts from the single concept of 'force' to the combination of 'will' and 'power'.

---

continuation of former page

[39] The following comparisons with Wolff are based on Christian Wolff, *Vernünfftige Gedancken von Gott, der Welt und der Seal des Menschen, auch allen Dingen überhaupt* (1751 ([1]1720)), abbreviated below as *Deutsche Metaphysik*. The Latin terms are given by Wolff himself in a supplementary glossary: "Das erste Register, darinnen einige Kunst-Wörter Lateinisch gegeben werden" (Wolff, *Deutsche Metaphysik*, [673]-[677]).

[40] Wolff, *Deutsche Metaphysik*, section 115 and section 120 respectively.

The fact that the examples appear to relate only to the second approach, the self-determination of fundamental force, must be attributed to poor formulation. This can be seen more clearly from a comparison with Wolff's system, which will be given in the following section. The examples indicate the need not only of an explanation of how a force comes to act, but also of how it is possible for its action to be impeded. On the other hand, the fact that the examples are restricted to mental forces suggests that an answer for material forces is considered optional, and even of lesser importance. Moreover, the examples suggest that the answer for mental forces will have to explain at least how 'mental force' is constituted of 'will', 'power' and perhaps a third element.

6.3.3 Comparison with Wolff's Concept of Force

To what extent is the formulation of the question really Wolffian? One must at least say that other elements have been added in two ways. In the first place, the identity of force and substance is not Wolffian. As we have seen, the formulation of the question centers on the idea that the determination of something *real and permanent* underlies the action of a force, and this something is the *substantial and fundamental force*. Thus the fundamental force is said to be a substance. This concept does remind one inevitably of Leibniz and Wolff, but it differs from their ideas considerably. Wolff in fact wanted to differentiate clearly between force and substance: for him, forces are *elements* of substances.

In the second place, the composition of mental force in the question is not entirely in accordance with Wolff's opinions. He does however make a corresponding distinction between force ("vis") and power ("potentia"). For Wolff, power is only the *possibility* of doing something, while force contains an actual *inclination*, the "effort to do something": "power only makes change possible; force makes it actual," he says.[41] In contrast, Wolff considers will, and also desire, memory and imagination, as a variety of changes in the soul, based on *a single underlying force.*[42] In the words of the Academy, the will appears rather to be *part of* the mental force.

Nevertheless, the Academy's identification of force and substance can be aligned with Wolff's system, by regarding this identification as a generalization of Wolff's view on living force. To do so, we will have to look at the way Wolff relates the concept of force to the "determination of something permanent." He does this in two places, the first in relation to the *self-determination of substance*, the other to the *self-determination of living force*.

---

[41] "Durch das Vermögen ist eine Veränderung bloß möglich; durch die Kraft wird sie würcklich" (*ibidem*, section 117).
[42] "(...) therefore only one force exists in the soul, from which all its changes originate (...)" ("(...) also ist in der Seele nur eine einige Kraft, von der alle ihre Veränderungen herkommen (...)") (*ibidem*, section 745).

Wolff defines 'substance' (also the 'self-existent thing', or 'simple thing') as that which has the "source of its changes within itself."[43] One example is the human mind, which produces its own thoughts and tendencies. At the same time, substance is something that exists continually and therefore knows neither creation nor destruction.[44] He now solves the hoary problem of how change is possible, i.e., how 'being' and 'not being' can combine in 'becoming', by interpreting the changes of a substance as "alterations to its limits."[45] A substance is comprised of an essence, which is immutable in itself, and the 'limits' of this essence, which naturally refers not to its spatial magnitude and form but to its *degree*.[46] Thus we see that the force of a substance is located in the substance itself and that it acts on the degree of an essence that is in itself invariable. Wolff does not say a word about the essence changing its own limits, which implies that, in contrast to Leibniz's "forma," the limits are changed by a third element—*outside* the invariable essence and *within* the variable limits. This in turn implies the presence of a force residing in the substance.

From this we can see that Wolff's answer to the question of how a substance can change itself entails shifting the problem from the substance itself to its essence. The cause of the change is simultaneously shifted from something internal to the essence to something external to it—although it is still internal to the substance. The question has now become, "How can limits be imposed externally on an *invariable* and *infinite* essence?" Wolff finds an analogy for the relationship between essence and limitation in the geometric relationship between abstract space ("Ausdehnung") and form ("Figur").[47] Space as such contains an infinity of forms, without itself being a form. Every three-dimensional figure can be seen as the limitation of a space that is infinite in both magnitude and form. Yet the essence, which is infinite space, is not altered by this limitation. Thus the concept of space is not the greatest common denominator of all possible forms, but neither is it a function that links the separate forms to one another. It is something undetermined which is the prerequisite for every possible spatial determination of form.

The presence of a third element naturally entails considerable problems. What is its origin, and how does it relate to the two other elements? How do the three elements comprise a unity? It is remarkable that Wolff does not elaborate on this for substances, but does do so for composite things, as I will demonstrate below. He also does not explain what the action of a force actually entails. The force that the substance contains, and which it exerts in itself, is taken as a given. He explains only the way that the essence of the substance can change, without addressing the question of the action itself.

---

[43] "Quelle seiner Veränderungen in sich hat" (*ibidem*, section 114).
[44] At least, not in a natural way. Of course, creation and destruction can always occur by means of a miracle, such as the creation or any other manifestation of God's omnipotence.
[45] "Abwechslungen seiner Schranken" *(ibidem*, section 107).
[46] *Ibidem*, section 106.
[47] *Ibidem*, section 54.

The second place where Wolff links the concept of force to the determination of something permanent relates to material bodies. 'Material force' is identical to moving force. A body is a composite thing, so it is not a substance. Its essence consists of the way it is composed (its "structura"), while that of which it is composed is called 'matter'.[48] In fundamental terms, matter consists of elements, i.e., substances, but being a derivative it is only suffering: passive and resistant. No active force can arise from matter.[49] Moreover, since no force either active or passive can arise from the essence, yet a body as a whole has the capacity to transmit motion in a collision, a body must be associated not only with matter and essence but also with force, which is the ground of that motion.[50] This force is manifest as a "constant effort to move matter."[51] In this case, where we are speaking of composite things, a force does not act upon the limits of an invariable essence, because the essence is already something limited. It does however act on matter. Moreover, while the body as a whole is not a substance, the moving force is a substance, a 'self-existent and permanent' thing that consequently has limits—its speed—and can change them.[52] However, the fact that speed is the limitation of moving force does not mean that the moving force itself is in motion. Motion is a category of composite bodies alone, and not of substances in themselves.[53]

It is hardly surprising that Wolff's concept of moving force is problematical. Chapter 3 has revealed the difficulties Leibniz encountered as he sought to link the fundamental force of individual substances with the phenomenal living force of matter. Wolff's system, which was very much oriented to Leibniz's thought, did not enable him to solve these problems. Wolff's analysis of substance as a totality of essence, limitation and force, and his confusing concept of moving force as both acting substance *and* the result of the actions of substances—even though substances in the strict sense cannot act—is certainly complex, but it clarifies little. Two of the issues that arise from it are, how can the moving force, which is a substance itself, act on matter, which is purely passive, and, how does the moving force relate to the forces of the elements of matter, i.e., the substances that make up a composite body?

Wolff tries to solve both problems by locating the moving force in its *fundamental* sense in the elements of matter, but in a *phenomenal* or physical sense in a 'fluid matter', a sort of ether that flows through the spaces between actual matter (that is: mass).[54] But this is a vain attempt, because he is confusing moving force and external force, as can be seen from the examples that he gives of such forces in the

---

[48] *Ibidem*, section 59.
[49] *Ibidem*, section 622. This is part of the problem of the relationship between phenomenal and metaphysical reality in the monadology of Leibniz and Wolff, that is, how are time, space, motion etc. founded in a reality that is not itself characterized by space and time? Their answer relies heavily on the concept of prime matter ("materia prima"). See also Chapter 3.
[50] *Ibidem*, section 623.
[51] "steten Bemühung die Materie zu bewegen" *(ibidem*, section 624).
[52] "für sich bestehendes und fortdaurendes" *(ibidem*, section 660).
[53] *Ibidem*, section 606 and section 693.
[54] *Ibidem*, section 697 and section 698, respectively.

same section: gravity, elasticity and magnetism. Moreover, what is the relationship between ether and actual matter? What is worse, he does not address the question of what precisely it means to say that the moving force is fundamentally in the elements of matter at all. He has evidently reached his intellectual limits, because he looks for an escape route:

> And from this it is clear that in explaining the events of the visible world it is not necessary to appeal to the fundamental force which is in these elements (...) We can limit ourselves to those forces that can be explained by the motion of subtle fluid matter in the empty space of bodies.[55]

The critique above is harsher than would have been appropriate if we were simply considering Wolff's philosophy in itself. However, this criticism of Wolff provides the key to the meaning of the Academy's competition question, which is what matters here. The obscure and uncompleted parts of Wolff's system can be detected in the competition question, partly in the question itself and partly in the suggestions added to it.

For Wolff, the action of force does not imply the determination of an invariable force at a fundamental level, but it does entail this on a phenomenal level, i.e., for moving forces. The competition question seems to assume Wolff's foundation of moving force, but then generalizes it to forces of all sorts, and even to fundamental forces. However, where Wolff only says that the moving force is located fundamentally in the elements—that is: in substances—without saying anything about how this may be so, the question identifies his concept of substance with the concept of force. Just as the essence is limited—by an inner force—in substance, the questions supposes that the fundamental force is also limited, either by itself or by something else. Although the fundamental force is not defined as infinite anywhere in the competition question, the fact that limitation was thought to be necessary, and the origin of this idea in Wolff's metaphysics, make it probable that this was implied.

Considered in this light, the Academy's question appears to identify the weak point in Wolff's concept of force, which is, the foundation of phenomenal moving forces in substance and in a fundamental force. Nevertheless, the formulation of the question poses less strict requirements for an answer than Wolff would have done: the fundamental force need not be determined by the substance or the force itself. The possibility that this determination may derive from another substance is explicitly left open.

---

[55] "Und hieraus erhellet, daß man in Erklärung der Begebenheiten in der sichtbahren Welt nicht nöthig hat sich auf die ursprüngliche Kraft, die in denen Elementen ist (...) zu berufen, sondern nur bey denjenigen Kräften verbleiben darf, die sich durch die Bewegung einer subtilen flüßigen Materie in dem leeren Raume des Cörpers erklären lassen" *(ibidem,* section 700).

6.4 THE COMPETITION ESSAYS

6.4.1 General

The manuscripts that the Academy received in response to its competitions were generally preserved in the archives. However, the manuscripts of the printed essays were usually not returned by the printer, and most must be presumed to be lost. Moreover, an author would sometimes ask that his manuscript should be returned. Authors whose essays were not published by the Academy regularly published them themselves.

For the competition for the year 1779, this means that sixteen of the unknown number of submissions have been entirely or partially preserved. Fifteen of these could be traced in the archives of the Academy (*Akademiearchiv* I–M722 to I–M736).[56] As explained above, the winning essay is not among them. However, the essay by August Wilhelm Rehberg, which was in fact awarded the Accessit, is in the archive. In a deviation from the usual procedure, his essay was not printed, and the manuscript therefore remained in the Academy.

Thirteen essays are available only as manuscripts. One, the winning essay by Pap de Fagaras, is available only in its published form, and two others are available in both manuscript and published form. The authors of the latter two are also known. One of them was Rehberg, the winner of the Accessit, who decided to publish his essay himself. The other was written by Michael Hißmann, who has already been discussed above in connection with the background of the competition.

Of the thirteen essays that have been preserved only in manuscript, three present material problems. For one manuscript, I–M732, only the title page and the foreword were in the archives. The first line, however, speaks volumes: "Where nature begins, our understanding has reached the end of its powers." Another manuscript, I–M736, is incomplete although it is still by far the longest essay, at 192 pages. These contain only parts A to H, which fortunately show enough of the author's argument for our purposes. A third, I–M725, was unreadable. This leaves a total of fourteen essays for analysis.

6.4.2 Selection

My purpose in studying the essays, as has been said, is to ascertain the extent to which developments in the concept of mechanical force had penetrated to the wider public following the turbulent years of disputes about *vis viva* and the principle of least action, but before the publication of Lagrange's *Méchanique analitique* and Kant's transcendental foundation for natural sciences. This immediately implies the most important selection criterion: only essays which locate the meaning and

---

[56] I am very grateful to members of the staff of the Berlin *Akademiearchiv*, which was then known as the *Zentrales Archiv*, who made copies of the relevant competition essays.

function of the concept of force in the field of mechanics, or if necessary a wider field of physics, would be suitable. Using this criterion, we are left with seven essays: I–M726, I–M729, I–M731, I–M733, I–M734, I–M736, and the winning essay. Of these seven, four are written in German, one in French and two, including the prize-winning essay, are in Latin. Some information regarding the name, origin and dating, is available for the seven selected essays from accompanying letters, from notes jotted down on the manuscripts (mainly the dates they were received and read) or from the printed version (in the cases of the winning essay and essays I–M731 (Rehberg) and I–M734 (Hißmann).

The Prize-Winning Essay:
General: the competition winner was Josephus Pap de Fagaras, a pastor at Háh-Wáros in Transylvania. Pap de Fagaras had written for other competitions. In his letter of thanks of 23 June 1779, he reports that he was awarded a prize by the *Societas Batavo Flessinganae* in 1772.[57] In 1782 he was to be awarded another prize, for his answer to a competition by the Netherlands Scientific Society on the topic of analogy.[58]
Title: "Dissertatio de vi substantiali, ejus notione, natura, et determinationis legibus. Auctore Josepho Pap de Fagaras, A.L.M. Philosophiæ doctore, et ecclesiæ reformatæ saxopolitanæ V.D.M. in Hah-Waros, urbe Transylvaniæ."
Dimensions: 20.3 by 25 cm.
Size: the essay is printed on pages [5] to 60 in the official publication: *Dissertation sur la force primitive, qui a remporté le prix proposé par l'Académy royale des sciences et belles-lettres pour l'année 1779*. A Berlin, Chez George Jacques Decker, imprimeur du roi. 1780. (60 pages).[59]

Essay I–M726
General: the author probably lived in Frankfurt. In an accompanying letter, posted on 12 October 1777, he gave his postal address as "Mr. Sattler in Franckfourt."
Title: "Beantwortung der Preisfragen [sic!], welche von der K. Preussischen Akademie der Wissenschaften und schönen Künste V[60] der Classe der Speculativischen philosophie ausgesezet worden."
Dimensions: 17 by 27.7 cm.
Size: A two-page accompanying letter, 1 title page, 27 pages of text.

---

[57] *Staatsbibliothek zu Berlin - Preußischer Kulturbesitz, Nachlaß Formey.*
[58] De Bruijn (ed., 1977) provides more information regarding the latter prize contest in *Inventaris van de prijsvragen*, 53-54, prize contest 36.
[59] Both the copy in the *Bibliothek der Akademie der Wissenschaften* and that in the *Staatsbibliothek zu Berlin, Preussischer Kulturbesitz* (which now covers both the eastern and western parts of the library) appeared to be missing during my stay there in November 1988. I would like to thank Cornelia Buschmann for providing a copy of the exemplar in the university library of Rostock.
[60] The meaning of this sign, which in the manuscript carries a small 'º' above it, is not very clear.

Essay I–M729
General: the author comes from Amiens; posted on 24 November, 1778.
Title: none.
Dimensions: 17.8 by 23.3 cm.
Size: front page, 30 pages of text.

Essay I–M731 (Accessit winner)
General: the author is August Wilhelm Rehberg, presumably from Leipzig, because he had his essay published there in the same year. Received before 8 April 1779.
Title: "Abhandlung über das Wesen und die Einschränckungen der Kräfte."
Dimensions: 16.9 by 21.8 cm (book: 9.3 by 15.5 cm).
Size: title page, 83 pages of text (book: title page, a two-page introduction, 84 pages of text).

Essay I–M733
General: the author comes from an Italian-speaking district (he cites the question in Italian.) Received 3 December 1778.[61]
Title: "Dissertatio de viribus."
Dimensions: 18 by 24.5 cm.
Size: front page, 2-page introduction, 69 pages of text, 21 figures on 3 pages.

Essay I–M734
General: the author is Michaël Hißmann from Göttingen, with an appointment as Professor to the *Göttinger Sozietät*.[62] It was posted in late 1778 or early 1779. Hißmann was an acquaintance of Merian (see page 179 above). The manuscript was published a few years later in Hißmann's journal *Magazin für die Philosophie und ihre Geschichte*. Göttingen und Lemgo: Meyerschen Buchhandlung 1783 (6), 3-110.
Title: "Versuch über das Fundament der Kräfte. Zur beantwortung der von der Königlichen Akademie der Wissenschaften in Berlin, für das Jahr 1779 aufgegebenen Preisfrage."
Dimensions: 18.5 by 23 cm.
Size: title page, a five-page introduction, and 94 pages of text.

---

[61] There is a jotted note from the recipient on the manuscript: "Reçu le 3 dc 1779." This can not mean that it was received on 3 December 1779, because this would be after the prize had been awarded, yet the manuscript was certainly received on time. The names of the members of the jury are listed on the cover sheet, showing that it was read, and thus must have been received on time, that is: before 31 May 1779. It is also unlikely that the 3 refers to the month, i.e., that the essay was received in March 1779, because essays I–M728 and I–M730, which were also received on time, bear notes in the same handwriting: "Reçu le 4 dc 1779" and "Reçu le 7 dc 1779" respectively. Therefore the person receiving the essays has presumably written 1779 instead of 1777 or, more likely, 1778. Another explanation is that the date refers to the time the essays were delivered from the jury to the archive—in one bundle presumably, but the archivist recorded the details at spare moments over the course of a week.

[62] Buschmann, "Philosophischen Preisfragen," 219. Buschmann remarks on the remarkably large participation in the Berlin philosophical prize contests from the circles centering on the *Göttinger Sozietät* and the *Georgia Augusta* (*ibidem*, note 118).

Essay I–M736
General: no information about the author is known. The manuscript has been only partially preserved, with the title page and the last part of the text missing.
Title: unknown.
Dimensions: 17.5 by 21.5 cm.
Size: 192 pages of text, in which the right hand side of each page is used for text, and the left for section headings.

6.5 OVERVIEW OF THE SELECTED ESSAYS

6.5.1   The Prize-Winning Essay (Pap de Fagaras)

"Dissertatio de vi substantiali, ejus notione, natura, et determinationis legibus" ("Dissertation on force, its concept, its nature and the laws of its determination").

It is hardly surprising that the prize-winning essay was written from a more Leibnizian-Wolffian point of view than the other essays that were submitted. The essay consists of three chapters, about the identity of mental force ("vis animae") and fundamental force, about the possibility of reducing all material phenomena to mental force and about the regularities that derive from this fundamental force. The first chapter proposes that changes exist in two forms, as motion ("motus"), and as observation ("perceptio"). The mental force is, in line with Leibniz and Wolff, defined as the "continuing drive to produce manifestations," and is at the same time the fundamental and permanent force.[63] The author cannot say precisely how it relates to the mind.[64] He does however claim that it is one, yet contains all the various faculties of the mind, just as the single force of elasticity manifests itself in various effects.[65] With this conclusion to Chapter 1, Pap de Fagaras has already demonstrated how important mechanics is as a metaphor for his ideas about the mind. However, in Chapter 2 he also addresses the problem of how *motion*, and thus all material change, can also be traced back to mental force. Here he changes his style, realizing that his understandings are second hand. Most of his sentences now begin with the words: "In my opinion (...)" ("video" + accusative with infinitive). However, his argument is essentially the same as that of Leibniz and Wolff.[66] The fundamental principles of mechanical philosophy, such as motion, space and time, cannot provide a foundation for the impenetrability, cohesion and inertia of matter. However, his conclusion requires a certain leap, although it is a leap that is common to the Wolffian tradition: "I am led inescapably to the proposition that what is real,

---

[63]   "continuum repræsentationes producendi nisum" (Pap de Fagaras, "Dissertatio de vi substantiali," 16).
[64]   *Ibidem*, 22.
[65]   *Ibidem*, 23.
[66]   For Leibniz's argument in favor, see section 3.3 above.

permanent and substantial in bodies consists of simple elements."[67] The alternative, according to him, is that extension and such like are *realities*.[68] Pap de Fagaras then concludes Chapter 2 by seeking to elucidate the simple elements, while remaining close to Wolff. He proposes, for example, that they must have an active force ("vis agendi"), since nothing can exist as the pure possibility of suffering. This active force is an urge ("nisus"), and not only a capacity ("potentia"): it generates an action continuously even if this is infinitely small. However, Pap de Fagaras does not want to go beyond a certain analogy between the active force of matter and mental force: he cannot say whether perceptions and representations underlie the active force of matter. That would first require an answer to the problem of how simple existents affect matter, and vice versa; and, secondly, how thoughts can arise from motion and motion from thoughts.

Pap de Fagaras's approach takes no account of the events described in the previous chapters. He appears not to be aware of changes in the modern concept of mechanical force. The fact that the competition was awarded to this essay justifies, in retrospect, the ridicule of the Paris Academy. The essay was, as Buschmann expressed it, a piece of "far-reaching insignificance."[69] However, this is not so much because he attempts to provide an ontological basis for material phenomena with the help of Wolffian forces, but rather because of his lack of understanding of more recent developments in mechanics.

6.5.2 Essay I–M726

"Beantwortung der Preisfragen [sic!], welche von der K. Preussischen Akademie (...) ausgesezet worden."
("Response to the Competition Questions [sic!], that have been announced by the Royal Prussian Academy (...)").
In his accompanying letter, the author indicates that, although he is a very busy man, he has great ambitions: he wants to introduce a whole new system of natural philosophy, that could "light a wholly new lantern in the schools." In his opinion, the reason that we have not yet gone beyond nature's outer shell, is primarily that we have an incorrect concept of what nature is. This is also the motto of his essay, which he has derived from Cicero: "Close examination is required so that we know what 'nature' is, leaving no room for any misunderstanding arising from the term."[70]

---

[67] "(...) in eam vel invitus deducor sententiam, forte id, quod in corporibus reale, perdurans, & substantiale est, (...) elementa simplicia esse" (Pap de Fagaras, "Dissertatio de vi substantiali," 27-28).
[68] This is put forward as a postulate in essay I–M729.
[69] "weitgehender Bedeutungslosigkeit" (Buschmann, "Philosophischen Preisfragen," 217).
[70] "Illud excutiendum, ut sciatur, quid sit natura, ne relinquatur aliquid erroris in verbo." The author gives the source as *Cicer. de offic*. However, the passage is not found in Cicero's *De officiis*, but rather in *Tusculanae disputationes* I, 36, 88, with the word 'carere' ('to miss', 'to lack') instead of 'natura' (*Loeb Classical Library*, 1956 edition). The author appears to have modified the quotation to suit his subject. My thanks to Prof. Heesakkers for finding the source.

Accordingly, he begins with a definition of 'nature', taking as a starting point that nature "harmonizes with both the greater and lesser worlds, i.e., the universe and human beings, or macrocosm and microcosm."[71] Nevertheless, the definition itself seems to be plucked from the air: 'nature' is "extended throughout the macrocosm, and limited according to the species and the degree of intensity in human beings and in all living creatures—in short, [it is] pneumatic forces."[72] All natural phenomena are thus based on 'pneumatic forces'. But what is a 'force'? It is that characteristic of nature, which does not itself have parts but can generate parts, and which varies only in degrees of intensity.[73] From this point on, the writer confuses his definitions: he first liberally treats the terms "vis," "qualitas," "virtus," and "proprietas" as synonyms, and then defines 'quality', and also 'matter', as a force. 'Quality' is the 'primary force', and 'matter' is the 'secondary force' in every natural thing.[74] After these definitions, any intuitive distinction the reader may have felt before between the terms 'characteristic', 'quality', 'force', and 'matter' has been lost in confusion. It is in any case clear that the concept of substance is central to the author's understanding of nature, but that 'substance' and 'quality' (or 'force', or 'characteristic', or 'matter') are not differentiated in absolute terms: any quality can shade into a substance. His argument does not explain what this implies in the case of the pneumatic forces, which are themselves qualities of nature.

The pneumatic forces provide the answer to the question of what the fundamental and substantial force is. His formulation illustrates of the confusion in his thinking: these fundamental forces are nothing other than "those pneumatic forces, that within certain limits hold themselves in equilibrium and determine themselves evenly, or that, in agreement with the laws of motion, [that are according to] the ideas of the Creator, determine themselves in the greater and lesser [world], and so realize themselves."[75] We can conclude that the author considers that the laws of motion are based on the characteristics of substances, that these characteristics are identical to forces, and can in some way shade into substances.

---

[71] "die im grossen Welt-bau ausgedehnte, in Menschen, u[nd] allen lebenden Geschöpfen aber der Gattung, u[nd] Grad der Heftigkeit nach sich selbst einschränkend — oder samlende Luft-Kräften" *(Akademiearchiv* I–M726, 4).
[72] "die im grossen Welt-bau ausgedehnte, in Menschen, u[nd] allen lebenden Geschöpfen aber der Gattung, u[nd] Grad der Heftigkeit nach sich selbst einschränkend — oder samlende Luft-Kräften" *(ibidem).* The text also includes Latin passages that may be derived from another author, since they are much more compact and more clearly worded than the German. The term 'natura naturata' might then indicate affinities with Spinozan ideas.
[73] *Ibidem*, 6.
[74] *Ibidem*
[75] "als die inner gewissen Schranken im Gleichgewicht sich selbst haltende, u[nd] sich selbst reg[e]lmässig disponierende, oder die im grossen u[nd] kleinen die Schöpfers-Ideen denen Gesätzen der Bewegung nach sich selbst disponierenden u[nd] so sich selbst ausführende Luftkräfte" *(ibidem,* 13).

6.5.3   Essay I–M729

(No Title).
The charm of this essay lies in the original way in which the author seeks a basis in substance for the characteristics of nature, combined with his unassuming approach to this audacious task. Nature, he says, is composed of elements. The various characteristics of matter, such as extension, form, gravity and hardness, recur in these elements.[76] However, the elements are only secondary, not primary, causes. They themselves stem from four principles: Space, Force, Time and Motion. These principles are necessary existents and constitute the primary causes of matter. For example, the principle of Space gives us extension and form, and the principle of Force gives us hardness and gravity. Considered in themselves, they are immaterial, but in combination they are in fact the foundation of all material characteristics. As for the necessity of these principles, the author says only that they "originate (...) from one whole of which they are the parts."[77] Each of these principal existents has its own "distinct and separate identity."[78] Logically, the author sets out to first discuss the characteristics of each principle separately, so as to be able to derive the necessary effects. At this point there is a degree of tension and ambiguity with regard to the ranking of the four principles. Although they were first presented to the reader as equal and independent from one another, in the separate discussions it emerges that they cannot be determined individually, but only in relation to one another. Space and Time cannot exist without one another, because together they are necessary prerequisites for every possible thing. Motion cannot exist without Space and Time, and Force works in Space and cannot act without Motion. Thus there is a hierarchy within the necessity of the principles.[79]

Later, in the sections entitled "Réflexions" that deals with the synthesis of the principles, these relationships take on a different character. Force and Space together comprise the first *cause*, Time and Motion the first *effect*. At the same time, this effect is necessary for the combination of Force and Space in the first cause. Thus cause and effect are united in the elements.[80] The argument here degenerates to playing with paradoxes.

However, the principles are also ranked in a third way, in the "Conclusion." There, the principle of Force is seen—without any preparation—as the divine Will, and the principle of Motion is the divine Intelligence![81]

Because of these three contradictory sets of relationships it is difficult to form any clear idea or what the author finally means to say. It is clear that, for this author, the foundation of nature as a whole, including space, time and motion, is located in

---

[76]   *Akademiearchiv* I–M729, 2.
[77]   "(...) tirent leur origine (...) d'un Tout dont ils sont les membres" *(ibidem,* 3).
[78]   "propriété distincte et séparée" *(ibidem).*
[79]   *Ibidem,* 22.
[80]   *Ibidem,* 23-24.
[81]   *Ibidem,* 31.

substances. It is therefore not surprising that mechanical force is also traced back to a substance. This substance is divisible, and is nevertheless conserved in the effects of the force. The author compares this to dividing and subdividing a material volume, and also with an 'elastic power'.[82] It is infinitely compressible and expandable. Although it acts continuously, it can also be passive, where it retreats within itself because of a predominance of external forces. This produces unlimited possible combinations embodying different relationships between power and resistance.[83] This dynamic theory of matter makes the essay a good example of the attempt to retain a substantial metaphysics as the foundation of mechanics.

6.5.4   Essay I–M731 (Rehberg)

"Abhandlung über das Wesen und die Einschränckungen der Kräfte"
("Treatise on the Essence and Limitations of Force").
Although the Accessit was awarded to Wilhelm August Rehberg's essay, this essay was not published as a supplement to the winning essay as was customary. Rehberg must have decided quite quickly to publish the essay through other channels, because his essay was published by the *Weygandsche Buchhandlung* in Leipzig in the same year.

In his answer, Rehberg begins with mechanics, but later goes beyond it. All changes in the corporeal world can be reduced to movements, and thus all forces are moving forces. In Rehberg we can still find a clear formulation of the unity of forces that act *in* a body, and those that act *on* it. The action of the force of a moving body is either to *change* the motion of another body, or to *continue* the motion in its own body. Thus the motion of a body is also the measure of the force of that motion, since forces can only be measured by their action.[84]

This does not mean that Rehberg's concept of the measure of force is unambiguous; he supports Kästner and d'Alembert who, he says, had proven that this measure depends on the definition of the word 'force'.[85] It is however simplest to take the quantity of motion as the measure of force.

But Rehberg also addresses the question or force in a more direct way. How are forces transferred? Is it by the transfer of substance from one body to another? Or should we perhaps locate the cause for each individual case in God, as in Malebranche's occasionalism? Finally: could the actions due to force be based on the imaginations of material particles, just as the movements of a human body are initiated because the act is conceptualized in the imagination? Rehberg's treatment of the strengths and weaknesses of the various points of view need not delay us here.

---

[82] "vertu élastique" (*ibidem*, 14).
[83] "puissance & résistance" (*ibidem*, 15).
[84] Rehberg, *Abhandlung über das Wesen*, 7 (I–M731, 5).
[85] Rehberg, *Abhandlung über das Wesen*, 8-9 (I–M731, 6-7). He refers to d'Alembert's *Traité de dynamique*, page 16 ff., and Kästner's *Höhere Mechanik*. From the page numbers it would appear that he is using the first edition of d'Alembert's *Traité*.

His own answer will suffice: he considers that "we must suppose that there is a material principle of motion in a body, whose activity harmonizes with the imaginations of the soul."[86]

Although the proposed harmony sounds Leibnizian, Rehberg did not consider the object of mechanics to be something purely phenomenal. On the contrary, the science of mechanics offers a pattern for the science of the human mind. Rehberg follows Mendelssohn in arguing for an analogy between the laws that govern the effects of our ideas and the laws of mechanical forces.[87]

It is matter itself that performs actions. But in Rehberg's opinion it is not possible to determine whether the material particles influence one another or each particle determines itself and unites all forces in itself.[88] Neither does it matter for mechanics since, as he says, mechanics "calculates only actions, and is therefore unconcerned about what subject the force may be in."[89]

Although Rehberg provides a clear analysis of some of the metaphysical problems of physics, he does not engage with physics closely enough to make such an analysis fruitful. His argument is largely devoted to elaborating Mendelssohn's idea of a mathematics of qualities (in this case of the power of imagination ("Vorstellungskraft")), and explaining the curtailment of the infinite power of imagination in each substance by other, finite, substances. He fails, however, to show that the three laws of Newton and gravitation, for example, are in harmony with the principle of the power of imagination.[90]

The two-fold distinction between the foundation of mechanics and praxis that he makes puts him outside mechanics, and thus also outside the discussion about the foundation of mechanics. The question he fails to ask is: what is the implication for the foundation of mechanics, if we say that mechanics is interested only in actions and not in substances?

---

[86]   "(...) [ein] materielles Principium der Bewegung im Körper annehmen zu müssen, dessen Thätigkeit mit den Vorstellungen der Seele harmonirt" *(ibidem,* 16).
[87]   Rehberg, *Abhandlung über das Wesen,* 17-27. See also Mendelssohn's competition essay for 1763, in which he proposes a mathematical approach to qualities such as goodness and beauty (Mendelssohn, *Abhandlung über die Evidenz).*
[88]   *Ibidem,* 30 (I–M731, 14).
[89]   "(...) nur Wirkungen berechnet, und sich wenig darum bekümmert, in welchem Subjekte die Kräfte seyn mögen" *(ibidem,* 28 (I–M731, 12)).
[90]   He does however make some sharp observations, saying for instance that "this [mathematical] method enabled [Descartes] to exorcise the Spirit of Hypothesis, which he himself possessed to such a great extent" ("durch diese [mathematische] Methode half [Descartes] den Geist der Hypothesen verdringen, den er selbst in so hohem Grade hatte") *(ibidem,* 36-37 (I–M731, 20)).

6.5.5 Essay I–M733

"Dissertatio de viribus"
("Dissertation on forces").

The author has understood the competition as a search for the foundations of mechanics. His answer is therefore more relevant than others to the issue considered in this chapter, the variety of opinions about the concept of mechanical force. The author himself teaches natural philosophy and mathematics, presumably at a university in an Italian or Italian-speaking city. He apologizes for his untidy handwriting and the lack of clarity in the essay, which he attributes to lack of time because of a heavy teaching load and to periodic states of physical and mental exhaustion.[91]

Despite these hindrances, he submits an essay of 69 pages. The essay provides a systematic structure for mechanics, in which metaphysics and epistemology are harmoniously combined with geometrical-mathematical proofs. The topics considered are, in order, the existence of the various mechanical forces; the principles of composition and resolution; the principles of static equilibrium; action and effect; the action of one substance on another; and forces that determine themselves versus those that are externally determined. I will consider the first chapter, which deals with the existence of forces.

There is a curious contrast between his frequent emphasis on the importance of observation and experiment and his own opinions, which are often dogmatic and rationalistic in character. He begins by explaining from observation the primary characteristics of matter, such as extension, impenetrability and inertia, and then tries to establish the basis of these characteristics.[92] In the end however, he creates more confusion than clarity. For example, in the sections dealing with inertia, he first states that the force of inertia is the passive potential that resides in a body, which Newton is therefore correct to call the innate passive force ("vis insita passiva"). He then concludes that 'inertia' means only that an external cause is required for any change in state. The inertial force (in contrast to other, external, forces) cannot cause any *change* in state, only *continuation* in a state.[93] One cannot avoid feeling disappointed here: while a positive existence is attributed to inertial force it is defined only in a negative way. What is the difference between this and the generally accepted view, as expressed by d'Alembert?

After the sections on the primary characteristics of matter, the author makes a transition to the causes of changes in states of motion. He introduces this transition

---

[91] *Akademiearchiv* I–M733, 3.
[92] One notable historiographical detail is that the author points to Joseph Ballus, *Demonstratio motus naturalis corporum* (Patavium [Padua] 1633) for the first correct formulation of the principle of inertia, prior to Galileo and Descartes (*ibidem*, 7). But in historiography, Giovanni Baliani is regarded as a secondary figure, overshadowed by Galileo (see Dijksterhuis, *Val en worp*, 321-331).
[93] *Akademiearchiv* I–M733, 7-8 (sections 10-16).

with a subtle point: whether any reality can be attributed to such changes in state.[94] Does not the relativity of motion undermine the reality of these changes? Since a change in state must nevertheless be traced back to either the body itself or the observer, or a combination of both, the reality of changes in state cannot be denied in absolute terms. There must therefore be causes, that is, forces. He defines these as "any magnitudes that are able to change the state of bodies whenever they are able to act without impediment."[95] The examples he gives are gravity, elasticity and rigidity ("tenacitas"). His choice of terms is striking: if one is looking for actual causes, is it appropriate to define 'force' as a magnitude? However, his use of this term does not imply that he considers that forces are purely mathematical. Forces are real, as he will later demonstrate.

First, however, he deals in more detail with the problem of the meaning of impediments. Does a force act even if it is impeded? If so, how is this in accordance with the necessary connection between cause and effect? Ingeniously, he notes that, if one argues along these lines, the true cause of the change in state would be the removal of the impediment rather than the force itself.[96] He explains the working of an impediment to a force by saying that a force always exerts the same drive ("nisus") or pressure ("pressio"), whether it acts ("agere") or not. So a distinction must be made between the force itself ("vis" or "potentia"), its action ("actio") and the possible resulting change in state ("status mutatio").[97]

For example, the action of gravity consists of the continual production of infinitely small movements, just as Leibniz said that dead force continually generated new "impeto."[98] Thus the action is always there but it only accumulates to a discernible change in state, consisting of living force, if there is no impediment. If there is an impediment, this annuls the small movements at the very moment they begin. In section 22, the author presents an analogy from mathematics, in the generation of a surface from the motion of a line. If we let the line represent a force, and the surface generated represents a change in state, then the motion of the line in generating the surface represents the action.

This explanation of impediments to forces is enough to suggest that the metaphysical status of force will be linked to that of substances. In the final sections of his first chapter the author does indeed defend the thesis that forces are something

---

[94] *Ibidem*, 8-9 (section 17).
[95] "(...) quantitates eas quaecumque tandem sint, quae aptae sunt ad mutandum statum corporum, quoties non impeditae actionem exercere possunt" (*ibidem*, 9 (section 18)). The Latin term 'quantitas' is translated here as 'magnitude' rather than 'quantity', the latter being reserved for physical existents.
[96] *Ibidem*, 9-10 (section 19).
[97] *Ibidem*, 11-12 (sections 20-22).
[98] This a rather free interpretation, but is justified in the historical context. What the author literally says is "the action of gravity is nothing more than the continual and sequential application of force at each point in time or space" ("(...) action (...) nihil est aliud quam continua, et successiva applicatio potentiae cuique puncto temporis, sive spatii (...)") (*ibidem*, 10 (section 21)).

real and have their own existence, just as bodies do.[99] They are not attributes, modifications, or products of bodies, they are external to the bodies.[100] The author makes an idiosyncratic attempt to link the conservation of force to this thesis, in such a way that the total quantity of force does not change, but forces can be transferred from one body to another. Living force appears to be confused here with external force. His concept of force may not be consistent, but it does provide another illustration of the continuing attempt to base the concept of mechanical force on a metaphysics of substances.

6.5.6 Essay I–M734 (Hißmann)

"Versuch über das Fundament der Kräfte"
("Essay on the Foundation of Forces").
Hißmann is exceptional among the authors who entered. I have already cited from a letter in which he said that the first requirement for winning was that the essay should propound a Leibnizian-Wolffian viewpoint. In the same letter he said that his own essay "contains much that cannot be considered true; simply because that is what they want in Berlin" and also: "I would prefer Locke and Condillac's approach to philosophy."[101] Nevertheless, this self-depreciation must be taken with a grain of salt, or why did he consider the essay of sufficient value to warrant its inclusion a few years later in his own journal of philosophy?[102]

For our purposes, the sincerity of his essay is hardly important. Whether it presents his true views or not we can suppose that he has designed the argument to be as plausible and persuasive as possible in the eyes of the jury. Its value for my argument is all the same, whether the essay represents *actual* opinions or opinions that were *plausible* at the time.

In the first place, Hißmann points to the ambiguity of the concept of force as both the potential and the principle of a change or event. A potential must still be activated, and thus requires some additional basis. A principle, in contrast, is in "a continuing expression and generation of effects."[103] The original force naturally can not be a potential and must therefore be interpreted as a principle.[104] He then defends the existence of forces by means of what he calls 'induction': 'force' is that with

---

[99] *Ibidem*, 12 (section 26).
[100] *Ibidem*, 11 (section 24).
[101] Letter dated 29 September 1780 from Hißmann to Merian (*Akademiearchiv* I-VI-10, 80-81).
[102] Hißmann was the founder and editor of the *Magazin für die Philosophie und ihre Geschichte* (Göttingen-Lemgo). His letter to Merian accompanied the third number (1780). He wrote to Merian that the aim of the journal was to publish German translations of articles from academic journals. His own article appeared in number 6 (1783), 3-110.
[103] "unaufhörliche Äußerung und Folgenerzeugung" (Hißmann, *Versuch über das Fundament*, 15 (I-M734, 10)).
[104] *Ibidem*, 16 (10).

which the effect is *usually* linked, and whose absence would be *surprising* if the effect were present.[105]

There is something quite remarkable going on here, something that will be evident throughout this essay. Metaphysical speculation and empirical criteria are used side by side, although they cannot easily be harmonized. For example, he attacks not only Hume but also the now relatively unknown English philosopher Henry Home, who says that experience allows us only to observe that one thing precedes another, and not any causal relationship. Hißmann recognizes that 'force' is no more than the "continuing association between two objects," although he adds that experience *does* yield such a force.[106] Despite this, he later tries to determine force metaphysically and concludes that substance both *has* force and *is* force.[107]

There are two keys to a better understanding of this tension. The first is the epistemological principle that Hißmann formulates in relation to the fundamental force: "we can conclude its existence only from its effects."[108] The second is his metaphysical proof, based on this principle, which states that since substances differ in their reactions only because of their force, and since no two substances react identically, each substance must have a unique force. Nevertheless, the same epistemological principle makes it impossible to say what this force is *in itself*. It manifests itself in a variety of actions *in relation to* other substances or forces—consider the number of possible effects of fire for example. If we attempt to deal with the force in itself, we cannot simultaneously examine its action, and it is then unknowable.

What is interesting about Hißmann's essay is that he rephrases the question of what something is in itself as a question about what something is in its contextually determined actions. That is, the focus of the question is shifted from individual substances to the *relationships between* substances. He does this by identifying substance and force.[109] Substance is not yet subsumed in its relationships, yet it can only be recognized through those relationships. Ontologically speaking, force and substance may be things in themselves, but in the process of knowing substance is subsumed in its action, in its relationship to other substances.

6.5.7 Essay I–M736

(Title unknown).
The author provides an elegant, step-by-step metaphysical treatment of the problem, in which he does not question the possibility of seeking the foundation of forces but does in places demonstrate that he is well-informed concerning contemporary

---

[105] *Ibidem*, 20 (15).
[106] "beständige Verknüpfung von ein paar Objekten" *(ibidem, 33-37 (28-32))*.
[107] *Ibidem*, 59 (53).
[108] "wir müssen erst aus seinen Wirkungen auf sein Daseyn schliessen" (*ibidem*, 58 (52)).
[109] *Ibidem*, 58 (52-53).

epistemological considerations. For example, in section 11, in which he explains that the substance of nature must be sought in nature itself and not in God, he recognizes that "our knowledge is based entirely on the relationships between things, and we cannot form any separate, individual picture of the substances (...)"[110] This does not prevent him penetrating to substances, even if access is, at least formally, limited to induction from natural phenomena to "characteristics (...) that are no longer depending [on other things], but rather must be immediate properties of substances."[111]

Naturally the actual reasoning assumes much more than phenomena alone. The principle of sufficient basis (sometimes in a negative form: the lack of any basis for denying a thesis); the impossibility of infinite regression; the principle of the simplicity of substance and the necessity that it should underlie nature; and analogical reasoning, they all are recognizable elements derived from rationalistic metaphysics. Step by step he builds up a dynamic concept of matter. Force and matter are both rooted in the same substance. Force must in fact be explained first as an expression or characteristic of a substance. He discusses this in a section entitled: "Elements Possess Force" ("Die Elemente haben Krafft").[112]

He bases his argument on the idea that phenomena constitute a causal chain in which every change originates in a previous one. But such a chain is not necessarily causal. There is also the possibility of an occasionalistic cause, which for the author is the same as saying that the cause is not located in the bodies themselves but in God. Moreover, necessity of the chain is only hypothetical, i.e., that the chain can only exist if the successive elements in the chain have specific characteristics. But the latter objection can be overcome if we accept the possibility of other worlds, thus of other chains of phenomena. As for the first objection, the author replies: "Since we have no reason at all to believe that God would have made nature so dependent on Himself, we have no choice but to accept as most probable the view that force inheres in nature and is an essential component of it (...)"[113] Then, from the impossibility of infinite regression, it follows that forces are finally rooted in elements, which are the substances of nature: "ultimately there must be bodies, the force of which is not grounded in the force of another body, but rather (...) they possess their own distinctive forces."[114]

This shows the fundamental position of the author clearly: science deals with relationships; metaphysics is possible, but only as an inductive science. The ground

---

[110] "unsere Kenntnisse sich bloß auf Verhältniße der Dinge gegen einander gründen, und wir uns von dem Substantiellen (...) keine individuelle Vorstellung machen können" (*Akademiearchiv* I–M736, 44-45 (section 11)).
[111] "Beschaffenheiten (...) die nicht mehr abhängig sind, sondern unmittelbare Eigenschaften der Substanzen sein müßen" (*ibidem*, 45 (section 11)).
[112] *Ibidem*, 49-59 (section 13).
[113] "Wenn wir gar keinen Grund haben zu glauben daß Gott die Natur von sich so abhängig gemacht haben sollte, so bleibt uns immer nichts anders übrig als mit der höchsten Wahrscheinlichkeit zu vermuthen, daß die Krafft in der Natur liege und einen wesentlichen Bestandtheil derselben ausmache (...)" (*ibidem*, 52 (section 13)).
[114] "so muß es endlich Körper geben, deren Krafft nicht mehr in der Krafft eines anderen Körpers ihren Grund hat, sondern die (...) eigenthümliche Krafft besitzen" (*ibidem*, 53 (section 13)).

of relational reality is a substantial reality, in which characteristics are absolute. Nevertheless, relational reality cannot be derived from these absolute characteristics, but does constitute the source through which the absolute characteristics can be known.

These views influence the ways in which nature must be studied. The variety of natural forces, for example, must be explained as the result of the composition of elementary forces. Reality is thus seen as a *mixture*, and this includes life itself: "The organic structure of the body differs from a mixture only by virtue of a higher degree of composition from diverse components."[115] Composition from material elements is therefore identical to a composition from forces. As the author later says: "Force and matter are modes of one and the same substance. To say that the [varying] forces of bodies arise from the diverse compositions of elements is the same as saying that the composite forces arise from the composition of individual distinctive forces."[116]

In several sections he seeks to clarify the implications of this for mechanics. For example, he explains how moving force that underlies motion is also the origin of mass, inertia and even form.[117] However, he does not elaborate this to any great extent, presumably because the actual derivation is beyond the scope of his metaphysical argument.

Thus the author of essay I–M736 seeks to establish a dynamic theory of matter in which force and matter are regarded as two aspects of the ultimate substances in nature. However, he is not arguing that matter is the result of the composition of forces, but rather that both matter and force are finally rooted in substances. Thus he continues to treat mechanics as a science determined by premises concerning substances.

6.6   CONCLUSION: DIVERGENCE OF METAPHYSICS AND MECHANICS

The Berlin competition for 1779 offers a surprising insight into the concept of force, at least by contrast to the radical changes in the concept of mechanical force described in previous chapters. The rationalistic and speculative formulation of the question, on the one hand, and the derision of mathematically oriented scientists on the other hand, show how wide the distance between philosophy and mechanics had become in the space of a few years. Where mechanics had closed ranks with increasing certainty against a metaphysical idea of force, in which force is founded in

---

[115] "Der organische Bau der Körper unterscheidet sich von der Mischung durch nichts anders, als durch einen höheren Grad der Zusammensetzung ungleichartiger Bestandtheile" (*ibidem*, 55-56 (section 13)).
[116] "Krafft und Materie sind Modi einer und eben derselben Substanz, und wenn man sagt, daß die Kräffte der Körper aus der verschiedenen Zusammensetzung der Elemente entstehen, so heißt das so viel als ob man sagte, die zusammengesetzten Kräffte entstehen aus der Zusammensetzung der einzelnen unter sich verschiedenen Kräffte" (*ibidem*, 133-134 (section 24)).
[117] *Ibidem*, 107-114 (section 20).

substances, philosophy tried to harmonize laws and understandings derived from mechanics with a substantial foundation. However, it would not be fair to regard the philosophical search for the foundation of forces as outmoded. While it is true that it was already derided from some quarters and that not long after, when Kant's transcendental philosophy had made its mark, many more thinkers would reject it as 'dogmatic', it is nevertheless preferable to see the whole process as an external symptom of the internal conflict between mechanics and metaphysics that was already present. Mechanics no longer had any way to deal with metaphysical questions, while the metaphysicists were developing the basis of their thought much more slowly, since they had been deprived of their sparring partner, mechanics. Seen in this light, the Academy's question appears to be intended mainly to strengthen the weak spot in the most commonly accepted metaphysics, Wolff's system. The weak spot was the foundation of phenomenal moving forces in individual substances, and thus in a fundamental force. However, the fact that the possibility that substances might determine one another was left open, rather than supposing that a substance can only act in itself as Leibniz and Wolff had said, can be regarded as a move in the direction of a more relational concept of force.

The essays themselves provide a variety of foundations for force. While a considerable proportion did not locate the foundation in Leibnizian-Wolffian substances, they did generally take the concept of substance as a starting point, in accordance with the wording of the question. This starting point was seldom questioned. The prize-winning essay by Pap de Fagaras, for example, was entirely in the tradition of Leibniz and Wolff, although the author did not dare to equate the fundamental forces in matter with mental forces. In the essay from Frankfurt, the author had force, quality and substance shade into one another. The essay from Amiens posits Space, Time, Motion and Force as necessary existents, the composition of which gives rise to material substances. While Rehberg recognized that the search for the foundation is not important for mechanics, because mechanics is only interested in actions, he was nevertheless looking for what we would, in metaphysical terms, call a material principle. The essay from Italy, whose author was without a doubt the most expert in mechanics of all the participants, defended the parallel and independent existence of force and matter. In essay I–M736 the author supposes that both force and matter are rooted in primary substances, so that both must be treated as aspects of the primary substance. Hißmann was the only author to express the epistemological criticism that substances can only be known through their actions. Nevertheless, he claims that force is identical to substance and is ontologically prior to action.

Thus the competition essays, with their wide variety of answers, show that a substantial foundation for mechanical forces was still an acceptable possibility around 1780. Only one essay, that of Hißmann, presented epistemological objections to this, but even Hißmann maintains in the end that an ontological foundation is necessary. The essays thus confirm the analysis given above of the formulation of the competition question, since they also reflect a growing gap between the disciplines of metaphysics and mechanics.

In this chapter, this growing gap has been shown to be felt within the discipline of metaphysics. That raises the question of whether similar effects can be observed in the way that mechanics was actually practiced. The next chapter will address this question using Lagrange's *Méchanique analitique*, published a few years later, which presented classical mechanics in its modern, analytic form.

Annex to Chapter 6

The Original Formulation of the Berlin Academy's Competition Question for 1779

Dans toute la nature on observe des effets; il y a donc des forces. Mais ces forces, pour agir, doivent être déterminées, cela suppose qu'il y a quelque chose de réel et de durable, susceptible d'être déterminé, et c'est ce réel et durable qu'on nomme force primitive et substantielle. En consequence l'Académie demande:

Quelle est la notion distincte de cette force primitive et substantielle qui lorsqu'elle est déterminée produit l'effet? Ou en autres termes: quel est le Fundamentum virium? Or, pour concevoir comment cette force peut être déterminée, il faut ou prouver qu'une substance agit sur l'autre, ou démontrer que les forces primitives se déterminent elles-mêmes.[118]

Im ersten Falle, verlangt die Akademie, ferner einen deutlichen Begrif von der ursprünglichen leidenden Kraft, und wie sie von einer andern Kraft leide. Im andern Fall aber müßte deutlich gezeigt werden: Wie und wodurch diese Kräfte eingeschränkt werden? und woher es kommt, daß dieselbe Kraft bisweilen eine gewisse Wirkung hervorbringe, die sie einandermal nicht mehr thun kann? Wie es zugeht, daß man, zum Beispiel, deutlich begreift, was ein andrer erklärt, da man eben diese Sache ohne Unterricht nicht würde begriffen haben? Warum man Vorstellungen, die man vergessen hat, nicht wieder hervorbringen kann, da man sie doch das erstemal hervorgebracht hat, und es immer ein Grundsatz ist, das Wollen und Vermögen zusammen vereinigt, nothwendig die Wirkung hervorbringen. Oder endlich, wenn die ursprüngliche Kraft alles durch sich selbst hervorbringt, worinn der eigentliche, wahre Unterschied der zwey Fälle bestehe, daß man in dem einen eine sehr künstliche, von einem großen Tonsetzer verfertigte Musik deutlich fasset, wenn man sie spielen hört; oder daß man die Auflösung einer sehr schweren mathematischen Aufgabe, die ein andrer gegeben hat, fasset, obgleich man in dem andern, so sehr man sich bestrebte, weder jene Musik hätte setzen, noch die Aufgabe auflösen können.[119]

---

[118] Harnack, Geschichte der Akademie II, 308.
[119] Rehberg, *Abhandlung über das Wesen u[nd] die Einschränkungen der Kräften*, 3-4. I have modernized the spelling of 'Grundsaz', 'Tonsezer' and 'sezen'.

CHAPTER 7

LAGRANGE'S CONCEPT OF FORCE

En réduisant les principes on les étendra.
Jean d'Alembert[1]

7.1 INTRODUCTION

Lagrange was not only the person who laid the analytic foundation for variational calculus, he was also willing to elaborate on the idea that the principle of least action could be the fundamental principle for all mechanics, including both statics and dynamics. In the 1750s he was enthusiastically hailed as the defender of this new approach to mechanics by Euler and Maupertuis. But in 1788 the same Lagrange wrote the classic work *Méchanique analitique*,[2] in which the principle of least action appears only as a derivative theorem, subordinated to the principle of virtual velocities. Thus Lagrange would seem to personify the transformation of the principle of least action from a teleological principle to a mathematical theorem.

Lagrange's involvement in the early development of the principle of least action, coupled with the great significance of his principle work both as a synthesis of the mechanics of the previous century and, because of its entirely analytic method, as a renewal of mechanics, make the *Méchanique analitique* exceptionally suitable for an examination of how the change in the metaphysical foundation of force was reflected within mechanics itself. To what extent did Lagrange realize that force no longer had a substantial foundation, and that the laws of motion should therefore be based on different foundations? What did he substitute for this substantial foundation? And what role did mechanical principles play in this?

But before I begin to answer these questions, I should first say how it all began.

---

[1] D'Alembert, "Méchanique" (1765), 224.
[2] The spelling here, which is no longer usual, is that used in the first edition. In the second and later edition, the title was spelt as *Mécanique analytique*.

7.2 LAGRANGE'S MATHEMATICAL REDUCTION OF MECHANICS

7.2.1 *The 'Peak of Perfection'*

Lagrange was born in 1736 in Turin in a well-to-do family, which suffered financial ruin during his youth. He was later to regard this misfortune as his own good fortune because if he had had enough money he would probably never have made mathematics his vocation.[3] Although many revolutionary things happened in society during his later life, his own life was remarkably undisturbed, if not dull. Perhaps the most interesting event in his life was in 1763, when he accompanied Marquis Caràccioli to Paris to take up his new post as ambassador. In Paris Lagrange met many prominent people of his time, such as d'Alembert, Fontaine, Clairaut and others. An illness forced him to stay for some time in Paris, during which he developed a close friendship with d'Alembert that was to last until the latter's death in 1783. This friendship has yielded us an exchange of letters in which Lagrange expresses himself more fully than he does elsewhere. From 1766 to 1787 he worked for Frederick the Great in Berlin, and from 1787 until his death in 1813 at the French *Académy des Sciences*.

Lagrange's interest in mechanics arose out of his interest in mathematics, especially variational calculus. Probably inspired and challenged by Euler's *Methodus inveniendi*, he worked on both the analytic foundation of the method of maxima and minima and on its application in mechanics.[4] At the end of his first letter to Euler of 28 June 1754 he writes that he has made several observations in the field of the maxima and minima found in natural phenomena.[5] However, we do not know exactly what observations he made. Euler did not reply, presumably because the main part of Lagrange's letter reported a mathematical discovery that had already been made by Leibniz.[6] Lagrange later discovered this himself, but did not permit it to discourage him.

In the following year he wrote to Euler again, when he had found a solution to a problem that Euler had outlined in his *Methodus inveniendi*.[7] The problem was how

---

[3] Delambre, "Notice sur la vie et les ouvrages de Lagrange" (1867), x.
[4] See section 5.3.4.
[5] "Haberem fortas[s]is alia tibi mittenda (...) observationesque nonnullas circa maxima, et minima, quae in naturae actionibus insunt (...)" (Letter from Lagrange to Euler, 28 June [1754] (*EOO* IVa.5, 361-366, this passage at 362).
[6] Lagrange hoped that his discovery of an analogy between the development of the power $(a+b)^m$ and the mth derivative of $(a \times b)$ would impress Euler, who was the most prominent mathematician of his day. However, this analogy had already been discovered in 1695 by Leibniz, and both published in an article in 1710 and described in a letter to Johann I Bernoulli (which was published around 1745). Lagrange only learned this a month after his letter (Juškevič and Taton, in *EOO* IVa.5, 365n.1). The fact that Euler only responded after Lagrange's second letter may well be attributed to his knowledge of Leibniz's prior claim to the discovery, which meant that he would have to share this rather deflating news with Lagrange.
[7] Letter from Lagrange to Euler, 12 August 1755 (*EOO* IVa.5, 366-375).

one could justify equating two differential terms. Euler could only provide a geometric argument that was not in accordance with the analytic character of his treatment.[8]

Lagrange had probably found his solution because he had perceived that what Euler calls the 'differential' and refers to with one symbol 'd', is in fact used in two different senses.[9] In his own solution Lagrange explicitly distinguishes these two meanings and allocates different symbols to them. The first meaning, for which he uses the symbol 'δ', is the change resulting from an infinitesimal movement of the curve C which Euler was later to call 'variation'. The second, 'd', is an infinitesimal movement *along* the curve, that is, it is the 'ordinary' differential (see Figure 7.1).

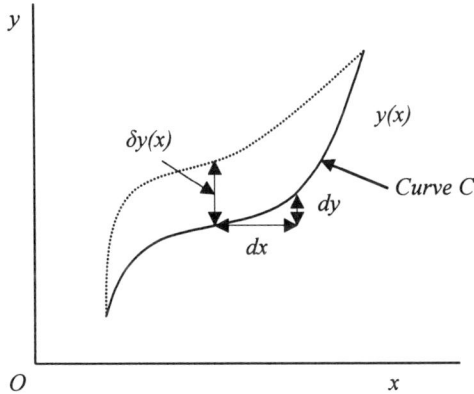

Figure 7.1 Variation δ and differential d of $y(x)$

Using what was later known as δ-calculus, Lagrange was able to develop an ingenious and easy to use method for determining the extremes of indefinite integrals.[10] This time Euler did reply. Full of enthusiasm, he told Lagrange that he had taken the theory of maxima and minima almost to the 'peak of perfection'.[11] He especially praises the fact that Lagrange has provided a fully analytic solution. He locates the

---

[8] "Desideratur itaque Methodus a resolutione geometrica et lineari libera, qua pateat in tali investigatione maximi minimive, loco *Pdp* scribi debere -*pdP*" ("Thus what is required is a method, apart from a solution achieved by geometry and drawing lines, which will show that in such an investigation of the maximum and minimum, one may write -*pdP* instead of *Pdp*") (Euler, *Methodus inveniendi*, 52 (Chapter 2, section 39). The word 'lineari' constitutes a problem in translating this passage. Delambre's translation is nonsense: "indépendante de la Géometrie et de l'Analyse" (Delambre, "Notice sur la vie," xvi). Stäckel translates it as "unabhängig (...) von der geometrischen Lösung," and simply omits 'lineari' (Stäckel ed., *Abhandlungen über Variationsrechnung*, 66), as do Juškevič and Taton, when they translate the fragment in Lagrange's letter to Euler, as "libre de considérations géométriques" (Juškevič and Taton, in *EOO* IVa.5, 370 and 374).

[9] Fraser, "J.L. Lagrange's Changing Approach to the Foundations of the Calculus of Variations" (1985), 163.

[10] The term 'indefinite integral', expressed as $\int f dx$, does not refer to a determined integral of f—i.e., a number—but rather the *set of primitive functions F* of f.

[11] "(...) ad summum fere perfectionis fastigium erexisse videris (...)" Letter from Euler to Lagrange, 6 September 1755 (*EOO* IVa.5, 375-378, this passage at 375).

essence of the generalization that characterizes Lagrange's method, in comparison to his own polygonal method, in the fact that Lagrange deals with simultaneous variation over the whole curve, while his own method only permitted him to vary one point at a time.[12] However, it is noteworthy that, amidst all this praise, he does not raise the significance of Lagrange's method for the further development of the principle of least action. That was not to come until six months later.

7.2.2   The Early Meaning of the Principle of Least Action for Lagrange

In the following years Lagrange worked on two articles that were eventually to be published in the *Miscellanea Taurinensia*.[13] The first is purely mathematical and contains an elaboration of his method of variational calculus. The second is an application of a limited form of the principle of least action to many mechanical and hydrodynamic problems. This is the article for which Lagrange, in his early years, won wide fame as the defender of the principle of least action. In spite of this reputation, the term 'principle of least action' is not mentioned in the articles at all. He begins the second article with a reference to the extremal principle that Euler had presented in "Additamentum II" of his *Methodus inveniendi*. Then he generalizes it to a general principle: $\Sigma M \int u ds$ is always a minimum or a maximum, for the case of a random number of bodies that not only act on one another but are also subject to central forces that are a random function of distance.[14] Although he says that this principle can easily solve *all* problems in dynamics, and that $\Sigma M \int v ds$ is *always* a maximum or minimum, he does not provide any basis for these absolute formula-

---

[12] Euler realised the importance of this for the future power of what he was later to call variational calculus: "(...) quam ob causam etiam non dubito, quin tua analysis, si penitius excolatur, ad multo profundiora mox sit perductura" ("and for this reason I have no doubt that your [method of] analysis, when it is further refined, will take [us] to much deeper matters") *(ibidem)*. It is however remarkable that Euler evidently identifies the analytic character of Lagrange's method with general variational calculus—although his own method could just as well be translated into an entirely analytic expression. In 1764 he was to repeat this position in the summary that introduces "Elementa calculi variationum" and "Analytica explicatio methodi maximorum et minimorum" (Carathéodory, "Einführung in Eulers Arbeiten über Variationsrechnung," XXXI). According to Carathéodory, the fact that Lagrange's method was used rather than Euler's polygonal method until the end of the nineteenth century should be attributed not the more analytic character of the former, but to the resulting simplification and generalization of the procedure used to formulate the differential equations of variational calculus. This simplification was due to Lagrange's use of the general variation of curves *(ibidem*, XXXII).

[13] Lagrange, "Essai d'une nouvelle méthode pour déterminer les maxima et les minima des formules intégrales indéfinies" and "Application de la méthode exposée dans le mémoire précédent à la solution de différents problèmes de dynamique" (both from 1760–1761).

[14] "Principe général — Soient tant de corps qu'on voudra $M, M', M'',...$, qui agissent les uns sur les autres d'une manière quelconque, et qui soient de plus, si l'on veut, animés par des forces centrales proportionnelles à des fonctions quelconques des distances; que $s, s', s'',...$, dénotent les espaces parcourus par ces corps dans le temps t, et que $u, u', u'',...$, soient leurs vitesses à la fin de ce temps; la formule

$$M \int u ds + M' \int u' ds + M'' \int u'' ds + ...$$

sera toujours un maximum ou un minimum" (Lagrange, "Application de la méthode," 365).

tions. This leaves the status of the principle unclear: he does not say whether it is an item of evidence, a hypothesis, or a metaphysically or experimentally derived theorem. He does not even use the term 'least action', or make any reference to Maupertuis.

Now one can understand that these 'omissions' did not at that time raise any questions regarding Lagrange's intentions, and certainly not for Euler. Just as in Bach's violin sonatas and partitas the chords are not played but must be heard, would not the readers have felt the teleological interpretation resonating with the mathematical formulation of the principle? Lagrange had also not given any grounds for doubts at any point in his previous correspondence with Euler and others. Moreover: is it reasonable to expect such a doubt, given that no-one knew the direction Lagrange's hierarchy of mechanical principles was later to take? Again, given that the stimulus in those years had come largely from Euler,[15] one might well suppose that Lagrange went through a process in which he was first taking his cues from Euler's and Maupertuis's teleological ideas, and later thought better of it and removed all the non-mathematical elements from his articles.[16]

Nevertheless, it is not plausible that Lagrange initially considered the principle of least action as a teleological principle. This can be shown by considering the history of the composition of the two articles, between 1756 and 1760, in more detail. Since no outlines for the articles have been preserved, it is not possible to see this directly. However, indirect evidence from his correspondence with Euler, Maupertuis[17] and the priest and mathematician Paolo Frisi[18] does support the probability that Lagrange was always thinking of Euler's more limited formula, with $\Sigma mvds$ as the integrand. The references to 'Maupertuis's principle' and the 'principle of least action' were no more than names adopted on the authority of Euler and Maupertuis, without Lagrange being aware of their teleological connotations.

---

[15] Lagrange's correspondence with d'Alembert did not begin until 1759, and even then the principle of least action seems never to have been mentioned. They did not meet in person until 1763 (see page 207).
[16] Fraser, "J.L. Lagrange's Early Contributions to the Principles and Methods of Mechanics" (1983), 233-234.
[17] The reference here is to a letter from Maupertuis to Lagrange on 4 January 1757, which was only discovered in 1986. The letter's existence had been known previously, but it had been dated before 4 May 1756 because of the incorrect dating of Lagrange's letter to Frisi in 1756 instead of in 1757 (see note 18). The letter is published in Taton, "Sur quelques pièces."
[18] The correspondence between Lagrange and Paolo Frisi, which was not yet available when Serret published Lagrange's oeuvre (1867-1892), was published separately in Favaro, "Sette lettere inedite di Giuseppi Luigi Lagrange al P. Paolo Frisi" (1895-1896). Favaro dates the letter that is important here on 4 May 1756. The discovery of a letter from Maupertuis to Lagrange dated 4 January 1757 (see note 17) prompted René Taton to study Lagrange's letter to Frisi more closely. He discovered that the date 1756 was a mistake in the manuscript—it should read 1757 (Taton, "Sur quelques pièces," 12-16). Paolo Frisi (1728-1784) was professor of mathematics and physics in Pisa from 1756. He was also well known in Berlin, where he had won the competition for the department of mathematics for 1756 (see also *Die Registres*, 224 and 224n.2(361)).

I will attempt to reconstruct the sequence of events.

1. Lagrange probably only heard the term 'principle of least action' in Euler's response to a treatise that Lagrange had sent him early in 1756. At any rate it was only then that he linked it with the dynamic extremal principle that he had addressed in the treatise.[19] Before this, the correspondence between Euler and Lagrange had dealt only with the mathematical theory of maxima and minima, and with the *Methodus Inveniendi*, which makes it unlikely that Lagrange had already heard of Maupertuis's minimal principle.[20]

2. Although there is no direct evidence of the contents of this treatise, and it is entered in the 'Régistres' as a treatise on the principle of least action,[21] we can be reasonably confident that this term was not used in the article itself, and that Lagrange did not even know of Maupertuis's formulation. Why? The extremal principle that it deals with is a generalization of Euler's extremal principle from the second "Additamentum" to his *Methodus inveniendi*.[22] If Lagrange had known about Maupertuis's formulation of the principle he could hardly have limited himself to Euler's formulation without providing some justification, and he would also have stated emphatically that, in his opinion, both a maximum and a minimum of the relevant quantity are possible. In that case, Euler's response could hardly have been entirely enthusiastic. Surely he would have made some critical comment on this point. But nothing of this sort can be seen in his letter. Euler simply conveys Maupertuis's thanks for Lagrange's defense of his principle, adding his promise of an appointment as non-resident member of the Academy and the question of whether he would like to come to work in Berlin.[23]

---

[19] This treatise and the accompanying letter have been lost. We only know its contents from a letter by Euler of 24 April 1756 (*EOO* IVa.5, 386-390), in which he responds to it. We also know that the treatise was presented in the Academy on 6 May 1756 by Euler himself (*Die Registres*, 223). The treatise and its covering letter must therefore have been sent before 24 April 1754, when Euler wrote his response. Since Euler had also given it to Maupertuis to read before he wrote his response, it is probable that the treatise and letter were sent as early as March 1754. The loss of both the letter and treatise may have occurred because Maupertuis took them with him on his definitive departure from Berlin at the end of May 1756 (Brunet, *Maupertuis* I. *Étude biographique*, 168-169).

[20] The previous correspondence, consisting of four letters, seems to have been preserved in its entirety. Two letters from Lagrange to Euler, of 28 June 1754 (*EOO* IVa.5, 361-366, see pages 362 and 364) and 12 August 1755 *(ibidem*, 366-375, see pages 366 and 369-370) are of special importance here.

[21] *Die Registres*, 223 (6 May 1756).

[22] That is, that $\int mvds$ is an extreme. See section 5.3.4.

[23] "[Maupertuis] tibi pro suscepto principii minimae [a]ctionis patrocinio maximas agit gratias (...)" ("[Maupertuis] thanks you very much for your defence of the principle of least [a]ction (...)") (letter from Euler to Lagrange, 24 April 1756, in *EOO* IVa.5, 386-390, this passage at 387). Lagrange was in fact nominated as a non-resident member by Euler—on behalf of the absent president Maupertuis—on 26 August 1756. The Academy accepted the nomination on 2 September of that year (*Die Registres*, 225; letter from Euler to Lagrange, 2 September 1756 (*EOO* IVa.5, 394-396, see page 394)).

3. Considering both points, it is plausible that Lagrange did not know about Euler's and Maupertuis's more far-reaching intentions when he wrote his treatise in the spring of 1756.[24]

4. In the period between receiving responses from Euler and Maupertuis and the publication of his papers in the *Miscellanea Taurinensea*, Lagrange used both expressions frequently and attributes great importance to them. The first occasion is in his animated answer to Euler of 19 May 1756, in which he applauds the principle of least action as "the universal key to all mechanics, both statics and dynamics."[25] In his next letter to Euler, of 5 October 1756, he speaks of a new way of applying the mathematical method of maxima and minima to Maupertuis's principle of least action.[26] He refers to the paper again in a letter to Paolo Frisi, saying that its subject is the "Principio Maupertuisiano."[27]

5. Although it is probable that Lagrange had taken the expression 'principle of least action' from Euler's letter of 24 April 1756, the same cannot be said for its context. He can also not have learnt this from Maupertuis, since the latter only wrote to him on 4 January 1757, when Lagrange had already been using the expression for eight months. It is most probable that he learnt about its context from Maupertuis's *Œuvres*.[28] However, what is remarkable in this is that, while Lagrange formally supports the principle of least action, he does not make any substantial changes in the application of the principle, compared to his early 1756 treatise. In fact, wherever he uses the term, he is actually drawing on Euler's dynamic extremal principle rather than Maupertuis's general minimal principle.

---

[24] This conclusion does not contradict that of Fraser who, on the basis of various correspondences between Lagrange's "Application de la méthode" of 1760 and Eulers "Harmonie entre les principes" of 1752, concludes that Lagrange must have known this work and been inspired by it, although he does not explicitly refer to it until the "Recherches sur la libration de la lune" (1764). However, there is no indication that Lagrange knew it as early as 1756 when he wrote his outlines (Fraser, "Lagrange's Early Contributions," 203-204 and 208-209).

[25] "De principio minimae quantitatis actionis ego ita sentio, (...), omnium tam staticorum, quam dynamicorum problematum universalem veluti clavem haberi posse (...)" (Letter from Lagrange to Euler, 19 May 1756 (*EOO* IVa.5, 390-394, this passage at 391).

[26] *Ibidem*, 396.

[27] Letter from Lagrange to Paolo Frisi, 4 May 1757 (and not 1756!) (Favaro, "Sette lettere inedite," 141-143, this passage at 142; Taton, "Sur quelque pièces," 16-18, this passage at 18). There are two later mentions in his letters of 28 July and 4 August 1759 to Euler, in which he refers to the subject of the papers as "de applicatione principii minimae quantitatis ad mechanicam universam" ("on the application of the principle of least action to universal mechanics") (*EOO* IVa.5 411-414, this passage at 411). In the letter of 4 August 1759 the term 'universam' is replaced by 'totam' ('the whole of') (*ibidem*, 414-417, this passage at 414).

[28] From Maupertuis's letter it is clear that Lagrange had read d'Arcy's criticism, probably in his "Réplique à un Mémoire de Mr. de Maupertuis sur le principe de la moindre action," which was printed in 1756 in the *Mémoires de Paris*. But it is unlikely that Lagrange would have drawn his enthusiasm from reading something so sharply critical. In his letter of 4 January, Maupertuis writes that he has sent one copy of the 1756 edition of his *Œuvres* to Ansaldi, a professor in Turin. He doubts whether he has in fact received it, and says that he will send two more exemplars, one of them for Lagrange. However, this does not exclude the possibility that Lagrange had already read his *Œuvres* by that time.

6. The complete lack of any reflection on the use of the term, linked to his ambitious plans to make it the foundational principle of mechanics, show that Lagrange never considered the principle of least action as a teleological principle, but only as a mathematical extremal principle.

On the basis of this reconstruction it is hardly strange that in the final version of "Application de la méthode" the teleological nature of the principle is never mentioned. For Lagrange that was of no essential importance. The development from the correspondence to the published form of the article therefore seems to have been formal rather than with respect to content: it is no more than shedding the coat in which the mathematical principle had been dressed. This does raise another question however: why does Lagrange suddenly present the principle naked, where he had previously dressed it up in a speculative garb? As late as 1759 he was still speaking of the subject of the article in these terms, and intended to include the term 'principle of least action' in the title.[29]

As has been said, Lagrange himself gives no explanation for this sudden change. We could however look at what he later said about it. In his biography of Lagrange, Delambre says that Lagrange described his first steps towards the theory of maxima and minima in a reminiscence just two days before his death. Lagrange is said to have recalled that he was inspired by a question from Euler regarding a metaphysical foundation, independent of geometry and analysis, for both the method of maxima and minima and the principle of least action. According to this account, Lagrange achieved all this in a completely analytic way, proving in the process that the principle was only a result of a much more general principle.[30] However, there is a double confusion in this account: the mathematical problem of the foundation of variational calculus is conflated with the metaphysical problem of the foundation of the principle of least action, and the "Application de la méthode" of 1760 with the *Méchanique analitique* of 1788. One has to conclude either that Delambre has misunderstood what he heard, or at least projected back the final result to the beginning of the process, or else that Lagrange later saw his involvement in solving a metaphysical problem as a flirt that he would rather gloss over.

The only point in his later publications at which Lagrange refers to the teleological function of the principle does not take us any further. In his *Méchanique analitique* of 1788 he criticizes Maupertuis's examples as too specific and too arbitrary to serve as a basis for a general principle. A better method, "more general and rigorous, and the only one worthy of consideration by geometricians," is Euler's method, developed in the second "Additamentum" to his *Methodus inveniendi*.[31] What Lagrange here calls 'least action' therefore refers only to Euler's method. From this it should be clear that what Lagrange tells us contributes nothing to

---

[29] See note 27.
[30] "plus générale et plus rigoureuse, et qui mérite seule l'attention des géomètres" (Delambre, "Notice sur la vie," xvi).
[31] Lagrange, *Méchanique analitique*, 229 (II.I.17).

understanding the *crucial change* in his ideas. His remarks refer only to his later understanding, around 1788 and later.

An alternative explanation could be sought in the way his articles were published, after a long process which he can hardly have found satisfactory. First, the Seven Years' War made communication with Berlin impossible from 1756 to 1759. Nevertheless, Lagrange expected to be able to publish his articles in Berlin and, presumably for that reason, took no action for some time. His new appointment at the Turin Academy was also keeping him very busy.[32] In 1759 communications were restored and, in two successive letters to Euler of 28 July and 4 August, Lagrange expressed his wish that his articles should be published in Berlin. Then came a second blow: Maupertuis's death on 27 July of that year and the deplorable state of the royal finances because of the war prompted Euler to inform Lagrange that publication in Berlin was impossible. Moreover, he said nothing about the treatise that Lagrange had sent in 1756.[33] This sudden coolness and indifference must have been a blow to Lagrange.[34] He waited two months before answering, and then told Euler that he would write to him "at some other time" about his new discoveries regarding variational calculus and the principle of least action.[35] Three years later, in a letter dated October 1762, he told him that the definitive versions had appeared, in a much abbreviated form, in the *Miscellanea Taurinensea*.[36]

Given this disappointing history, it is plausible that Lagrange altered the title and terminology of the article after the death of Maupertuis, perhaps thinking also that Euler had betrayed him, so as to free it from any specific metaphysical interpretation now that he could not be sure of protection.

---

[32] In his letter to Frisi of 4 May 1756, Lagrange writes that he has already got the two articles almost completely in order, but cannot find time to finish them because of the intense activity entailed by his chair at the Artillery school in Turin (Taton, "Sur quelques pièces," 17; Favaro, "Sette lettere inedite," 142). First Euler and later Maupertuis himself had by then assured him that they would publish his work—including the memoir he had already sent—in the *Mémoires* of the Berlin Academy (letter from Euler to Lagrange, 2 September 1756 (*EOO* IVa.5, 395); letter from Maupertuis to Lagrange, 4 January 1757 (Taton, "Sur quelques pieces," 9)).
[33] Letter from Euler to Lagrange, 2 October 1759 (*EOO* IVa.5, 418-423, this passage at 418).
[34] See also Taton and Juškevič, [Introduction to *Correspondance*], 42-43.
[35] Letter from Lagrange to Euler, 24 November 1759 (*EOO* IVa.5, 429-432, this passage at 430).
[36] Letter from Lagrange to Euler, 38 [28?] October 1762 (*ibidem*, 446-448, this passage at 447). Another event is also explained by this disappointment, namely a certain hostility on Lagrange's part towards Euler. In the letter in which Euler tells him the sad news, he also writes about an analytic solution that he has developed himself, inspired by Lagrange's treatise of early 1756, but that he does not intend to publish until Lagrange has published his own, so as not to steal his thunder. Euler's articles ("Analytica explicatio methodi maximorum et minimorum" and "Elementa calculi variationum") were in fact not published until 1766, in the *Nouveaux Comm. Petrop.*, although they had been presented in the Berlin Academy on the 9th and 16th of September 1756, respectively. If Taton and Juškevič are right in their suggestion that Lagrange took the rejection personally (see note 34), it is quite possible that Lagrange developed a personal resentment towards Euler out of disappointment. This could plausibly explain his strange refusal to take a position in Berlin in 1764: "(...) I feel that Berlin would not suit me, while Monsieur Euler is there" (Lagrange's letter to d'Alembert, 13 November 1764, *Œuvres de Lagrange* XIII, 20-23, this passage at 23).

7.3 LAGRANGE'S 'TRUE METAPHYSICS'

7.3.1 A Puzzling Remark

The significance of the principle of least action for the early Lagrange, as outlined above, confirms the usual interpretation of him. He is usually described—and often praised—as a man who was averse to metaphysical speculations. Nevertheless, in 1762, in the same letter in which he told Euler that the two articles mentioned above, had been printed, he also mentioned his pursuit of a metaphysical foundation. He does this in the last paragraph, tucked in inconspicuously between the report that his articles had been published and a greeting to Euler's son Albert. The reference is to the textbooks that Lagrange had developed as a lecturer at the newly established academy of Turin. He writes:

> I myself have also composed elements of mechanics and differential and integral calculus, for my students, and I believe that I have developed the true metaphysics of their principles, so far as this is possible.[37]

Nothing has been preserved from these courses except the lecture notes made by one of his students, covering the introduction to what is probably an earlier version of the course in mathematics.[38] We therefore have no textual evidence that would clarify the meaning of "the true metaphysics" ("la vraye métaphisique"). The passage quoted above is generally interpreted as a reference to a transition that would become visible in Lagrange's work soon after. Fraser, for example, reads it as a first indication of Lagrange's shift from the principle of least action to the principle of virtual velocities.[39] Pulte considers that the quotation may refer to Lagrange's search for a single principle for both statics and dynamics, a role for which the principle of least action seemed to be inadequate.[40] Such interpretations are quite possible, although the way Lagrange refers to the value of metaphysics in other places, and some of his own treatments that he described as metaphysical, give us no reason to expect much of the metaphysical level of these foundations.

7.3.2 Lagrange's Concept of Metaphysics

It is peculiar that, although Lagrange was very interested and knowledgeable in metaphysics, history, religion, linguistics, medicine and botany, he seldom gave his opinions on these fields, even in his correspondence.[41] One example of this attitude

---

[37] "J'ai aussi composé moi même des elemens de Mécanique et de Calcul differentiel et integral à l'usage de mes ecoliers, et je crois avoir developpé la vraye metaphisique de leurs principes, autant qu'il est possible" (letter from Lagrange to Euler, 24 November 1759 (*EOO* IVa.5, 429-432, this passage at 430-431)).
[38] Taton and Juškevič, [Notes to *Correspondance*], 432n.14 and 15.
[39] Fraser, "Lagrange's Early Contributions," 233.
[40] Pulte, *Das Prinzip der kleinsten Wirkung*, 257-258 and 258n. 275.
[41] Sarton, "Lagrange's Personality (1736–1813)" (1944), 477-478; Itard, "Lagrange," 569.

of rejection of metaphysics, combined with tacit understanding, can be found in his correspondence with d'Alembert, who was perhaps the only scholar with whom he corresponded on genuinely friendly terms. The exchange in question relates to the Berlin competition for 1779 that was described in the previous chapter. D'Alembert, who was constantly encouraging Lagrange to venture outside his narrow scientific domain, writes jestingly to him from Paris: "I am sure that you were not consulted. Everyone is amused about this program, and the Academy could not suppress a laugh when Monsieur de Condorcet read it out."[42] In his answer, Lagrange was able to remain neutral with a pointed relativization: "You are quite right to believe that I have not had any part in the metaphysical program. This science, if it is one, is not to my taste. It seems to me that every country almost has its own metaphysics, like its own language, and the question that has been put forward is German and Leibnizian metaphysics."[43]

There is evidence that at one point at the beginning of his career Lagrange had a less skeptical and relativizing picture of metaphysics. The title of his article "Note sur la métaphysique du calcul infinitésimal" (1760) suggests that its contents would be metaphysical, although it is in fact purely mathematical. In the article, he asks whether the asymptote of a curve is identical to the tangent in infinity. He says it is not, with the argument that the asymptote is only the *limit* of the tangent and is not itself a tangent. He compares this to a problem in differential calculus, namely the way a polygon with infinitesimal sides approaches a curve. If the curve and polygon are supposed to be the same, as Leibniz did, this introduces an error, however small it may be. The only reason the answers are correct is that this error is compensated by another error (also the equation of infinitesimal magnitudes to zero).

Lagrange's understanding of metaphysics here is related to the mathematically infinite; in precise terms, metaphysics is related to the transition from the finite to the infinite. From this we could derive something like a lower limit for the significance that Lagrange could attribute to metaphysics in relation to mechanics. For Lagrange, a metaphysical foundation could be relevant to those aspects of mechanics that relate to the transition from the finite to the infinite. With this background, one could read Lagrange's reference to the 'true metaphysics' in relation to mechanical principles as a reference to the problem of infinitesimal variational displacements and movements.

---

[42] "Je suis bien sûr que vous n'avez pas été consulté. Tout le monde se moque de ce programme, et l'Académie n'a pu s'empêcher d'en rire quand M. de Condorcet l'a lu" (letter from d'Alembert to Lagrange, 22 September 1777 (*Œuvres de Lagrange* XIII, 330-332, this passage at 332)).

[43] "Vous avez bien raison de croire que je n'ai eu aucune part au programme de Métaphysique. Cette science, si c'en est une, n'est nullement de mon gibier. Il me semble que chaque pays a presque sa Métaphysique particulière comme sa langue, et la question proposée est de Métaphysique allemande et leibnitzienne" (letter from Lagrange to d'Alembert, 27 January 1778 *(ibidem*, 334-336, this passage at 336)).

7.3.3 Recognition and Rejection of Metaphysics

In the light of the interpretation above, one must suppose that Lagrange's remark in his letter to Euler had a rhetorical function.[44] Its position, as a throwaway remark in the concluding section of the letter and immediately after Lagrange had coolly informed Euler that he would write to him at some later date about variational calculus and the principle of least action, provides the key. Assuming that the picture given in section 7.2.2 is correct, and Lagrange removed the metaphysical language from the final version of the "Application de la méthode" out of disappointment with Euler, it is no great step to read his remark about providing his own metaphysical foundation as an ironic rebuke to Euler. What Lagrange wants to say, and expresses very indirect with the subtle modifier "so far as this is possible" ("autant qu'il est possible") is that there is little possibility of providing a metaphysical foundation for mechanical principles, and Euler should also not make the attempt.

The reader may well have read the two previous sections with increasing puzzlement. If my goal is to show that the development of mechanics in the eighteenth century also entailed a development in metaphysics, why is it necessary to refute, at some length, the metaphysical pretensions of one of the most important shapers of analytic mechanics? Would it not serve the argument much better to cherish the little metaphysics that Lagrange seemed to recognize in his younger years, and magnify it wherever possible?

But, if I may question the reader in my turn, is it not clear that, in the case of Lagrange, the *explicit* metaphysics says little, but rather conceals the *implicit* metaphysics in his mechanics? His pragmatic rejection of the teleological interpretation of the principle of least action reveals how little he cares for any specifically avowed metaphysics. His remark to Euler about the true metaphysics, then, is no more than an ironic commentary on the latter's metaphysical ambitions. Much more interesting, in contrast, is what Lagrange *does not* say, and what he does *not* write about. However, an analysis of Lagrange's implicit metaphysics will require seven-league boots, to cross the three decades that separate his letter to Euler from his *Méchanique analitique*.

7.4 LAGRANGE'S ANALYTIC FOUNDATION OF MECHANICS

7.4.1 "Recherches sur la libration de la lune" (1764)

Soon after the publication of his "Application de la méthode" in 1762, Lagrange removed the principle of least action from his foundation of mechanics and replaced it with the principle of virtual velocities. D'Alembert had already given this principle a foundational role as an equilibrium theorem in his *Traité de dynamique*,

---

[44] See the quotation on page 215.

and in doing so he was an important source of inspiration for Lagrange.[45] However, in his own formulation Lagrange uses the concept of force, which d'Alembert so detested, to make d'Alembert's labored kinematic formulation easier to understand and apply. The principle of virtual velocities, for the simple case of a system of masses that is subject to two forces, can then be formulated as 'forces F_1 and F_2 are in equilibrium with one another if they are inversely proportional to their virtual velocities v_1 and v_2, taken in the direction of the forces' (see Figure 7.2).[46]

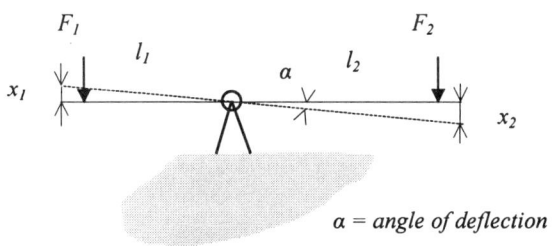

$\alpha = $ angle of deflection

$$F_1 \times x_1 - F_2 \times x_2 = 0 \quad \Leftrightarrow \quad F_1 \times l_1 \times \alpha - F_2 \times l_2 \times \alpha = 0 \quad \Leftrightarrow \quad F_1 \times l_1 = F_2 \times l_2$$

Figure 7.2. The equilibrium of forces according to the principle of virtual velocities

Lagrange first accords the principle a fundamental place in his competition essay "Recherches sur la libration de la lune" (1764), dealing with the irregular movements of the moon. Perhaps the fact that d'Alembert was on the jury for this competition encouraged Lagrange to give the principle of virtual velocities such a prominent place, but I will not consider this further. It is more interesting at this point to see how Lagrange justified the shift. He does not do so explicitly, except by emphasizing in his discussion of the principle that the derivation of various other principles depends on the principle of virtual velocities. After introducing the

---

[45] D'Alembert, *Traité de dynamique*, 50-51 (see also section 4.4.3.2). Strangely enough, there is no mention of the principle of virtual velocities in the extensive correspondence between d'Alembert and Lagrange. Lagrange has also not given us any insight into this transition in other places (Fraser, "Lagrange's Early Contributions," 220 and 233-235).

[46] This formulation is a paraphrase of Lagrange, *Méchanique analitique*, 18. Lagrange also gives a general formulation on page 20. In mathematical form the principle then reads: $Pdp + Qdq + Rdr + \ldots = 0$ (*ibidem*, 26). A general description of 'virtual velocities' is: "by *virtual velocities* is meant that speed that a body in equilibrium is disposed to receive if its equilibrium is destroyed, that is, the velocity that the body would actually assume in the first instant of its motion" ("On doit entendre par *vitesse virtuelle*, celle qu'un corps en équilibre est disposé à recevoir, en cas l'équilibre vienne à être rompu, c'est-à-dire la vitesse que ce corps prendrait réellement dans le premier instant de son mouvement") (*ibidem*, 17-18). A second definition of 'virtual velocities' is given apart from the general formulation of the principle: the *infinitesimal distance dx* that an element in a mass system moves, if an infinitesimal arbitrary motion is given to that system (*ibidem*, 20). Notice that Lagrange makes a silent transition from *velocity* to *distance*.

principle, he cites three of these dependant mechanical principles in a scholium.[47] The first is the static minimal principle ('law of rest'), as formulated by Maupertuis in 1740,[48] and consequently the second, the principle of least action, since Euler showed that it can be derived from the law of equilibrium.[49] The third is the principle of the conservation of living force, which as d'Alembert demonstrated in his *Traité de dynamique*, can be derived from the principle of virtual velocities. Lagrange recapitulates d'Alembert's derivation at length, but in a version of his own.[50]

However, the principle of virtual velocities does not hold the scepter on its own: it is only a static principle, and needs a dynamic principle if it is also to be applicable in dynamics. Lagrange introduces this principle, 'd'Alembert's principle', *en passant* during his derivation of the conservation of living force, without formally designating it as the second foundational principle. The general formula that results from the combination of the two principles, and which rather remarkably is not given a name of its own, "(...) contains the solution to all problems regarding the motion of bodies."[51]

The "Recherches sur la libration" thus shows that the degree of generality of principles was an important ranking criterion for Lagrange as early as 1764. The principle of virtual velocities is prior to the principle of least action and is therefore more fundamental.[52] However, if we look more closely we see that the principle of virtual velocities was no more able to satisfy his hopes of finding a 'universal key' for both statics and dynamics than was the principle of least action. In the case of the latter, Lagrange only achieved results in dynamics, whereas the principle of virtual velocities, in itself, is only valid for statics. He certainly saw this problem, although in his presentation he seems to want to draw our attention away from it.

In 1780, some years before the publication of his main work, Lagrange was again to write a treatise on the problem of the irregular movements of the moon.[53] He used the same method as in "Recherches sur la libration" and expected just as much from it. However, his justification of the method was more extensive. Although he no longer mentions the principle of least action, he gives another proof for the law of the conservation of living force.[54] On this occasion he shows the distinction between

---

[47] Lagrange, "Recherches sur la libration," 10-12.
[48] See section 5.3.1 above.
[49] Lagrange is referring here to Euler's "Harmonie entre les principes généraux" (see above, Chapter 5, page 167).
[50] For d'Alembert's derivation see Chapter 4, page 129.
[51] "(...) renferme la solution de tous les Problèmes qui regardent le mouvement des corps" (Lagrange, "Recherches sur la libration," 12).
[52] Lagrange does not ask himself whether this derivation could also be reversed, which is an argument for the suggestion that he wrote the piece with d'Alembert in mind but it is not an argument against my interpretation of the *way* he ranks principles, using an axiomatic and deductive ranking based on mathematical principles.
[53] Lagrange, "Théorie de la libration de la lune et des autres phénomènes qui dépendent de la figure non sphérique de cette planète" (1780).
[54] In which he introduces a mathematical variable V which would later be called 'potential energy': $V = \Sigma\{m\int(Pdp + Qdq + Rdr + ...)\}$ *(ibidem, 24)*.

the static and dynamic principle more clearly by introducing them in various sections rather than postulating the latter in the middle of an explanation of the first, as in "Recherches sur la libration." This increased clarity is also a sign of a much more acute awareness of incompleteness, as he writes: "the combination of these two principles remains to accomplish, and is perhaps the only step that remains, after Monsieur d'Alembert's discovery, to perfect the theory of dynamics."[55]

7.4.2   The Méchanique analitique

In the *Méchanique analitique* (1788), published five years after the death of Euler and d'Alembert, Lagrange takes the method he used in "Recherches sur la libration" and "Théorie de la libration" and systematically works out its implications for all mechanics. He also clarifies the relationship between the principle of virtual velocities and principles such as the principle of least action and the conservation of living force: the latter are merely derivative theorems, "general results of the laws of dynamics," and are therefore exiled to a remote corner of mechanics.[56]

The book is divided into two parts, dealing with statics and dynamics respectively. In contrast to his earlier work, Lagrange introduces the principle of virtual velocities as the fundamental principle, the general law, for both parts of mechanics. The transfer of this principle from the field of statics to dynamics is now presented as an *extension*. This will be discussed further below. The principle truly belongs to statics, and he discusses it in detail in that part. He makes his formulation of the principle of virtual velocities somewhat more compact by introducing the expression 'the moment of a force'. This is the product of a force and the virtual velocities (read: displacement) of the physical point on which that force acts, projected in the direction of that force.[57] However, the application of the principle to various fields in statics comes only after an explanation of the way in which it is used, and an introduction in which he provides historical and systematic reflections on the existing principles of statics.

This introduction is largely a distillation of mathematical formulations of mechanical principles from the literature and of what Lagrange regards as valid proofs of these.[58] The first of these principles is the principle of the lever formulated by Archimedes, but still without a satisfactory proof. The second principle, also known to the Greeks, is the principle of the 'composition of forces', now called the

---

[55] "Mais la combinaison de ces deux principes est un pas qui n'avait pas été fait, et c'est peut-être le seul degré de perfection qui, après la découverte de M. d'Alembert, manqait encore à la Théorie de la Dynamique" *(ibidem,* 11).
[56] "Ces principes doivent être regardés plutôt comme des résultats généraux des lois de la Dynamique (...)" (Lagrange, *Méchanique analitique,* 225).
[57] *Ibidem,* 26. See also note 46.
[58] Lagrange's 'Whig history' has left a considerable mark on the historiography of eighteenth century mechanics, especially through the work of Ernst Mach. Mach's original inspiration for *Die Mechanik, historisch-kritisch dargestellt* (1883) came from Lagrange's historical introductions in the *Méchanique analitique.* See also Chapter 1 above, pages 10-11.

principle of the superposition of forces. This principle is sufficient for the derivation of all of the laws of statics.[59] His third principle is the principle of virtual velocities, originating not with the ancient Greeks but as recently as the previous century.[60] Although this principle is "very simple and very general," two requirements for a fundamental principle, it is not sufficiently evident.[61] Therefore it requires a proof, which prior to Lagrange had usually been given by reducing it in one way or another to one of the two other principles. Lagrange seeks to provide this proof in a more direct way, on the basis of the 'principle of pulleys'.[62] Evidently he thought that the assumption made in this principle, that the weights in a pulley system will descend as far as possible, is evident!

However, the lack of evidence is more than outweighed by an important advantage of the principle: the possibility of translating it into a general mathematical formula with which all equilibrium problems can be solved.[63] Lagrange is even convinced that all possible general static principles are in essence nothing more than the same principle of virtual velocities.[64] This means that the principle of virtual velocities, while it may not be possible to regard it as the "primary principle" ("principe primitive") because of the lack of evidence for it, can at least be designated as the "general expression of the laws of equilibrium."[65] In basing statics on the principle of virtual velocities Lagrange is thus making a compromise: it has the advantage of being mathematically expressable, but it is less than ideal in terms of his pursuit of an *axiomatic foundation* for statics, and thus for all mechanics, in which the axioms would be *self evident truths*.[66]

The section of the *Méchanique analitique* that deals with dynamics is ordered in the same way as the part on statics. For his general laws, Lagrange introduces the law of inertia and the principle of composite motion, as d'Alembert had also done.[67] However, Lagrange's use of the concept of force makes it necessary to define the effect of force. He introduces two different measures, for accelerating force and for the force of impact. The measure is grounded in the effect produced, i.e., in the

---

[59] Lagrange, *Méchanique analitique*, 2-10, and 10-17 respectively.
[60] *Ibidem*, 17-23.
[61] *Ibidem*, 20, 19 and 21 respectively.
[62] This proof is based on a thought experiment with a pulley system. The suggestion is that the principle of virtual velocities can be based on what is in effect a generalised principle that the centre of gravity tends to the lowest point. The latter is called the 'principle of pulleys' here, and was later elaborated further. The arbitrariness of this suggestion can be seen from the fact that d'Alembert had earlier used the argument in the opposite direction: his principle was supposed to explain that the centre of gravity in a mechanical system tends to the lowest point. See *ibidem*, 21-22; Lagrange, "Sur le principe des vitesses virtuelles" (1797); d'Alembert, *Traité de dynamique*, 93-95.
[63] Lagrange, *Méchanique analitique*, 20.
[64] *Ibidem*, 20.
[65] "expression générale des lois de l'équilibre"*(ibidem*, 21).
[66] As was noted in section 7.3, something like this may have been what Lagrange meant, ironically, by his reference to 'true metaphysics'.
[67] For the way in which d'Alembert introduced these principles in his *Traité de dynamique* see section 4.4.

resulting speed and in the distance traveled in a given time, respectively.[68] Thus for accelerating force the measure is the ratio between speed and time, while the measure for the force of impact is the quantity of motion that it can produce.[69] Yet, although the concept of force plays a central role in the *Méchanique analitique*, what Lagrange is really concerned with is reducing mechanical interactions to analytical equations.[70] We must reduce forces, distances, durations and velocities to 'ordinary' mathematical quantities.[71] As he says:

> In mechanics, one must assume as known the simple effects of forces; the art of this science is nothing more than deducing the composite effects that must result from the combined and modified action of these same forces.[72]

The multivalence of the concept of force therefore presents no real problems. The formulas simply have to be adapted sometimes, according to whether one is dealing with an accelerating force or a force of impact.

In the second part, the principle of virtual velocities is extended, with a maneuver, to become a principle of dynamics:

> (...) to apply the formula for a system's equilibrium to the movement of a system of bodies, one needs only to introduce forces that result from the variations in the movement of each body, and which must be destroyed.[73]

The method, which bases dynamics on the conditions of static equilibrium, is analogous to d'Alembert's method, but Lagrange uses the concept of the destruction of *forces* rather than of *velocities*, which while it makes the method less direct, is also more practical.[74]

Immediately after formulating this general dynamic principle, and before applying it to dynamic problems, Lagrange shows that a number of *general characteristics of motion* can be derived from it. Of these, the conservation of living force and

---

[68] Lagrange, *Méchanique analitique*, 209.
[69] *Ibidem*, 210 and 213 respectively. In the previous section, article 4 *(ibidem*, 213) Lagrange says that the measure of forces of impact requires a new principle, because the forces are in this case unknown, but he then introduces a measure that is analogous to the measure of accelerating force. However, in this case the measure is not related to acceleration, but to imposed velocity ("mouvement imprimé").
[70] *Ibidem*, 208.
[71] *Ibidem*, 231. When Lagrange refers here to 'reducing' such magnitudes, he is pointing to the need to introduce a *measure* for the various magnitudes, by taking a known force, distance, etc. as a unit and relating the others to it, "(...) thus the forces, spaces, times, and speeds will be no more than simple ratios, that is, ordinary mathematical quantities" ("(...) les forces, les espaces, les temps, et les vitesses ne seront que des simples rapports, des quantités mathématiques ordinaires") *(ibidem)*.
[72] "(...) pour appliquer au mouvement d'un système de corps la formule de son équilibre, il suffira d'y introduire les forces qui proviennent des variations du mouvement de chaque corps, et qui doivent êtres détruites" *(ibidem*, 231).
[73] "Il faut, dans la Mécanique, prendre les effets simples des forces pour connus; et l'art de cette science consiste uniquement à en déduire les effets composés qui doivent résulter de l'action combinée et modifiée des mêmes forces" *(ibidem*, 224-225).
[74] *Ibidem*, 224. For d'Alembert's treatment of this transition see section 4.4, especially page 126. Despite Lagrange's reformulation of the principle, it is known as 'd'Alembert's principle'.

the principle of least action are interesting here.[75] Providing that the constraints[76] are not dependant on time and $Pdp + Qdq + Rdr + ...$ is a *total differential*—as Lagrange adds, the latter condition is always met in the case of central forces—living force is conserved and action is minimal.[77]

These derivations are not intended as proofs of the *truth* of the principle of virtual velocities or of d'Alembert's principle, but for its *universality*. The issue in question is to determine the relationship of the various laws, characteristics and principles to one another: it is only by actually deriving various sorts of more limited mechanical principles from a single principle that the latter is shown to be more universal.

It is precisely the desire to find axiomatic truths that must have motivated Lagrange to replace the principle of least action with the principle of virtual velocities. One has to add the conservation of living force to the principle of least action as an extra condition, to make it suitable for dynamics, whereas both the conservation of living force and the principle of least action can be *derived* from the principle of virtual velocities. Moreover, the principle of virtual velocities is, with a maneuver, valid for both statics and dynamics, but the two other principles are limited to dynamics.[78]

What has been said here shows that Lagrange sought a unified, axiomatic and deductive mechanics in both his early and in his late work. Therefore the reduction of the principle of least action from the universal key to a derivative theorem does not indicate his conversion from the school of Euler and Maupertuis to that of d'Alembert, but rather of the universalizing ambitions that always marked his quest.[79]

7.5 CONCLUSION: FORCE AND STRUCTURE IN LAGRANGE'S THOUGHT

The previous section has briefly surveyed the meaning of the analytic character of Lagrange's *Méchanique analitique*. I have emphasized its logical structure and paid little attention to the contents of the fundamental concepts of 'force', 'space', 'time' and 'mass'. Lagrange says little about these anyway. Pulte correctly remarks that

---

[75] *Ibidem*, 267-274 and 274-281 respectively.

[76] French: 'disposition'; German: 'Zwangsbedingung'. An example of a constraint is the connection of two masses by a weightless rod of length l. We can express this mathematically as the condition that for the coordinates of the positions of mass 1 and mass 2, $x1$ and $x2$ respectively, $(x1-x2)2 = l2$.

[77] *Ibidem*, 268. The formula $Pdp + Qdq + Rdr + ...$ represents the sum of the products of forces P, Q, R... with displacements dp, dq, dr... in the direction of the forces. Lagrange calls this the "intégrable," rather than the total differential, which amounts to the same thing: the integration here is independent of the path, and so depends only on the beginning and end points.

[78] For the different ideas on both heads Lagrange still had in 1760, when the principle of least action promised to be the key for all mechanics, see note 27.

[79] For a contrary opinion, see Fraser, "Lagrange's Early contributions," 234. Fraser interprets the change in status from universal to derivative as a rejection of the metaphysical status of the principle of least action. D'Alembert's influence would then be very plausible. In my interpretation, Lagrange was only concerned with universality, and for that, he did not need d'Alembert.

Lagrange's work lacks any reflection on the fundamental concepts and their application. His interpretation of this is that mathematics is a universal language for Lagrange, meaning that the problem of foundations needs not be addressed. For example, the theme of the impenetrability of matter is not found in Lagrange's work, although it played a large role in the formulation of mechanical principles for the people who influenced him most, that is: Newton, d'Alembert and Euler. Therefore Lagrange's mechanics cannot be categorized as Leibnizian, Newtonian or Cartesian, simply because he does not provide such metaphysical reflections.[80]

Nevertheless, it would be going too far to characterize Lagrange's analytic mechanics as being, in metaphysical terms, 'mathematical instrumentalism'. Perhaps Lagrange himself thought that his analytic approach had freed mechanics from metaphysical premises, but he was considering only *explicit* premises, and not the premises that were *implicit* in his own analytic approach. This metaphysical 'blind spot' can be revealed, using the broader concept of metaphysics applied in this book, as the hypothetical *a priori* of reality and of our possibility to know it. This can be done by locating Lagrange's approach to the concept of force in the historical development outlined in previous chapters. It can then be seen that there is an important difference, as compared to the beginning of the eighteenth century, with regard to the metaphysical aspect of reality on which the mathematization of the concept of force is based.

We have seen that Lagrange regarded mechanics as an analytic science, which for him meant: an axiomatic and deductive system of purely mathematical statements. Lagrange draws his fundamental concept of the mutual relationships between the elements of matter from Newtonian force. Coming after the shift to a structural metaphysics, teleological or otherwise, by his direct teachers in mechanics (d'Alembert, Maupertuis and Euler), Lagrange might appear to mark a return to Newton's substantial concept of force. That is however only apparent. Like Newton, he regards force as a mathematical magnitude, and like Newton he rejects the idea that force requires an explanation before it may be used in a theory. Thus far, he does not differ from the eighteenth century Newtonians. But whereas for Newton and the Newtonians the idea of causality was always in the background, Lagrange has brushed it aside. To be more precise, and taking the example of gravity: where Newton placed the force of attraction, in ontological terms, somewhere in an indeterminate no man's land between being a phenomenon and being a mechanical or other type of ontological explanation, but did not deny the possibility or usefulness of seeking such explanations, for Lagrange an explanation has become entirely irrelevant.

It makes little difference here that Lagrange formally defines force as 'cause'.[81] In fact this reference to a causative character is a misplaced addition, a reminder of earlier times when force was still included in mechanical theories as a cause. But

---

[80] Pulte, *Das Prinzip der kleinsten Wirkung*, 236-238.
[81] Lagrange, *Méchanique analitique*, 1.

when Lagrange says that two forces are equal, he does not mean that two *causes* are equal, but two *effects*.[82] In Lagrange's mechanics, force is not a cause but an effect. The transformation from cause to effect, which d'Alembert made explicit (as shown in Chapter 4) has become implicit in Lagrange's mechanics. It is taken as self-evident.

Thus Lagrange incorporates the concept of force as effect in his mechanics. In his first principles, force is given a mathematical form, not by relating effect to cause, but by relating the effects to *one another*. This places the concept of force in a new context: it is 'effect', but is nevertheless also something that in an ontological sense is prior to effect. Thus the mathematization of force is no longer based on reality as substance, but as structure. Where the concept of force had previously been anchored to the ground by the weight of its substantial foundation, Lagrange allows it to fly by taking it as a structural concept.

This means that Lagrange's analytic mechanics is in fact founded on a metaphysics, one in which structure, and not substance, is the central concept. Lagrange himself did not recognize this change in the metaphysical foundation. He undertook the search for principles on analytic grounds, and did not realize that his emphasis on principles entails a transition to the structural aspect of reality.[83]

One could question whether this shift should be attributed to Lagrange. Is it not inherent in the mathematical approach itself? This could be argued on a philosophical level, but historically speaking it is clear that a mathematical approach can continue for a long time in parallel with assumptions about substances as efficient causes. While it had been recognized since the beginning of the eighteenth century that forces cannot be observed by the senses and must therefore be derived from their effects, in thinking about physics the concept of force nevertheless continued to be contrasted to phenomena in the same way as the efficient cause relates to its effects. Therefore it also continued to provide the connection between various substances, and metaphysical understandings about substances were applicable to the effect of force. While d'Alembert had abandoned the position that principles concerning effects can be derived from our understanding of substances and efficient causes, his concept of force was still linked to the old way of thinking—not so much by his terminology, but rather because he tries *to reduce* force to its effects. Lagrange however has no need at all to clarify the relationship between cause and effect. In his case, reflection on this point is absent yet the reader does not feel its absence. It is only with Lagrange that the emphasis moves entirely to the *structure of the effects*, and force itself is concealed in its *appearance* as effect.[84]

---

[82] See pages 221-222.
[83] This could be the reason for Duhem's agreement with Fourier's accolade to Lagrange's work as a "philosophical mechanics" (Duhem, *L'évolution de la mécanique*, 23).
[84] The evolution of mathematics itself is not considered here. Mathematics was originally qualitative in nature. In Pythagorean number theory, ratios were associated with psychic relationships such as harmony and jealousy. Therefore the Pythagorean motto, 'everything is number', is not simply the reduction of reality to quantity. With this in mind, one might, as Lanczos suggests, draw a parallel

continued on next page

This also puts Lagrange's relationship to the teleology of Maupertuis and Euler in a different light. It seemed at first that his rejection of their teleological views entailed a rejection of their metaphysical program. However, as we have seen in Chapter 5, Maupertuis and Euler had two aims in mind with their program, the transition from a physics based on efficient causes to one based on structure, and the determination of that structure on the basis of teleological considerations. The first of these was concealed under the second, and when Lagrange rejected the second, he thought he had to reject the idea of metaphysics as a whole. Thus the irony of Lagrange's *Méchanique analytique* is that, although Lagrange himself rejected metaphysical speculations, this work completed a metaphysical transformation of mechanics, and so gave Newton's mechanics of force a new, structural, foundation.

---

continuation of former page
between the transition from geometry to analysis in eighteenth century mathematics and the transition in ideas about force (see Lanczos, *The Variational Principles*).

CHAPTER 8

METAPHYSICS CONCEALED

> The phrases of last century in this
> Linger to play tricks –
> Vis Viva and Vis Mortua and Vis
> Acceleratrix: –
> Those long-nebbed words that to our text
>   books still
> Cling by their titles,
> And from them creep, as entezoa will,
> Into our vitals.
> But see! Tait writes in lucid symbols clear
> One small equation
> And Force becomes of Energy a mere
> Space-variation.
> James Clerk Maxwell[1]

8.1   EXPOSITION: A PARADOX IN THE HISTORY OF SCIENCE

This book began with an expression of astonishment. At the beginning of the nineteenth century, and possibly earlier, a certain distinction between natural philosophy and natural science was unmistakable. The contemptuous reactions of Goethe's contemporaries to his theory of color and the growing opposition that Schelling's and Hegel's *Naturphilosophie* met with in Germany after 1830 were symptoms of the autonomy that natural science had achieved, which entailed that research in natural science should be evaluated using inherent norms.[2] This growing separation was already apparent in mechanics, as has been shown in the last two chapters. At one extreme we have the essay competition concerning the foundation of force, which the speculative philosophy department of the Berlin Academy made so bold as to announce in 1779, and which was already the target of merciless derision at that time, from most of the practitioners of mechanics. On the other extreme stands

---

[1] From the poem "Report on Tait's Lecture on Force" (1876). In: L. Campbell and W. Garnett: *The Life of Sir James Clerk Maxwell*. London 1882 (reprint New York: Johnson Reprint 1969), 346-348.
[2] See Snelders, *Wetenschap en intuïtie* (1994).

Lagrange's classical work on analytic mechanics, which was marked by the absence of considerations from natural philosophy.

This divorce between natural philosophy and natural science would have been unthinkable a century earlier. Although Newton's *Principia mathematica* was a very important work in the development of natural science, and especially of mechanics—it is even frequently regarded as the completion of the scientific revolution—nevertheless, when read with modern eyes, both it and the rest of the seventeenth century works of natural science could equally well be classed as natural philosophy or natural science. It is significant that natural science was originally a branch of natural philosophy—in England the name 'natural philosophy' still reflects this origin.

The astonishment that was referred to arises from the fact that this differentiation can hardly be explained on the basis of the position that historians have, until recently, allocated to the science of the century following Newton's *Principia mathematica*. If we consider only the history of mechanics, we can see a conflation in which later developments are projected backwards, so that the seventeenth century became, metaphorically, rather overweight, and there was little left to be done in the eighteenth century except to elaborate on what had in principle already been determined. Other developments in eighteenth century science were therefore measured in comparison with the great masters of the seventeenth century, and could only be considered as support for, or reaction against, the process of realizing the potential of their thinking.

Against this view there has been an impression—growing in strength in the course of this study—that the eighteenth century had a much greater significance for the development of mechanics and certainly for the definition of its proper domain. The fact that the practitioners of mechanics were also very much involved in a number of metaphysical controversies in the period between Newton and Lagrange, and that their scientific work was sometimes influenced by these controversies, suggests that the methodological limitation of mechanics that was later to be regarded as self-evident was only accepted, and indeed developed, in the course of the eighteenth century.

8.2   Development: The Significance of Metaphysical Premises

The argument that has been developed in this book embodies a historical approach that is, in one respect, new. This was scarcely mentioned in the introductory chapter because that would assume too much of what was still to come. Moreover, historical methodology is best worked out in practice. I would like to return here to the internal structure of my approach to define my position more precisely in comparison to the historiographical tradition. This will also give me the chance to show that my goal has been not so much to treat my theme exhaustively, as to approach it in a way that is partly new. This approach can be seen as an accumulation of four sequential specifications for reevaluating eighteenth century mechanics.

First specification. The development of mechanics in the eighteenth century should be given more attention, and should not be regarded only as the formalization of Newton's mechanics. This point was discussed in detail in the introductory chapter. The historiographic tradition has maintained the latter position—'just formalization'—throughout the period between Lagrange's *Méchanique analitique* and the 1950s. Truesdell may be regarded as an important revisionist, because he showed clearly that it was not only the *formalization* of the available *mathematical* principles that remained to be completed after Newton, but also the *mathematization* of *mechanical* principles.

Second specification. The development of mechanics in the era after Newton is a continuation of the Scientific Revolution of the seventeenth century, i.e., the Scientific Revolution was not completed—not even in principle—by Newton's *Principia mathematica*. Questions concerning the foundation and true character of mechanics were still being intensively discussed in the eighteenth century, showing clearly that not only the mathematization, but also the foundation and definition of mechanical principles were still open questions.

Third specification. Eighteenth century debates about mechanical principles are closely linked to theological and metaphysical controversies, such as the question of human freedom. A practitioner of mechanics—a category that lacks its own name—considered himself to be simultaneously a natural philosopher. Mechanical controversies and developments must therefore be studied in relation to these other controversies and developments.

Fourth specification. The series of disputes in mechanics in the course of the eighteenth century reflects, *in itself,* a metaphysical development, a development in fact of the metaphysical basis of mechanics. The *vis viva* dispute and the early years of the principle of least action cannot be entirely explained historically unless we also consider developments in metaphysics. Only at this level is a sensible interpretation of the change in the meaning of living force and the principle of least action possible.

What is new about my approach can be found mainly in the fourth specification. The relationships referred to in the third specification have been given an important place in recent historiography, to the extent in some cases that changes in actual contents are treated as subordinate to their context. However, in my approach the aim is not to compare and evaluate mutual influences, but to make the simultaneous presence of various levels in the development of mechanics visible. Mathematics and experiment alone do not constitute the heart of mechanics, they are combined with generally self-evident opinions that are metaphysical in origin. The invisibility of these opinions is no reason to suppose that they are not important. On the contrary, only by making them visible can we achieve a historical understanding of the power and significance of mechanics. This also provides us with the test of the approach: the extent to which making the metaphysical premises of a scientific discipline visible also provides a better insight into its historical development.

8.3 CLIMAX: THE CONCEPT OF MECHANICAL FORCE BETWEEN SUBSTANCE AND STRUCTURE

'Mechanics' is, by the broadest definition, the science of the motions of matter and the interaction of matter. However, this is not the end of the story. In addition to differences in the type and sophistication of the mathematics that is used, differences in the choice of fundamental principles and concepts, and differences in the way the conclusions of mechanics are tested against reality, there are also differences in which *aspect of reality* is considered to be the proper object of mechanics. Few, apart from some obdurate positivists, would dispute the idea that the practice of mechanics entails the abstraction of particular facets of reality. Nobody nowadays would equate matter with extended substance, as Descartes thought he could.[3] If anyone still wanted to consider extension as the primary concept in mechanics, it would be labeled as an *abstraction* from matter. Abstraction amounts to the *distillation of one aspect* or, formulated subjectively, the view from a particular perspective.

The above does not imply any position on how the aspect in question relates to the essence of reality, it addresses the question of what a particular science allows us to know about reality. Thus it speaks not so much about reality itself as about the science concerned in its orientation to reality: its intentionality. What I have called metaphysics in this book is based on this intentionality: it is the distillation of the aspect of reality, chosen *a priori*, to which mechanics relates.

The key distinction in the historical connection between metaphysics and mechanics, as I have analyzed it here, is based on a difference as to the *object* of mechanics—substance versus structure—and on a difference in how ideas about that object are incorporated in mechanics—as explicit foundation or as implicit premise. The grounding of the concept of force in a metaphysics of substance amounts to saying that the ground of the separate forces is in some way located in the separate material substances, because force is regarded either as a characteristic of separate substances, or a modification of substance, or something that is analogous to substances. The scholastic concept of impetus and the Leibnizian concept of living force are based on just such a ground. If the concept of force is grounded in a structural metaphysics, this implies that the foundation of force lies in the structure of matter, i.e., in the system of external relationships between material elements. In modern mechanics the basis of these relationships consists of space and time, while its elements are point masses, the residue of the earlier substances.

Let us review the whole foregoing argument once more, but formulated in a more general and emphatic form, showing the meaning of the two antitheses more clearly and briefly than was possible in the separate chapters.

It is generally accepted, in the historiography of the Scientific Revolution, that Aristotelian-scholastic mechanics focused on the substantial aspect of reality and

---

[3] Descartes, *Principes de la philosophie* II.4: "la nature de la matiere (...) consiste (...) seulement en ce qu'il est une substance (...) qui a de l'extension."

explicitly based their mechanical conclusions on it. In the seventeenth century there was a shift, related to the mathematization of people's picture of the world, to focus on the structural aspect of reality. For example, Galileo provisionally bracketed out the question of *why* a body falls, because he thought that that question could only usefully be posed *after* one had some understanding of the *how* of the fall. However, the second chapter showed that the question of the *cause* of changes in the motion of matter, especially in collisions, nevertheless continued to be closely linked to the old considerations in terms of substance. Although Descartes based his laws of motion on a structural principal—his *Principia philosophiae* uses the metaphysical premise that God always maintains the same quantity of motion (that is: the scalar mv), so that whatever motion one body loses must at the same time reappear in other bodies—yet this principle is accompanied by considerations based on ideas about the substances themselves. Matter is in the first place extension, but since this cannot be used to explain what happens in collisions, Descartes feels forced to add other, qualitative, characteristics. Descartes's third and subsequent laws of motion, that relate to collisions between hard particles, are based on the belief that a heavier body always has more force than a lighter one, so a lighter particle will never be able to move a heavier one.

By the end of the seventeenth century this sort of reasoning was usually not made explicit, but it was still implicitly present. This can be seen, for example, in Newton's concept of force, as I have illustrated from his observations about inertial force and from his reduction of continuously acting forces to instantaneously acting forces. In his treatment of the transfer of motion between bodies it appears that he presumed that force is itself transferred from one body to another during the action of a force. The way in which this transfer was supposed to occur can be compared to the transfer of an infinitely divisible material, such as water. So for Newton, force is analogous to an infinitely divisible material substance. Another illustration of this is that the question of whether it is sensible to aggregate or integrate force over space and time was shown to hinge on the assumption that such an aggregate or integration could also be regarded as a quantity of force.

In Leibniz's mechanics we even found mechanical force being explicitly grounded in a metaphysics of substance: "it is a characteristic of substances to act." At the same time Leibniz realized that this primary concept of force had to be linked to the structure of matter. He did this by distinguishing between a real and a phenomenal aspect in relation to matter and force. The real, metaphysical aspect corresponds to what I have called the substantial aspect, while the phenomenal aspect corresponds to the structural aspect. His famous principle, "the cause is equal to the effect," was important in connecting these two levels. His attempt can therefore be seen as an admittedly ingenious but nevertheless doomed attempt to synthesize *structure,* the primary object of classical mechanics, and *substance,* which was the primary object of Aristotelian-scholastic metaphysics. This attempt to create a synthesis also had implications for his work in the field of mechanics, and more so than Leibniz himself realized: his distinction between dead and living force was a direct consequence of grounding the concept of mechanical force in this metaphysics of substance. The substantial grounding of the concept of force was also expressed in his thought in the qualitative equation of force itself with its integral over time or over distance. All three are on the same level, and so had the same quality from a metaphysical point of view.

Thus substantial roots of the concept of force were not left behind as outmoded during the transition from the seventeenth to the eighteenth century. It may well be correct to say that few, except for those carrying on the Aristotelian-scholastic tradition, such as the Jesuits, continued to explicitly base their arguments on substantial foundations. Nevertheless, as the example of Newton shows, this did not at all prevent the substantial foundation, through implicit premises, from continuing to have a decisive influence on the mathematization of the concept of force.

The substantial concept of force was one of the keys to the eighteenth century controversy about the true measure of living force. This dispute involved almost everyone who considered himself a scholar in the period up to about 1750, but after that time gradually declined in status and was finally derided as a mere dispute about words. This development can be better understood if we note that the problem arose from asking how forces, each of which was thought to be separately linked with a substance, are to be related to changes in motion, which relate to simultaneous interrelations between more than one substance.

The way d'Alembert solved the question, by transforming the concept of force from a cause to an effect, also entailed a shift in the concept of force, from a substantial to a structural concept. In fact shifting the concept of force from the level of substances to that of space-time relationships meant that the *metaphysical causes* of changes in motion were excluded from the field of mechanics. D'Alembert spoke instead about mechanical causes. However, these are not proper causes, but 'effects'. But since cause and effect are relative concepts (the one cannot exist without the other), the original sense of the concept 'cause' disappears from view along with the original sense of the concept 'effect'. There is no sense in calling something an effect, if we cannot say anything about the cause. So rather than speaking of 'effect', it is more appropriate to speak of a 'space-time structure'. With this transition, the concept of mechanical force has achieved its own form, distinct from the general concept of force.

His denial of the substantial ground of the concept of force led d'Alembert to conclude that force could only have a derivative function. Force is purely a mathematical magnitude that can be used to express the relationships between material elements in space and time in a simple way. The laws of the conservation of living force (mv^2) and of quantity of motion (mv) were, for him, nothing more than a consequence of the principles of mechanics. However, the fact that d'Alembert did not want to call them metaphysical principles means only that they were not based on the equality of the substantial cause and the empirical effect. It does not mean that they could not be necessarily derived from metaphysical principles of space and time. In fact d'Alembert based mechanical principles on metaphysical understandings of the structure of space and time. Therefore his instrumentalist view of force is not a question of principle but the consequence of a restrictive approach to the reality of mathematical magnitudes. For him, only the fundamental concepts, and not the mathematical functions derived from them, are real.

The transition from substance to structure as a basis for the concept of mechanical force, which for d'Alembert took the form of a transition from cause to effect, also opened up the possibility of a new, explicitly metaphysical, approach to the concept of force. As was discussed in Chapter 5, the early history of the principle of least action can be understood as an exploration of the nature of such an approach based on a metaphysics of structure. Maupertuis tried to do this by analogy to optical laws, in which he combined the *physical analogy* of light and mechanics that Newton had assumed with Fermat's supposed *harmony* of force and purpose. Euler too was convinced of this harmony. He tried to use it as a basis for developing an *indirect* way of calculating effects on the basis of minimal principles, as an alternative to the *direct* calculation of the effects on the basis of forces. The concept of action related to force as *acting*, rather than considering force as a *cause*, and in this it precisely matched d'Alembert's transformation of the concept of force. According to this principle, the effects do not follow from material substances but from a material structure. Thus Maupertuis's and Euler's work on the principle of least action was an attempt to design another type of metaphysics and apply it explicitly in mechanics. The fact that this metaphysics was a teleological one appeared to be of secondary importance. Their work on the principle of least action was primarily an attempt to ground mechanics explicitly in structure, rather than in substance.

8.4 Resolution: Separation and Transformation around 1780

D'Alembert, Maupertuis and Euler appeared to be practitioners of mechanics who were consciously seeking an alternative for its traditional foundation, especially for the foundation of the concept of mechanical force. Their attempts all explored structure as the foundation of the laws of motion. However, in the latter part of the century the mathematical approach seemed to be sufficient in itself. The development of analytic mechanics by Euler and especially Lagrange seemed to have made such reflections superfluous. Metaphysical approaches were for many people no more than a target for satire. Nevertheless, if we consider the mathematical foundation of analytic mechanics more closely, it is possible to see a hidden metaphysical foundation in which the transition from substance to structure is clearly recognizable.

The essays for the Berlin competition for 1779 show that this metaphysical foundation was no longer explicitly provided. The essays lead one to suspect that, around 1780, the opinions about the essence of force current among philosophers who were not specialists in mechanics had not been much influenced by the developments in mechanics. The essays that have been considered quite consistently adopt the suggestions of the academy as self-evident, and all seek the foundation of mechanical force in substances, in one way or another. Thus the concept of mechanical force was still seen following naturally from the earlier general concept of force. Although the level of the essays is generally not high (for example, mechanics was treated in merely geometric terms), the competition does provide an illustration of the increasing separation of metaphysics and mechanics as disciplines.

Lagrange's *Méchanique analitique*, in contrast, shows that the concept of mechanical force had taken on an entirely different character in the eighteenth century. Although Lagrange presented the book as a formalization of mechanics expressed in purely analytic mathematics, and he generally did not deal with metaphysical questions, there can be no doubt that his concept of force differs in metaphysical terms from that held at the beginning of the century. While the form had remained the same, to the extent that terms such as the *mutual equilibrium* and the *action* of forces were still being used, the content had been transformed. Force was no longer seen as simultaneously cause and effect. In other words: the effect was no longer the *transfer of a cause of motion*, and thus was no longer linked to force *as cause*. Effects are now *transformations* of other effects. While the term 'force' was still used, it refers now to an effect that is prior to another effect, and so is still formally called a *cause*, although it has actually become a *relationship between the elements* of a mechanical system. Force has become a structural concept, because it links masses in space and time.

But even as the object of the quest for the substantial character of force vanished from sight, another quest was put in its place: what is the structure of matter? The foundation of mechanics was also no longer concerned with the relationship between the effects and their substantial cause, but rather with the relationship between phenomena and the structure of mass, space and time. One consequence is that force can now be expressed as a function of mass, space and time, and even as the gradient of a function of potential, so that it almost disappears from view.

Thus the relation between mechanics and metaphysics in the era after Newton appears to have been a dual one. On an explicit level a rejection of the traditional metaphysical approach to the foundations of mechanics gradually (and, from our modern point of view, rather belatedly) took place. This was accompanied by a transition toward a metaphysics of structure that appeared to a range of scholars (sometimes implicitly, sometimes explicitly so) to fit better the new, rational mechanics. Thus at the implicit level what had happened was not a separation but a *transformation*.

8.5   Epilogue: History and Metaphysics in Modern Natural Science

From a long perspective, one can observe a general trend in natural science since the days of Galileo and Kepler. There is a tendency to marginalize metaphysics, leaving no place for metaphysical questions except in the realm of faith and subjectivity. In its place, mathematics became the form of thinking in which the practitioners of natural science feel most at home, the form that governs both experience and practice. That at least is how Dijksterhuis, for example, summed up the character of natural science after the seventeenth century.[4] It is also confirmed by a well-known

---

[4]   Dijksterhuis, *Mechanization*, 500-501.

saying of Kant: "I assert that the proportion of true science contained in each branch of natural philosophy is commensurate to the amount of mathematics it contains."[5] In that line of thought, it is an easy step to a strict separation between metaphysics and the natural sciences.

But the progress that has accompanied the mathematization of natural science in the West is also open to question. It is undeniable that metaphysics was often a fruitful source of inspiration for natural science, as can be seen from what has been said in this book about the concept of living force and the principle of least action. The same could also be said for Kepler's divine harmony, and about mathematical order in the thought of Stephen Hawking. Yet in recounting its own history, natural science has consistently disavowed the old metaphysical formulations and retained only the mathematical formulas. However, this disavowal also means that it has disavowed its own source of inspiration.

The difficult, often hostile relationship between physics and metaphysics has had a negative effect on physicists' image of themselves. All too often they have put forward indefensible simplifications of the proper nature of physics and its object. This has not necessary been done deliberately. The mathematization of a problem necessarily removes any aspects that cannot be expressed mathematically from sight. This makes it no easy task to make a historical or philosophical study of the adventure we know as 'science'. We cannot always simply accept what physicists *say they do:* sometimes one must include what they *do not say* they do, what they say they *do not do*, and even what they *do not say they do not do*.

One has also to ask whether this antithesis between physics and metaphysics might be more apparent than real; whether, in modern physics, mathematics and metaphysics might not be the same thing. In other words: is mathematics the metaphysics of modern physics? This would solve the whole issue of interpreting physics in a single blow, because we would then understand a problem completely once we have expressed it in mathematical formulas. But it quickly becomes clear that this identity cannot be maintained, once we realize that mathematics says nothing about physical existence. In mathematics, something exists if it entails no contradictions. But in physics we distinguish between purely mathematical quantities and physical existents. We have seen how large the differences were between the ways in which Newton and d'Alembert answered the question of the reality of force, in a physical sense, even though force had almost the same meaning for them in a mathematical sense.

It is also far from clear that the mathematical form of modern physics unambiguously determines a specific type of structural metaphysics as its basis. The structural metaphysics that was concealed in the analytic mechanics of the late eighteenth century is not necessarily required for other mathematical sciences, and is also not required for the further development of mechanics. A metaphysics of substance

---

[5] "Ich behaupte aber, daß in jeder besonderen Naturlehre nur so viel eigentliche Wissenschaft angetroffen werden könne, als darin Mathematik anzutreffen ist" (Kant, *Metaphysische Anfangsgründe*, 14 (A VIII)).

could also be paired with a mathematical form. In the late eighteenth century, for instance, the idea arose that a variety of qualities such as heat and electricity lay concealed in an equal variety of substances.[6] Lorentz, in a dispute with Einstein, defended the need for ether as a basis for the electromagnetic field.[7] And in modern quantum mechanics we have the theory of elementary particles, such as photons, quarks, leptons, baryons, mesons, etc.

It would be out of place to paint ideas of this sort as simply illegitimate, as if they derived only from a need for simplified conceptions.[8] Rather we would be well advised to consider them as something like a dialectic pair, in the sense that any consistent interpretation in terms of one form leads us to the other. A structure cannot exist without any underlying substance, and substances together comprise a system with a structure. This dialectic pairing might to some extent underlie a recurring phenomenon in the history of physics: the postulation and then rejection of one new substance after the other. If we suppose that structural metaphysics is the basis on which modern physics is built, it is a consequence that it spontaneously generates its complement, in the form of substances that carry the structure. Seen in this way, elementary particles, energy and ether all flow from the abstractness of the structural aspect of physics.

It is then interesting to ask what difference there is between such substances in modern physics and the substances in a scholastic or mechanistic physics that is grounded in a metaphysics of substance. A comparison between the nineteenth-century concept of energy and the Leibnizian concept of living force provides a good illustration.

In the nineteenth century, energy was usually regarded as a structural property: it was described as a *quantity that is conserved,* that is, a quantity that can be found in various *equivalent* forms, and whose total magnitude is a constant. A few people concluded from this immutability that energy is a substance. But what sort of substance? Its relationship to structure is completely different than in Leibniz's thought. For Leibniz, substance constituted the individual external ground of structure, and living force was the phenomenal expression of substantial force. Living force belongs first to the material elements, and second to the resulting system. In the nineteenth century, substance is the common inward basis of structure: energy is the substantial unity that embraces the multiplicity of its differentiated physical forms. Energy belongs first to the material system, and second to its elements.

The quest for the metaphysics that underlies a particular science cannot be resolved by pointing to the extent to which it has been mathematized. It requires an approach that is both historical and philosophical. Only a combined approach can deal with the fact that every scientific theory stands at the juncture between eternity and time, and of universality and fragmentation. In this book I have sought to provide a preliminary response to the question, as it applies to mechanics, using a

---

[6] Heimann and McGuire, "Newtonian Forces and Lockean Powers" (1971).
[7] See Berkson, *Fields of Force* (1974), 256-274; Nersessian, *Faraday to Einstein* (1984).
[8] See Bosch *et al., Substantie* (1966); Hutchison, "Dormitive Virtues" (1991).

philosophical approach. The converse approach, which would be a philosophical examination of the metaphysical nature of modern physics, will also not get very far unless it is supplemented with a historical approach. By combining the two we can avoid treating science in the abstract, either as purely the accumulation of universal knowledge or as merely the sum total of opinions at a given moment. Where we can avoid such abstractions, our understanding of the metaphysical foundation of a particular science will prove much more concrete than many are inclined to think: what science was about all along, suddenly becomes clear.

This brings us back to the first pages of this book. The question of what significance modern physics may have for problems that lie *outside* its own proper domain cannot be answered without first discovering what foundation underlies its findings *within* its own proper domain. And just as metaphysics seemed to be unable to come to grips with analytic mechanics, yet continued to exist concealed within it, so we will find that, whatever our opinions may be regarding the explicit meaning of modern physics in non-physical questions, physics has become part of our way of thinking, even where it is, like metaphysics, concealed.

BIBLIOGRAPHY

ABBREVIATIONS

Akademiearchiv	Archive of the Academy of the Berlin-Brandenburg Academy of Sciences and Humanities.
Comment. Petrop.	Commentarii academiae scientiarum imperialis Petropolitanae
DSB	Gillispie ed., *Dictionary of Scientific Biography*.
EOO	*Leonhardi Euleri opera omnia*.
En.x	Eneström x = Index of Euler's papers according to Eneström, *Verzeichnis der Schriften Leonhard Eulers*.
Gerh. *Math. Schr.*	[G.W. Leibniz], *Leibnizens mathematische Schriften*. C.I. Gerhardt ed.
Gerh. *Phil. Schr.*	[G.W. Leibniz], *Die philosophische Schriften von G.W. Leibniz*. C.I. Gerhardt ed.
Histoire de Berlin	*Histoire de l'Académie Royale des Sciences et Belles-Lettres [de Berlin]*. Including the presented *Mémoires* of the relevant year.
Mémoires de Paris	*Mémoires de l'Académie Royale des Sciences et Belles-Lettres [de Paris]*.
Ms.	Manuscript.
Rav.x	Index according to the Leibniz-bibliography of Ravier, *Bibliographie des œvres de Leibniz* (1937).

UNPUBLISHED SOURCE MATERIALS

The sources below can be found in the *Archive of the Academy of the Berlin-Brandenburg Academy of Sciences and Humanities*.

1. Régistres de l'Académie (I-IV-32). The part concerning the period 1746–1766 is printed in *Die Registres der Berliner Akademie* (E. Winter ed.).
2. Findbuch Preisschriften (D1/14). Subtitle: Die an die Akademie eingesandten Preisarbeiten 1745–1939; Verzeichnis der noch vorhandenen Manuskripte. Lore Ulbricht ed. Berlin 1956, verified and improved by R. Faber (without year).
3. Prize essays. Coded with 'I–M' and a sequential number. The essays for the competition for the year 1779 have the numbers 722 up to and including 736.

PUBLICATIONS

Aarsleff, Hans, "The Berlin Academy under Frederick the Great." In: *History of the Human Sciences* 1989 (2), 193-206.

Abro, A. d', *The Rise of the New Physics. Its Mathematical and Physical Theories.* New York: Dover ^{2}1952 (^{1}1939 under the title *Decline of Mechanism*).

Aiton, Eric J., *Leibniz: a Biography.* Bristol [etc.], Hilger 1985.

d'Alembert, Jean le Rond dit, varying articles in: Diderot ed., *Encyclopédie*.
"Discours préliminaire des editeurs." In: *Encyclopédie* I (1751), i-xlv.
"Explication détaillée du systeme des connoissances humaines." In: *Encyclopédie* I (1751), xlvii-li.
"Accélération." In: *Encyclopédie* I (1751), 60-62.
"Action." In: *Encyclopédie* I (1751), 119-120.
"Attraction." In: *Encyclopédie* I (1751), 846-856.
"Causes finales." In: *Encyclopédie* II (1755), 789.
"Cause, en méchanique & en physique." In: *Encyclopédie* II (1755), 789-790.
"Communication." In: *Encyclopédie* III (1753), 727-729.
"Corps." In: *Encyclopédie* IV (1754), 261-263.
"Cosmologie." In: *Encyclopédie* IV (1754), 294-297.
"Dynamique." In: *Encyclopédie* V (1755), 174-176.
"Élémens des sciences." In: *Encyclopédie* V (1755), 491-498.
"Équilibre." In: *Encyclopédie* V (1755), 873-874.
"Force." In: *Encyclopédie* VII (1757), 110-120.
"Gravitation." In: *Encyclopédie* VII (1757), 871-873.
"Gravité." In: *Encyclopédie* VII (1757), 873- 876.
"Impulsion." In: *Encyclopédie* VIII (1765), 635.
"Matiere." In: *Encyclopédie* X (1765), 189-191.
"Méchanique." In: *Encyclopédie* X (1765), 222-[2]26.
"Nature." In: *Encyclopédie* XI (1765), 40-41.
"Percussion." In: *Encyclopédie* XII (1765), 330-335.
"Puissance." In: *Encyclopédie* XII (1765), 555-556.
"Statique." In: *Encyclopédie* XV (1765), 496-497.
"Uniforme." In: *Encyclopédie* XVII (1765), 381.

—, *Traité de dynamique* (Paris: David ^{2}1758 (^{1}1743)). Facsimile New York: Johnson Reprint 1968.

—, "Memoire historique sur la vie & les ouvrages de M. Jean Bernoulli." In: *Mercure de France* March 1748, 39-79. Partly reprinted in facsimile in Costabel, *Signification* (1984), 76-78.

d'Arcy, Sir Patrick, "Réflexions sur le principle de la moindre action de Mr. de Maupertuis." In: *Mémoires de Paris pour l'année 1749* (Paris 1753), 531-538.

—, "Réplique à un Mémoire de Mr. de Maupertuis sur le principle de la moindre action." In: *Mémoires de Paris pour l'année 1752* (Paris 1756), 503-519.

Ariès, Philippe, *Centuries of Childhood.* (Original title: *L'enfant et la vie familiale sous l'Ancien Régime.* Paris: Plon 1960) Revised paperback-edition 1973, Harmondsworth: Penguin 1986.

Aristotle, *Metaphysics.* I have used the German translation by E. Rolfes. Leipzig: Felix Meiner ^{3}1928.

—, *Aristotle's Physics.* Hippocrates G. Apostle ed. English translation. Grinnell, Iowa: The Peripatetic Press 1980 (^{1}1969)).

Arithmos-Arrythmos. Skizzen aus der Wissenschaftsgeschichte. In honour of J.O. Fleckenstein. München 1979.

Aquinas, St. Thomas, *Commentary on Aristotle's 'Physics'*. Richard J. Blackwell ed. translation. New York: Yale University Press 1963.

Barroso Filho, Wilton and Claude Comte, "La formalisation de la dynamique par Lagrange (1736–1813)." In: Roshdi Rashed ed., *Sciences à l'époque de la Révolution Française*. Paris: Blanchard 1988, 329-348.

Bauerreis, Heinrich, *Zur Geschichte des spezifischen Gewichtes in Altertum und Mittelalter*. Dissertation Friedrich-Alexander-Universität Erlangen. Erlangen: Junge & Sohn 1914.

Beckner, Morton, "Teleology." In: Edwards ed., *Encyclopaedia of Philosophy* VIII (New York [etc.]: MacMillan [etc.] 1972=1962), 88-91.

Béguelin, Nicolas, "Essai d'une conciliation de la métaphysique de Leibnitz avec la physique de Newton, d'où résulte l'explication des phénomenes les plus généraux et les plus intéressans de la nature." In: *Histoire de Berlin pour l'année 1766* (XXII, Berlin 1768), 365-380. Read 30 January 1766.

—, "De l'usage du principe de la raison suffisante dans les loix générales de la mécanique. Cinquiéme mémoire sur les principes métaphysiques." In: *Histoire de Berlin pour l'année 1768* (XXIV, Berlin 1770), 367-383. Read 8 September 1768.

—, "Conciliation des idées de Newton et de Leibnitz sur l'espace et le vuide." In: *Histoire de Berlin pour l'année 1769* (XXV, Berlin 1771), 344-360. Read 12 October 1769.

Berg, Jan Hendrik van den, *Metabletica van de materie I. Meetkundige beschouwingen*. Nijkerk: Callenbach 1968.

Berger, Herman H., "Substantie, een metafysische beschouwing." In: Bosch a.o., *Substantie* (1966), 43-66.

Berghuys, Johannes J.W., "Zelfstandigheid van de materie." In: Bosch a.o., *Substantie*, 27-43.

Bergson, Henri, *Essai sur les données immédiates de la conscience* (1889). In: *Œuvres*, 1-157.
—, *Œuvres; Édition du centenaire*. A. Robinet ed.; H. Gouhier, Introduction. Paris: Presses Universitaires de France 1959.

Berkson, William, *Fields of Force: The Development of a Worldview from Faraday to Einstein*. London: Routledge and Kegan Paul 1974.

Bernoulli, Daniel, "Examen principiorum mechanicae, et demonstrationes geometricae de compositione et resolutione virium." In: *Comment. Petrop. Februari 1726* (I, Petersburg 1728), 126-142. Reprinted in —, *Werke* III, 119-135.

—, "De variatione motuum a percussione excentrica." In: *Comment. Petrop. 1737* (IX, Petersburg 1744), 189-206. Reprinted in —, *Werke* III, 145-159.

—, "Commentationes de immutatione et extensione principii conservationis virium vivarum, quae pro motu corporum coelestium requiritur." In: *Comment. Petrop. 1738* (X, Petersburg 1747), 116-124. Reprinted in —, *Werke* III, 160-169.

—, "Remarques sur le principe de la conservation des forces vives pris dans un sens general." In: *Histoire de Berlin pour l'année 1748* (IV, Berlin 1750), 356-364. Read 16 May 1748. Reprinted in *Werke* III, 197-206.

—, *Die Werke von Daniel Bernoulli III. Mechanik*. David Speiser a.o. eds. Basel [etc.]: Birkhäuser 1987.

Bernoulli, Johann I, *Discours sur les lois de la communication du mouvement. Pièce qui a merité les Eloges de l'Académie des Sciences (1724–1726)*. Paris 1727. Partly reprinted in facsimile: Costabel, *Signification*, 81-82 and 111-123.

Bertrand, L., "Examen des Réflexions de M. le Chevalier d'Arcy sur le principle de la moindre action." In: *Histoire de Berlin pour l'année 1753* (IX, Berlin 1755), 310-320. Presented in Berlin, 3 October 1754.

Beth, Evert W., *Natuurphilosophie*. Noorduijn's Scientific Series; 30. Gorinchem: Noorduijn 1948.

Biermann, Kurt-R., "Aus der Geschichte Berliner mathematischer Preisaufgaben." In: *Wissenschaftliche Zeitschrift der Humboldt-Universität zu Berlin, mathematisch-naturwissenschaftliche Reihe*. 1964 (13), 185-198.

Black, Jeremy, *Eighteenth Century Europe 1700 – 1789*. MacMillan History of Europe. London: MacMillan 1992 (1990).

Bois-Reymond, Emil du, "Maupertuis." In: *Sitzungsberichte der Königlich Preussischen Akademie der Wissenschaften zu Berlin* 1892 (1), 393-442.

Bongie, Laurence L., [Introduction to Condillac, *Les Monades* (1747)]. Oxford: The Voltaire Foundation at the Taylor Institution 1980, 9-108. In his introduction, Bongie attributes the prize essay to Etienne Bonnot de Condillac.

Bos, Henk J.M., et al., eds., *Studies on Christiaan Huygens. Invited Papers from the Symposium on the Life and Work of Christiaan Huygens, Amsterdam, 22–25 August 1979*. Lisse: Swets & Zeitlinger 1980.

—, "Mathematics and Rational Mechanics." In: Rousseau and Porter eds., *The Ferment of Knowledge* (1980), 327-355.

Bosch, J.W., Johannes J.W. Berghuys, Herman H. Berger, *Substantie*. Utrecht and Antwerpen: Spectrum 1966.

Breidert, Wolfgang, "Leonhard Euler und die Philosophie." In: Burckhardt a.o. eds., *Leonhard Euler* (1983), 447-457.

Briggs, J. Morton, "D'Alembert: Philosophy and Mechanics in the 18th Century." In: *University of Colorado Studies, Series in History* 1964 (3), 38-56.

—, "Alembert, Jean le Rond d'." In: *DSB* I (1970), 10-117.

Briggs, J. Morton jr., [Commentary to Costabel, "Newton's and Leibniz' dynamics" (1967).] In: Palter ed., *The Annus Mirabilis* (1970), 117-119.

Brunet, Pierre, *Maupertuis I. Étude biographique*. Paris: Blanchard 1929.

—, *Maupertuis II. L'œuvre et sa place dans la pensée scientifique et philosophique du XVIIIe siècle*. Paris: Blanchard 1929.

—, *Étude historique sur le principe de la moindre action*. Actualités scientifique et industrielles; 693. Paris: Hermann 1938.

Buchdahl, Gerd, [Commentary to Herivel, "Newton's achievement" (1967).] In: Palter ed., *The Annus Mirabilis* (1970), 136-142.

—, "Explanation and Gravity." In: Teich and Young, *Changing Perspectives* (1973), 167-203.

Burckhardt, J.J., Emil A. Fellmann and W. Habicht eds., *Leonhard Euler 1707–1783; Beiträge zu Leben und Werk. Gedenkband des Kantons Basel-Stadt*. Basel: Birkhäuser 1983.

Burtt, Edwin Arthur, *The Metaphysical Foundations of Modern Physical Science*. London and Henley: Routledge and Kegan Paul 1972 (=^{2}1932) (^{1}1924).

Buschmann, Cornelia, "Der Monadenstreit um 'ein Prämium von 50 Ducaten'." In: *Spectrum, Monatzeitschrift der Akademie der Wissenschaften der DDR* 1987 nr.12, 30-31.

—, " 'Die elendste Schrift, die je gekrönet worden'." In: *Spectrum, Monatzeitschrift der Akademie der Wissenschaften der DDR* 1988 nr.9, 28-29.

—, "Wolffianismus in Berlin." In: Förster ed., *Aufklärung in Berlin* (1989), 73-101.

—, "Die philosophischen Preisfragen und Preisschriften der Berliner Akademie der Wissenschaften im 18. Jahrhundert." In: Förster ed., *Aufklärung in Berlin* (1989), 165-228.

Calinger, Ronald S., "Euler's 'Letters to a Princess of Germany' as an Expression of his Mature Scientific Outlook." In: *Archive for the History of Exact Sciences* 1975–1976 (15), 211-233.

Capra, Fritjof, *The Tao of Physics*. London: Fontana 1979 (1975).

Carathéodory, Constantin, "Einführung in Eulers Arbeiten über Variationsrechnung." In: *EOO* I.24 (Bern 1952), viii-lxii.

Casini, Paolo, "D'Alembert, l'économie des principes et la 'métaphysique des sciences'." In: Emery and Monzani, *Jean d'Alembert* (1989), 135-152.

Cassirer, Ernst, *Leibniz' System in seinen wissenschaftlichen Grundlagen*. Darmstadt: Wissenschaftliche Buchgesellschaft ²1962 (1900).

—, *Substanzbegriff und Funktionsbegriff: Untersuchungen über die Grundfragen der Erkenntniskritik*. Darmstadt: Wissenschaftliche Buchgesellschaft 1969 (=1910).

Catelan, Abbé de, "Courte remarque de M. l'abbé D.C., où l'on montre à Mr. G.G. Leibnits le paralogisme contenu dans l'objection précédente." [Criticism to Leibniz's "Brevis demonstratio."]. In: *Nouvelles de la République des Lettres* Sept. 1686, 999-1003. Reprinted in Gerh. *Phil.Schr.* III, 40-42.

Châtelet, Gabrielle-Emilie le Tonnelier de Breteuil, Marquise du, *Institutions physiques de madame la Marquise du Chastellet adressées à Mr. son fils*. Amsterdam ²1742 (¹1740). Reprint Hildesheim: Georg Olms 1988. (Christian Wolff, *Gesammelte Werke*. III. Abt. Materialien und Dokumente; 28).

Christie, J.R.R., "Aurora, Nemesis and Clio." In: *British Journal for the History of Science* 1993 (26), 391-405.

Clagett, Marshall, *The Science of Mechanics in the Middle Ages*. Madison 1959.

Clercq, Peter de, "The 's Gravesande Collection in the Museum Boerhaave, Leiden." In: *Annali di storia della scienza* 1988 (3), 127-137.

—, *At the Sign of the Oriental Lamp. The Musschenbroek Workshop in Leiden, 1660 – 1750*. Dissertation. Rotterdam: Eramus Publishing 1997.

Cochius, Leonhard, "Examen de la question: si toute succession doit renfermer un commencement?" In: *Nouveaux mémoires de l'Académie Royale des Sciences et Belles-Lettres [de Berlin]* 1773: 325-346.

Cohen, H. Floris, *The Scientific Revolution. A Historiographical Inquiry*. Chicago: University of Chicago Press 1994.

Cohen, I. Bernard, "Newton's Second law and the Concept of Force in the *Principia*." In: *The Texas Quarterly* 1967 (10, 3), 127-157. Strongly revised in: Palter ed., *The Annus Mirabilis* (1970), 143-185, there followed by McGuire, [Commentary].

—, *The Newtonian Revolution. With Illustrations of the Transformation of Scientific Ideas*. Cambridge [etc.]: Cambridge University Press 1980.

Cohen, I. Bernard ed., *The Conservation of Energy and the Principle of Least Action*. New York: Arno Press 1981.

Cohen, I. Bernard and Alexandre Koyré, "Newton & the Leibniz-Clarke Correspondence." In: *Archives internationales d'histoire des sciences* 1962 (15), 63-126.

—, [Notes to Isaac Newton, *Principia.*] Cambridge, Mass.: Harvard University Press 1972.

Comte, Auguste, *Cours de philosophie positive* I. *Philosophie première: leçons 1 à 45.* II. *Physique sociale: leçons 46 à 60.* Paris: Hermann 1975 (1830–1842).

[Condillac, Etienne Bonnot de], *Les monades.* Prize essay, published anonymously in Justi a.o., *Dissertation* (1748). Laurence L. Bongie ed. Oxford: The Voltaire Foundation at the Taylor Institution 1980.

Corr, Charles A., "Christian Wolff and Leibniz." In: *Journal of the History of Ideas* 1975 (36), 241-262.

Costabel, Pierre, *Leibniz and Dynamics. The Texts of 1692* (Translation by R.E.W. Maddison from: *Leibniz et la dynamique: Les textes de 1692.* Paris: Hermann 1960). London: Methuen 1973.

—, "Le *De viribus vivis* de R. Boscovich ou de la vertu des querelles de mots." In: *Archives internationales d'histoire des sciences* 1961 (14), 3-21.

—, "'s Gravesande et les forces vives ou des vicissitudes d'une expérience soi-disant cruciale." In: *Mélanges Alexandre Koyré* I. *L'aventure de la science.* Paris: Hermann 1964, 117-134.

—, "Newton's and Leibniz' Dynamics." (translation from the French by J. Morton Briggs jr.) In: *The Texas Quarterly* 1967 (10, 3). Reprinted in Palter ed., *The Annus Mirabilis* (1970), 109-116. In the latter followed by: Briggs jr., [Commentary].

—, "L'affaire Maupertuis-König et les 'questions de fait'." In: *Arithmos-Arrythmos* (1979), 29-48.

—, *La signification d'un débat sur trente ans (1728–1758): la question des forces vives.* Paris: Centre National de la Recherche Scientifique [etc.] 1984.

—, [Introduction to *Correspondance d'Euler avec Maupertuis.*] In: *EOO* IVa.6 (Basel: Birkhäuser 1986), 1-28.

—, "D'Alembert et la querelle des forces vives: leçons d'un examen critique." In: Emery and Monzani eds., *Jean d'Alembert* (1989), 377-393.

Delambre, Jean Baptiste Joseph, "Notice sur la vie et les ouvrages de M. le Comte J.-L. Lagrange." In: Lagrange, *Œuvres de Lagrange* I (Paris 1867), ix-li.

Dellian, Ed, "Newton, die Trägheitskraft und die absolute Bewegung." In: *Philosophia naturalis* 1989 (26), 192-201.

Descartes, René, *Principia philosophiae* (Amsterdam: Ludovicum Elzevirium 1644). Facsimile in: *Œuvres de Descartes* VIII-1.

Non-authorised translation and revision as:

—, *Les principes de la philosophie* (Paris: Henry le Grass 1647). Facsimile in: *Œuvres de Descartes* IX-2.

— *Œuvres de Descartes.* 11 Volumes. Charles Adam and Paul Tannery eds. Paris: Vrin 1964–1974 (1897–1913).

Dhombers, Jean and P. Radelet-de Grave, "Contingence et nécessité en méchanique. Etude de deux textes inédits de Jean d'Alembert." In: *Physis Nuova Serie* 1991 (28), 35-114.

Diderot, Denis, "Leïbnitzianisme, ou Philosophie de Leïbnitz." In: Diderot ed., *Encyclopédie* IX (1765), 369-379.

Diderot, Denis and (up to Volume VII) Jean le Rond dit d'Alembert eds., *Encyclopédie, ou dictionnaire raisonné des sciences, des arts et des métiers, par une société de gens de lettres.* 35 Volumes. Paris: Briasson [etc.] 1751–1780. Facsimile Stuttgart: Frommann 1966–1967.

Dijksterhuis, Eduard Jan, *Val en worp; een bijdrage tot de geschiedenis der mechanica van Aristoteles tot Newton*. (*The Motions of Free Fall and of Projectile Motion: A Contribution to the History of Mechanics from Aristotle to Newton.*) Groningen: Noordhoff 1924.

—, *The Mechanization of the World Picture*. Oxford: Oxford University Press 1961. (Translation of *De mechanisering van het wereldbeeld*. Amsterdam: Meulenhoff 1950.)

Dolby, R.G.A., "A Note on Dijksterhuis' Criticism of Newton's Axiomatization of Mechanics." In: *Isis* 1966 (57), 108-115

Dugas, René, "Le principe de la moindre action dans l'œuvre de Maupertuis." In: *La revue scientifique; revue rose illustrée* 1942 (80, 2), 51-59.

—, *Histoire de mécanique*. Neuchatel: Éditions du Griffons 1950. Duhem, Pierre, *L'évolution de la mécanique*. Paris: Hermann 1905.

École, Jean, "Des rapports de Wolff avec Leibniz dans le domain de la métaphysique." In: Heinekamp, *Beiträge* (1986), 88-96.

Ellis, Brian D., "Newton's Concept of Motive Force." In: *Journal of the History of Ideas* 1962 (23), 273-278.

Emery, Monique and Piere Monzani eds., *Jean d'Alembert, savant et philosophe: portrait à plusieurs voix*. Montreux: Gordon and Breach 1989.

Eneström, Gustaf, *Verzeichnis der Schriften Leonhard Eulers*. Jahresbericht der Deutschen Mathematiker-Vereinigung. Der Ergänzungsbände IV. Band. 1. Lieferung. Leipzig: Teubner 1910.

Engfer, Hans-Jürgen, "Teleologie und Kausalität bei Leibniz und Wolff. Die Umkehr der Begründungspflicht." In: Heinekamp ed., *Beiträge* (1986), 97-109.

Est-il utile de tromper le peuple? Concours de la classe de philosophie spéculative de l'Académie de Sciences et de Belles-Lettres de Berlin pour l'année 1780. Werner Krauss ed. Berlin: Akademie-Verlag 1966.

Euler, Leonhard, "De communicatione motus in collisione corporum sese non directe percutientum." In: *Comment. Petrop. pour l'année 1737* (IX, Petersburg 1744), 50-76. En.22. *EOO* II.8 (1964), 7-26.

—, *Methodus inveniendi lineas curvas maximi minimive proprietate gaudentes, sive solutio problematis isoperimetrici latissimo sensu accepti*. Lausanne and Genève: Bousquet 1744. En.65. *EOO* I.24 (Bern 1952).

—, "Additamentum I. De curvis elasticis." Annex to *Methodus inveniendi* (1744). *EOO* I.24 (Bern 1952), 231-297.

—, "Additamentum II. De motu proiectorum in medio non resistente, per methodum maximorum ac minimorum determinando." Annex to *Methodus inveniendi* (1744). *EOO* I.24 (Bern 1952), 298-308.

—, "De la force de percussion et de sa véritable mesure." In: *Histoire de Berlin pour l'année 1745* (I, Berlin 1746), 21-53. En.82. *EOO* II.8, 27-53.

[—], *Gedancken von den Elementen der Cörper, in welchen das Lehr-Gebäude von den einfachen Dingen und Monaden geprüfet, und das wahre Wesen der Cörper entdecket wird*. Berlin: Haude and Spener 1746. En.81. *EOO* III.2 (Genève 1942), 347-366.

[—], *Considérations sur les élémens des corps, dans lesquelles on examine la doctrine des monades et l'on découvre la veritable essence du corps*. Translation (with comments) of *Gedancken von den Elementen der Cörper* in: [Formey], *Recherches* (1747), 159-242.

—, "Différentes pièces sur les monades 1-25." Ms from 1747. En.854. Commentary on the essays that had been submitted to the Berlin competition for the year 1747 about monadology. Published posthumously in *Opera posthuma* II (1862), 805-813. Here I used the version in *EOO* III.2 (Basel 1942), 416-429.

—, "Recherches sur le mouvement des corps célestes en général." In: *Histoire de Berlin pour l'année 1747* (III, Berlin 1749), 93-143. Read Berlin, 8 June 1747. En.112. *EOO* II.25, 1-44.

—, *Rettung der göttlichen Offenbarung gegen die Einwürfe der Freygeister.* Berlin: Haude and Spener 1747. En.92. *EOO* III.12 (Zürich: Orell Füßli 1960), 267-286.

—, "Recherches sur les plus grands et plus petits qui se trouvent dans les actions des forces." In: *Histoire de Berlin pour l'année 1748* (IV, Berlin 1750), 149-188. Read Berlin, 19 December 1748. En.145. *EOO* II.5 (Lausanne: Orell Füßli 1957), 1-37.

—, "Réfléxions sur quelques loix générales de la nature qui s'observent dans les effets des forces quelconques." In: *Histoire de Berlin pour l'année 1748* (IV, Berlin 1750), 189-218. Read 6 February 1749. En.146. *EOO* II.5 (Lausanne: Orell Füßli 1957), 38-63.

—, "Découverte d'un nouveau principe de mécanique." In: *Histoire de Berlin pour l'année 1750* (VI, Berlin 1752), 185-217. Read Berlin, 3 September 1750. En.177. *EOO* II.5 (Lausanne: Orell Füßli 1957), 81-108.

—, "Recherches sur l'origine des forces." In: *Histoire de Berlin pour l'année 1750* (VI, Berlin 1752), 419-447. Read Berlin, 1 October 1750. En.181. *EOO* II.5 (Lausanne: Orell Füßli 1957), 109-131.

—, "Exposé concernant l'examen de la lettre de Mr. de Leibnitz, alluguée par M. le prof. Koenig, dans le mois de mars, 1751 des Actes de Leipzic, à l'occasion du principe de la moindre action." In: *Histoire de Berlin pour l'année 1750* (VI, Berlin 1752), 52-62. Read Berlin, 13 April 1752. En.176. *EOO* II.5 (Lausanne: Orell Füßli 1957), 64-73.

—, "Harmonie entre les principes généraux de repos et du mouvement de M. de Maupertuis." In: *Histoire de Berlin pour l'année 1751* (VII, Berlin 1753), 169-198. Read Berlin, 9 November 1752. En.197. *EOO* II.5 (Lausanne: Orell Füßli 1957), 152-176.

—, "Examen de la dissertation de M. le Professeur Koenig, inserée dans les actes de Leipzig, pour le mois de mars 1751." In: *Histoire de Berlin pour l'année 1751* (VII, Berlin 1753), 219-245. Read Berlin, in Latin, 21 December 1752. En.199A. *EOO* II.5 (Lausanne: Orell Füßli 1957), 194-213.

—, "Sur le principe de la moindre action." In: *Histoire de Berlin pour l'année 1751* (VII, Berlin 1753), 199-218. Read Berlin, 22 Februari 1753. En.198A. *EOO* II.5 (Lausanne: Orell Füßli 1957), 179-193.

—, "Essay d'une démonstration métaphysique du principe général de l'équilibre." In: *Histoire de Berlin pour l'année 1751* (VII, Berlin 1753), 246-254. En.200. *EOO* II.5 (Lausanne: Orell Füßli 1957), 250-256.

—, "Analytica explicatio methodi maximorum et minimorum." In: *Novi commentarii academiae scientiarum Petropolitanae 1764* (X, Petersburg 1766). Read Berlin, 9 September 1756. En.297. *EOO* I.25 (Bern 1952), 177-207.

—, "Elementa calculi variationum." In: *Novi commentarii academiae scientiarum Petropolitanae 1764* (X, Petersburg 1766). Read Berlin, 16 September 1756. En.296 *EOO* I.25 (Bern 1952), 141-176..

[—], *Lettres à une princesse d'Allemagne sur divers sujets de physique & de philosophie.* 3 Volumes. Berlin 1760. En.343, 344 and 417, respectively. It was generally known from the start that Euler was the author. Reprinted after the edition Petersburg 1768 in: *EOO* III.11 and 12 (Zürich: Orell Füßli 1960).

—, "De insigni paradoxo, quod in analysi maximorum et minimorum occurrit." In: *Mémoires de l'académie des sciences de St.-Pétersbourg* 1809/10 (III, Peterburg 1811), 16-25. Read 31 May 1779. En.735. *EOO* I.25 (Bern 1952), 286-292.

—, *Correspondance de Leonhard Euler avec A.C. Clairaut, J. d'Alembert et J.L. Lagrange.* Adolf P. Juškevič and René Taton eds. *EOO* IVa.5 (Basel: Birkhäuser 1980).

—, *Correspondance de Leonhard Euler avec P.-L.M. de Maupertuis et Frédéric II.* P. Costabel a.o. eds. *EOO* IVA.6 (Basel: Birkhäuser 1986).

—, *Leonhardi Euleri opera posthuma mathematica et physica; anno 1844 detecta.* 2 Volumes. Paulus Henricus Fuss and Nicolaus Fuss eds. Petersburg 1862. En.805. Reprinted in facsimile: Nendeln: Kraus 1969.

—, *Leonhardi Euleri opera omnia.* Charles Blanc a.o. eds. Basel: Birkhäuser

Favaro, Antonio ed., "Sette lettere inedite di Giuseppi Luigi Lagrange al P. Paolo Frisi." In: *Atti della R. Academia delle scienze di Torino* 1895–1896 (31), 138-150.

Fehér, Marta, "The Role of Metaphor and Analogy in the Birth of the Principle of Least Action of Maupertuis (1698–1759)." In: *International Studies in the Philosophy of Science* 1988 (2), 175-188.

—, "The Role Accorded to the Public by Philosophers of Science." In: *International Studies in the Philosophy of Science* 1990 (4, 3), 229-240.

Fellmann, Emil A., "Leonhard Euler." In: *Kindler Enzyklopädie. Die Großen der Weltgeschichte* VI (Zürich 1975), 496-531.

Ferola, R., "Some Ideological Dimensions of the Principle of Least Action." In: *Fundamenta Scientia* 1982 (3, 2), 161-176.

Feynman, Richard P., *The Feynman Lectures on Physics. Mainly Mechanics, Radiation and Heat.* 3 Volumes. Lectures held for 1st and 2nd year students at the *California Institute of Technology* (Caltech), 1961–1963. Reading, Mass. [etc.]: Addison-Wesley 1965.

Finster, Reinhard, *Gottfried Wilhelm Leibniz* (translated from the German by Gerd van den Heuvel). Baarn: Tirion 1992. Original German edition: Reinbek: Rowohlt Taschenbuch 1990

Fleckenstein, Joachim O., [Preface to Euler, *Commentationes mechanicae; principia mechanica.*] In: *EOO* II.5 (Lausanne: Orell Füßli 1957), vii-lii.

Forbes, Robert James, and Eduard Jan Dijksterhuis, *A History of Science and Technology; 'Nature Obeyed and Conquered'.* 2 Volumes. Harmondsworth: Penguin 1963 (Translation of *Overwinning door gehoorzaamheid; geschiedenis van natuurwetenschap en techniek.* I. *Van Thales tot Newton.* II. *Van Newton tot Lorentz.* Zeist: de Haan 1961.)

[Formey, Jean Henri Samuel], *Recherches sur les éléments de la matière.* S.l.: 1747. Contains on pp.159-242 a translation, commented by diverse authors, of [Euler], *Gedancken* (1746).

—, *Souvenirs d'un citoyen.* 2 Volumes. Berlin: Lagarde 1789.

Förster, Wolfgang ed., *Aufklärung in Berlin.* Berlin: Akademie-Verlag 1989.

Fourier, Jean-Baptiste Joseph, Baron, *Théorie analytique de la chaleur.* Paris: Didot 1822. *Œuvres de Fourier* I (Paris 1888).

Fraser, Craig, "J.L. Lagrange's Early Contributions to the Principles and Methods of Mechanics." In: *Archive for the Hstory of Exact Sciences* 1983 (28), 197-241.

—, "J.L. Lagrange's Changing Approach to the Foundations of the Calculus of Variations." In: *Archive for the History of Exact Sciences* 1985 (32, 2), 152-191.

—, "D'Alembert's Principle: The Original Formulation and Application in Jean d'Alembert's *Traité de dynamique* (1743)." In: *Centaurus* 1985 (28), 31-61.

Funk, Paul, *Variationsrechnung und ihre Anwendung in Physik und Technik.* Die Grundlehren der mathematischen Wissenschaften; 94. Berlin [etc.]: Springer 1962.

Furetière, Antoine, *Dictionnaire universel des arts et sciences*. 4th edition, revised by B. de la Rivière. The Hague 1727 ([1]1690).

Gabbey, Alan, "Force and Inertia in Seventeenth-Century Dynamics." In: *Studies in History and Philosophy of Science* 1971 (2, 1), 1-67.

Galilei, Galileo, *Dialogo sopra i due massimi sistemi del mondo Tolemaico, e copernicano*. Florence: Landini 1632. *Opere di Galilei* VII.
For the translation, use has been made of:
—, *Dialogue Concerning theTwo Chief World Systems: Ptolemaic & Copernican*. Stillman Drake ed. Berkeley and Los Angeles 1974(=[2]1967) ([1]1953).

—, *Discorsi e demonstrazioni matematiche intorno à due nuoue scienze; attenenti alla mecanica & i movimenti locali*. In Leida, Apresso gli Elsevirii 1638 (Leiden: Elsevier 1638). The first edition contains only the first four *Days*. The *Fifth Day* was published posthumously in 1674, the *Sixth Day* and some fragments only in 1718 (*Opere di Galileo Galilei* II, Florence). The whole is included in *Opere di Galilei* VIII, 41-448.
For the translation, use has been made of:
—, *Two New Sciences. Including Centers of Gravity & Force of Percussion*. Stillman Drake ed. Madison, Wisconsin: The University of Wisconsin Press 1974.

—, *Le opere di Galileo Galilei. Edizione nationale*. 20 Volumes in 21 Books. Antonio Favaro (1st edition) and Giorgio Abetti (2nd edition) eds. Florence: Barbèra 1929–1939 (1890–1909); Reprint 1968.

Garber, Daniel, "Leibniz and the Foundation of Physics: The Middle Years." In: Okruhlik and Brown eds., *The Natural Philosophy of Leibniz* (1986), 27-130.

Gillispie, Charles Coulston ed., *Dictionary of Scientific Biography*. 16 Volumes. New York: Scribner 1970–1980.

Goethe, Johann Wolfgang von, *Zur Farbenlehre* (1810). *Goethes Werke* II.2 (Weimar 1890).

Golinski, Jan V., "Science *in* the Enlightenment." In: *History of Science* 1986 (24), 411-424. Review essay of Hankins, *Science and the Enlightenment* (1985).

Görland, Albert, *Die Hypothese; Ihre Aufgabe und ihre Stelle in der Arbeit der Naturwissenschaft. In Briefen zweier Freunde*. Göttingen: Vandenhoeck and Ruprecht 1911.
—, *Prologik; Dialektik des kritischen Idealismus*. Berlin: Bruno Cassirer 1930.

Grant, Edward, *Physical Sience in the Middle Ages*. Cambridge: Cambridge University Press 1977 (1971).

's Gravesande, Willem Jacob, "Essai d'une nouvelle théorie sur le choc des corps." In: *Journal litéraire de la Haye* 1722 (12, 1), 1-54.
—, "Sup(p)lement à l'Essai sur le choc des corps." In: *Journal litéraire de la Haye* 1722 (12, 2) 190-197.
—, "Remarques sur la force des corps en mouvement, & sur le choc; précédées de quelques réflexions sur la maniere d'écrire de Monsieur le Docteur Samuel Clarcke." In: *Journal litéraire de la Haye* 1729 (13, 1), 189-297 and (13, 2), 407-432.
—, "Nouvelle expériences sur la force des corps en mouvement." In: *Journal historique de la République des Lettres* 1733 (3), 374 and further.
—, *Welzijn, wijsbegeerte en wetenschap*. Cees de Pater ed. Geschiedenis van de wijsbegeerte in Nederland; 13. Baarn: Ambo 1988.

Gueroult, Martial, "Métaphysique et physique de la force chez Descartes et chez Malebranche." In: *Revue de métaphysique et de morale* 1954 (5), 1-37 and 113-134.

—, *Leibniz: dynamique et métaphysique, suive d'une note sur le principe de la moindre action chez Maupertuis*. Collection analyse et raisons; 7. Paris: Aubier-Montaigne ^{2}1967 (^{1}1934 under the title of *Dynamique et métaphysique leibniziennes*).

Hakfoort, Casper, "Nicolas Béguelin and his Search for a Crucial Experiment on the Nature of Light (1772)." In: *Annals of Science* 1982 (39), 297-310.

—, "Christian Wolff tussen cartesianen en newtonianen" ("Christian Wolff between Cartesians and Newtonians"). In: *Tijdschrift voor de geschiedenis der geneeskunde, natuurwetenschappen, wiskunde en techniek* 1982 (5), 27-38.

—, "Science Deified: Wilhelm Ostwald's Energeticist World-View and the History of Scientism." In: *Annals of Science* 1992 (49), 525-544.

—, "Fysica en wereldbeeld van Descartes tot Hawking" ("Physics and World-View from Descartes to Hawking"). In: *De Gids* 1992 (150), 713-722.

Hamel, Georg, "Die Lagange-Eulerschen Gleichungen der Mechanik." In: *Zeitschrift für Mathematik und Physik* 1904 (50), 1-57.

Hankins, Thomas L., "Eighteenth-Century Attempts to Resolve the *Vis Viva* Controversy." In: *Isis* 1965 (56), 281-297.

—, "The Influence of Malebranche on the Science of Mechanics during the Eighteenth Century." In: *Journal of the History of Ideas* 1967 (228), 193-210.

—, [Introduction to d'Alembert's *Traité de dynamique*.] In: D'Alembert, *Traité de dynamique*, reprint 1968, ix-xxxvi.

—, *Science and the Enlightenment*. Cambridge, Mass.[etc.]: Cambridge University Press 1985.

Harman, Peter M. (Former name Peter M. Heimann)

—, *Metaphysics and Natural Philosophy: The Problem of Substance in Classical Physics*. Sussex: Harvester Press 1982.

—, "Force and Inertia: Euler and Kant's 'Metaphysical Foundations of Natural Science'." In: W.R. Shea ed., *Nature Mathematized: Historical and Philosophical Case-Studies in Classical Modern Natural Philosophy* (Dordrecht: Reidel 1983), 229-249.

Harnack, Adolf, *Geschichte der Königlich Preussischen Akademie der Wissenschaften zu Berlin*. 3 Volumes. Berlin 1900.

Heidegger, Martin, *Prolegomena zur Geschichte des Zeitsbegriffs*. Marburger Vorlesung Sommersemester 1925. Petra Jaeger ed. Gesamtausgabe Band 20. Frankfurt am Main: Klostermann 1979.

—, *Aristoteles, Metaphysik J 1-3; Von Wesen und Wirklichkeit der Kraft*. Freiburger Vorlesung Sommersemester 1931. Heinrich Hüni ed. Gesamtausgabe Band 33. Frankfurt am Main: Klostermann 1981.

Heilbron, John Lewis, *Electricity in the 17th and 18th Centuries: A Study of Early Modern Physics*. Berkeley [etc.]: University of California Press 1979.

Heimann, Peter M. and J.E. McGuire, "Newtonian Forces and Lockean Powers: Concepts of Matter in Eighteenth-Century Thought." In: *Historical Studies in the Physical Sciences* 1971 (3), 233-306.

Heinekamp, Albert ed., *Beiträge zur Wirkungs- und Rezeptionsgeschichte von Gottfried Wilhelm Leibniz*. Studia Leibnitiana Supplementa; 26. IV Int. Leibniz-Kongreß der G.W.-Leibniz-Gesellschaft 1983 (Hannover). Stuttgart: Steiner Verlag 1986

Helmholtz, Hermann L.F. von, "Ueber die physikalische Bedeutung des Princips der kleinsten Wirkung." In: *Journal für angewandte und reine Mathematik* 1886 (100), 137-166 and 213-222. Reprinted in —, *Wissenschaftliche Abhandlungen* III, 203-248.

—, "Rede über die Entdeckungsgeschichte des Princips der kleinsten Action." Read 27 January 1887. Published in: Harnack, *Geschichte der Akademie* II, 282-296.

—, "Zur Geschichte des Princips der kleinsten Action." In: *Sitzungsberichten der Akademie der Wissenschaften zu Berlin*, 10 March 1887, 225-236. Reprinted in: —, *Wissenschaftliche Abhandlungen* III, 249-263.

—, "Das Princip der kleinsten Wirkung in der Elekrodynamik." In: *Sitzungsberichten der Akademie der Wissenschaften zu Berlin*, 12 May 1892, 459-475. Reprinted in: —, *Wissenschaftliche Abhandlungen* III, 476-504.

—, "Nachtrag zu dem Aufsatze: Ueber das Princip der kleinsten Wirkung in der Elektrodynamik." Read 14 June 1894. Published in: —, *Wissenschaftliche Abhandlungen* III, 597-603.

—, *Wissenschaftliche Abhandlungen*. Arthur König ed. Leipzig: Barth 1895.

Herivel, John, "Newton's Achievement in Dynamics." In: *The Texas Quarterly* 1967 (10, 3). Reprinted in Palter ed., *The Annus Mirabilis* (1970), 120-135, followed by Buchdahl, [Commentary].

Hiebert, Erwin N., *Historical Roots of the Principle of Conservation of Energy*. New York: Arno Press 1981 (=1962).

Hißmann, Michaël, "Versuch über das Fundament der Kräfte." In: — ed., *Magazin für die Philosophie und ihre Geschichte aus den Jahrbüchern der Akademien angelegt*, Band VI (Göttingen 1983). Publication on his own of his prize essay (essay I–M734 from the Akademiearchiv.)

Hoenen, Petrus S.J., *Philosophie der anorganische natuur*. Philosophische Bibliotheek. Nijmegen: Dekker & vandeVegt ^{3}1947 (^{1}1938).

Hösle, Vittorio, "Über die Unmöglichkeit einer naturalistischen Begründung der Ethik." In: *Wiener Jahrbuch für Philosophie* 1989 (21), 13-29.

Hülsen, August Ludewig, *Prüfung der von der Akademie der Wissenschaften zu Berlin aufgestellten Preisfrage; Was hat die Metaphysik seit Leibniz und Wolf für Progressen gemacht?*. Altona: Hammerich 1796.

Hume, David, *Enquiries concerning Human Understanding and concerning the Principles of Morals* (1748). L.A. Selby-Bigge ed. following the posthumous edition from 1777. Oxford: Oxford University Press 1985 (=^{3}1975).

Hutchison, Keith, "Dormitive Virtues, Scholastic Qualities, and the New Philosophies." In: *History of Science* 1991 (29), 245-278.

Huygens, Christiaan, "Extrait d'une lettre de M. Hugens à l'auteur du Journal." In: *Journal des Sçavans*, March 18th, 1669, 19-24. Here I referred to the reprint in: —, *Œuvres complètes* XVI, 179-181.

—, "A Summary Account of the Laws of Motion, Communicated by Mr. Christian Hugens in a Letter to the *Royal Society* ..." In: *Philosophical Transactions*, April 12th, 1669. Here I referred to the reprint in: —, *Œuvres complètes* XVI, 429-433.

—, *Œuvres complètes de Christiaan Huygens*. 22 Volumes. The Hague: Nijhoff 1888–1950.

Iltis, Carolyn, "D'Alembert and the *Vis Viva* Controversy." In: *Studies in History and Philosophy of Science* 1970 (1), 135-144.

—, "Leibniz and the *Vis Viva* Controversy." In: *Isis* 1971 (62), 21-35.

Inventaris van de prijsvragen uitgeschreven door de Hollandsche Maatschappij der Wetenschappen 1753–1917. J.G. de Bruijn ed. Groningen: Tjeenk Willink/ Haarlem: Hollandsche Maatschappij der Wetenschappen 1977.

Itard, Jean, "Lagrange, Joseph Louis." In: *DSB* VII (1973), 559-573.

Jacobi, Carl Gustav Jacob, "Das Princip der kleinsten Wirkung." In: *Vorlesungen über Dynamik*. A. Clebsch ed. Berlin: Georg Reimer 1866, 43-51. Lecture read during the winter semester 1842–43 at the *Königsberger Universität*.

Jammer, Max, *Concepts of Force: A Study in the Foundation of Dynamics*. Cambridge, Mass.: Harvard University Press 1957.

Jespers, F.P.M., *De kracht in alles; Het mechanistisch en metafysisch systeem van Leibniz*. Assen: Van Gorcum 1997.

Jourdain, Philip E.B., *The Principle of Least Action*. Chicago: The Open Court 1913.

Juškevič, Adolf P., "Euler, Leonhard." In: *DSB* IV, 467-484.

—, "Lazare Carnot and the Competition of the Berlin Academy in 1786 on the Mathematical Theory of the Infinite." In: Charles Coulston Gillispie and Adolf P. Juškevič, *Lazare Carnot, Savant*. Princeton, N.J.: Princeton University Press 1971, 149-168.

Justi, Johann Heinrich Gottlob, a.o., *Dissertation qui a remporté le prix proposé (...) sur le système des monades avec les pieces qui ont concouru*. Berlin: Haude and Spener 1748.

Kant, Immanuel, *Gedanken von der wahren Schätzung der lebendigen Kräfte und Beurtheilung der Beweise, deren sich Herr von Leibniz und andere Mechaniker in dieser Streitsache bedient haben, nebst einige vorhergehenden Betrachtungen, welche die Kraft der Körper überhaupt betreffen*. (1747). Reprint in: *Gesammelte Schriften* I (1900), 1-182.

—, *Kritik der reinen Vernunft* (Riga: Hartknoch ^{2}1787 (^{1}1781)). Reference is made to the reprint of the edition by Raymund Schmidt (1926): Hamburg: Meiner Verlag 1978.

—, *Metaphysische Anfangsgründe der Naturwissenschaft* (Riga: Hartknoch 1786). *Theorie-Werkausgabe* (Wilhelm Weischedel ed.); IX. Unchanged pocket edition: Frankfurt am Main: Suhrkamp 1977.

—, *Über die von der Königl. Akademie der Wissenschaften zu Berlin für das Jahr 1791 ausgesetzte Preisfrage: Welches sind die wirklichen Fortschritte, die die Metaphysik seit Leibnizens und Wolf's Zeiten in Deutschland gemacht hat?* (1796). Friedrich Theodor Rink ed. Königsberg: Goebbels and Unzer 1804. Reprint in: *Gesammelte Schriften* XX (Berlin 1942).

—, *Gesammelte Schriften*. 23 Volumes. Berlin: Königlich Preussischen Akademie der Wissenschaften 1900–1955.

Karskens, Machiel, [Introduction to Leibniz's *Metafysische verhandeling*]. Dixit. Bussum: Het Wereldvenster 1981, 9-58.

Kemble, Edwin C., *The Fundamental Principles of Quantum Mechanics: With Elementary Applications*. New York: Dover ^{2}1958 (1937).

Klein, F. and A. Sommerfeld, *Über die Theorie des Kreisels*. New York [etc.]: Johnson [etc.] 1965 (Leipzig 1897).

Kleinert, Andreas, "D'Alembert et le prix de l'Académie de Berlin en 1746." In: Emery and Monzoni eds., *Jean d'Alembert* (1989), 415-431.

Kleijwegt, Marc, *Ancient Youth. The Ambiguity of Youth and the Absence of Adolescense in Greco-Roman Society*. Dissertation Leiden. Amsterdam: Gieben 1991.

Kneser, Adolf, *Das Prinzip der kleinsten Wirkung von Leibniz bis zur Gegenwart*. Wissenschaftliche Grundfragen, philosophische Abhandlungen; 9. Leipzig and Berlin: Teubner 1928.

König, Samuel, "De universali principio aequilibrii et motus in vi viva reperto, deque nexa inter vim vivam et actionem, utriusque minimo." In: *Nova acta eruditorum* (Leipzig, March 1751), 125-135, 162-176. Reprinted in: *EOO* II.5 (Lausanne: Orell Füßli 1957), 303-324. Partly translated in English as: *Vis Viva and the Principle of Least Action, Dissertation on a Principle of Universal Equilibrium and Motion Made Plain with the Use of Vis Viva*. In: Lindsay ed., *Energy* (1975), 149-157.

Krauss, Werner, "Eine politische Preisfrage im Jahre 1780." In: — ed., *Studien zur deutschen und französischen Aufklärung* (Berlin 1963), 68-71.

Kutschmann, Werner, *Die Newtonsche Kraft; Metamorphose eines wissenschaftlichen Begriffs*. Wiesbaden: Steiner 1983.

Kuhn, Thomas S., "Energy Conservation as an Example of Simultanuous Discovery." Originally from 1959; reprinted in —, *The Essential Tension* (1977), 66-104.

—, *The Structure of Scientific Revolutions*. Chicago [etc.]: The University of Chicago Press ^{2}1970 (^{1}1962).

—, "The Relations between the History and the Philosophy of Science." Isenberg lecture, March 1st, 1968. Printing in revised form in: —, *The Essential Tension* (1977), 3-20.

—, "Mathematical versus Experimental Traditions in the Development of Physical Science." In: *The Journal of Interdisciplinary History* 1976 (7), 1-31. Reprinted in: —, *The Essential Tension* (1977), 31-65.

—, *The Essential Tension. Selected Studies in Scientific Tradition and Change*. Chicago [etc.]: The University of Chicago Press 1977.

Lagrange, Joseph Louis, "Note sur la métaphysique du calcul infinitésimal." In: *Miscellanea Taurinensia* 1760–1761 (2), 17-18. *Œuvres de Lagrange* VII, 597-599. This note is a comment on: P. Gerdile, "De l'infini absolu considéré dans la grandeur."

—, "Recherches sur la méthode de maximis et minimis." In: *Miscellanea taurinensia* 1759 (1), 18-32.

—, "Essai d'une nouvelle méthode pour déterminer les maxima et les minima des formules intégrales indéfinies." In: *Miscellanea Taurinensia* 1760–61 (2), 173-195. *Œuvres de Lagrange* I, 335-362.

—, "Application de la méthode exposée dans le mémoire précédent à la solution de différents problèmes de dynamique." In: *Miscellanea Taurinensia* 1760–1761 (2), 196-268. *Œuvres de Lagrange* I, 363-468.

—, "Recherches sur la libration de la lune, dans lesquelles on tache de résoudre la question proposée par l'Académie Royale des Sciences, pour le prix de l'année 1764." *Œuvres de Lagrange* VI (1873), 3-61.

—, "Théorie de la libration de la lune et des autres phénomènes qui dépendent de la figure non sphérique de cette planète." In: *Nouveaux mémoires de l'Académie Royale des Sciences et Belles-Lettres de Berlin 1780*, 203-209. *Œuvres de Lagrange* V, 5-122.

—, *Méchanique analitique*. Paris 1788. The spelling of the title is that of the first printing, which is no longer common . Beginning with the second edition, the title was written as *Mécanique analytique*. Here I refer to the *Édition complète*, which is based on the 2nd editie (1811–1815), but includes the comments from Joseph Bertrand (3rd edition 1853) and Gaston Darboux (4th edition 1888/9). 2 Volumes. Paris: Blanchard 1965.

—, "Sur le principe des vitesses virtuelles." In: *Journal de l'École Polytechnique* an VI [1797]. *Œuvres de Lagrange* VII, 317-321.

—, "Le raison du principe suffisante." Read at Paris, May 1806. *Œuvres de Lagrange* VII.

—, *Correspondance de Lagrange avec Euler*. *Œuvres de Lagrange* XIV, 133-245. On pages 323-327 a list of contents of the letters can be found. In the present book I refer to the more complete and accessible edition in *EOO* IVa.5 (1980).

—, *Œuvres de Lagrange*. 14 Volumes. J.A. Serret and Gaston Darboux eds. Paris 1867–1892. Reprint Hildesheim: Olms 1973.

Lambert, Johann Heinrich, *Johann Heinrich Lamberts "Cosmologische Briefe" mit Beiträgen zur Frühgeschichte der Kosmologie*. Reprint of Lamberts *Cosmologische Briefe über die Einrichtung des Weltbaus* (Augsburg 1761). G. Jackisch ed. Berlin: Akademie-Verlag 1979.

Lanczos, Cornelius, *The Variational Principles of Mechanics*. Toronto: University of Toronto Press ^{4}1970.

Landau, L.D. and E.M. Lifshitz, *Course of Theoretical Physics* I. *Mechanics*. II. *The Classical Theory of Fields*. III. *Quantum Mechanics: Non-Relativistic Theory*. Translated from the Russian by J.B. Sykes a.o. Oxford [etc.]: Pergamon Press ^{2}1969 (^{1}1960); 1989=^{4}1975 (^{1}1951); 1991=^{3}1977 (^{1}1958).

—, *A Shorter Course in Physics* I. *Mechanics and Electrodynamics*. II. *Quantum Mechanics*. Translated from the Russian by J.B. Sykes and M. Hamermesh. Oxford [etc.]: Pergamon Press ^{2}1972–1974 (^{1}1969).

Latour, Bruno, *Science in Action: How to Follow Scientists and Engineers through Society*. Cambridge, Mass.: Harvard University Press 1987.

Laudan, Laurens L., "The *Vis Viva* Controversy; A Post-Mortem." In: *Isis* 1968 (59), 130-143.

Leibniz, Gottfried Wilhelm, "Brevis demonstratio erroris mirabilis Cartesii et aliorum circa legem naturalem, secundam quam volunt ad Deo eundam semper quantitatem motis conservandi, qua et omne mechanica abituntur." Rav.94. *Acta eruditorum* March 1686, 161-163. Gerh. *Math. Schr.* VI, 117-119. English translation as: —, "A Short Demonstration." In: Loemker, *Philosophical Papers*, 296-302 (nr.34).

—, "Discours de métaphysique." Ms. from 1686. Published posthumously by C.E. Grotefend (1846). Reprinted untitled in Gerh. *Phil. Schr.* IV (1881), 427-466.

—, "Réplique de M. L[eibniz] à M. l'abbé D[e] C[atelan] contenuë dans une lettre écrite à l'auteur de ces Nouvelles le 9 de Janv. 1687. Touchant ce qu'a dit M. Descartes que Dieu conserve toujours dans la nature la même quantité de mouvement." In: *Nouvelles de la République des Lettres* February 1687, 131-136. Rav.98 [Response to "Courte remarque" by De Catelan]. Gerh. *Phil. Schr.* III, 42-49.

—, "Réponse de M. L[eibniz] à la Remarque de M. l'abbé D[e] C[atelan] contenuë dans l'article I. de ces Nouvelles, mois de Juin 1687 où il prétend soutenir une loi de la nature avancée par M. Descartes." In: *Nouvelles de la République des Lettres* Sept. 1687, 952-956. Rav.100. [Reply to 2nd criticism by Catelan]. Gerh. *Phil. Schr.* III, 49-51.

—, "De causa gravitatis, et defensio sententiae autoris de veris naturae legibus contra cartesianos." In: *Acta eruditorum* May 1690, 228-239. Rav.105. Gerh. *Math. Schr.* VI, 193-203.

—, "De legibus naturae et vera aestimatione virium motricium contra cartesianos. Responsio ad rationes a D[e]n[is] P[apin] mense Januarii proximo in Actis hisce p.6 propositas." In: *Acta eruditorum* September 1691, 439-447. Ravier 112. Gerh. *Math. Schr.* VI, 204-211.

—, "Animadversiones in partem generalem Principiorum Cartesianorum." (1692). Published posthumously in: Gerh. *Phil. Schr.* IV, 350-392. English translation as —, "Critical Thoughts." In: Loemker, *Philosophical Papers*, 383-412 (nr.42).

—, "Essay de Dynamique." (1692, Ms. send to Pelisson). Published posthumously in: Foucher de Careil ed., *Œuvres de Leibniz* I (Paris 1859), 470-483. A copy of Des Bilettes appeared in: Costabel, *Leibniz* (1960), 108-131.

—, "De primae philosophiae emendatione et de notione substantiae." In: *Acta eruditorum* March 1694, 110-112. Rav.133. Gerh. *Phil. Schr.* IV, 468-470. English translation as —, "On the Correction of Metaphysics." In: Loemker, *Philosophical Papers*, 432-434 (nr.45).

—, "Specimen dynamicum pro admirandis naturae legibus circa corporum vires et mutuas actiones deligendis et suae causas revocandis." Part 1 in: *Acta eruditorum* April 1695, 145-157. Rav.139. Gerh. *Math. Schr.* VI, 234-246. Part 2 remained unpublished until print in *ibidem*, 246-254. English translation as —, "Specimen Dynamicum." In: Loemker, *Philosophical Papers*, 435-452 (nr.46).

—, "Système nouveau de la nature et de la communication des substances, aussi bien que de l'union qu'il y a entre l'âme et le corps." In: *Journal des Sçavans* June–July 1695, 294-306. Rav.141. Gerh. *Phil. Schr.* IV, 477-487. On pages 471-477 a draft of "Système" is included.

—, "[Remarques sur les objections de M. Foucher]." In: *Histoire des ouvrages des savans* Februari 1696, 274-276. Rav.145. Gerh. *Phil. Schr.* IV, 490-493.

—, "Éclaircissement du nouveau système de la communication des substances, pour servir à ce qui en est dit dans le Journal du 12 Septembre 1695." In: *Journal des Sçavans*, April 1696, 166-171. Rav.147. Gerh. *Phil. Schr.* IV, 493-498.

—, [Postscript to the letter to Basnage de Beauval from 3/13 January 1696, concerning the "Système nouveau"]. Gerh. *Phil. Schr.* IV, 498-500.

—, "Tentamen anagogicum. Essay anagogique dans la recherche des causes" (around 1696). Published posthumously in: Gerh. *Phil. Schr.* VII, 270-279.

—, "De ipsa natura sive de vi insita actionibusque creaturum, pro Dynamicis suis confirmandis illustrandisque." In: *Acta eruditorum* Sept. 1698, 427-440. Rav.154. Gerh. *Phil. Schr.* IV, 504-516. English translation as —, "On Nature Itself." In: Loemker, *Philosophical Papers*, 498-508 (nr.53).

—, "Eclaircissement des difficultés que Monsieur Bayle a trouvées dans le système nouveau de l'union de l'âme et du corps." In: *Histoire des ouvrages des savans* July 1698, 332-342. Rav.153. Gerh. *Phil. Schr.* IV, 517-524.

—, "Essay de dynamique sur les loix du mouvement, où il est montré, qu'il ne se conserve pas la même quantité de mouvement, mais la même force absolue, ou bien la même quantité de l'action motrice" (1698). Published posthumously in: Gerh. *Math. Schr.* VI, 215-230.

—, "Extrait du Dictionnaire de M. Bayle article Rorarius p. 2599 sqq de l'Edition de l'an 1702 avec mes remarques." Published posthumously in: Gerh. *Phil. Schr.* IV, 524-571.

—, *Leibnizens mathematische Schriften*. (7 Volumes). C.I. Gerhardt ed. Berlin, Halle: Schmidt 1849–1863. Reprinted Hildesheim: Olms 1971 (1962).

—, *Die philosophische Schriften von G.W. Leibniz*. (7 Volumes). C.I. Gerhardt ed. Berlin 1875–1890. Reprinted Hildesheim: Olms 1965.

—, *Philosophical Papers and Letters of Leibniz*. Leroy E. Loemker ed. Dordrecht: Reidel 1969 (^{1}1956).

Leibniz, Gottfried Wilhelm and Samuel Clarke, *A Collection of Papers which Passed between (...) M. Leibnitz and Dr. [Samuel] Clarke in 1715 and 1716 Relating to the Principles of Natural Philosophy and Religion.* London 1717. Gerh. *Phil. Schr.* VII, 344-440.

Lindberg, David C., "Conceptions of the Scientific Revolution from Bacon to Butterfield: A Preliminary Sketch." In: Lindberg & Westman, *Reappraisals* (1990), 1-26.

Lindberg, David C. and Robert S. Westman eds., *Reappraisals of the Scientific Revolution*. Cambridge [etc.]: Cambridge University Press 1990.

Lindsay, R. Bruce, *Energy: Historical Development of the Concept*. Benchmark Papers on Energy; 1. Stroudsberg, Penns.: Dowden, Hutchinson and Ross 1975.

Lindt, Rich., "Das Prinzip der virtuellen Geschwindigkeiten; Seine Beweise und die Unmöglichkeit seiner Umkehrung bei Verwendung des Begriffes 'Gleichgewicht eines Massensystems'." In: *Abhandlungen zur Geschichte der mathematischen Wissenschaften* 1904 (18), 145-195.

Loemker, Leroy E. ed., [Commentary on *Philosophical Papers and Letters of Leibniz*]. Dordrecht: Reidel 1969 ([1]1956).

Lorenz, Martina, "Der Beitrag Christian Wolffs zur Rezeption von Grundprinzipien der Mechanik Newtons in Deutschland zu Beginn des 18. Jahrhunderts." In: *Historia Scientiarum* 1986 (31), 87-100.

Lunteren, Frans H. van, "Hegel and Gravitation." In: Rolf-Peter Horstmann and Michael John Petry eds., *Hegels Philosophie der Natur* Klett-Cotta 1986, 45-53.

—, *Framing Hypotheses; Conceptions of Gravity in the 18th and 19th Centuries*. Dissertation. Utrecht 1991.

—, "Eighteenth-Century Thought on the Nature of Gravitation." In: Michael J. Petry ed., *Hegel and Newtonianism*. Dordrecht: Kluwer Academic Publishers 1993, 343-366.

Mach, Ernst, *Die Mechanik, historisch-kritisch dargestellt* (1883). Darmstadt: Wissenschaftliche Buchgesellschaft 1973 (=[9]1933). Original title: *Die Mechanik in ihrer Entwicklung historisch-kritisch dargestellt*.

Maffioli, Cesare S., "Italian Hydraulics and Experimental Physics in Eighteenth-Century Holland. From Poleni to Volta." In: Maffioli and Palm eds., *Italian Scientists* (1989), 243-275.

Maffioli, Cesare S. and Lodewijk C. Palm, *Italian Scientists in the Low Countries in the XVIIth and XVIIIth Centuries*. Amsterdam: Rodopi 1989.

Mairan, Jean Jacques d'Ortus de, "Dissertation sur l'estimation et la mesure des forces motrices des corps." In: *Mémoires de Paris* 1728, 1-70. Partly reprinted in facsimile in: Costabel ed., *Signification*, 131-149.

Mason, Stephen F., *A History of the Sciences*. New York: Collier [2]1962 ([1]1956). Original title: *Main Currents of Scientific Thought*.

Maupertuis, Pierre Louis Moreau de, [N.b.: the spelling used for Maupertuis' works is that of the bibliography by Tonelli in *Œuvres* I (1768), xxiv*-xxxviii*].

—, *Discours sur les différentes figures des astres avec une exposition abrégée des systèmes de M. Descartes et M. Newton*. Paris 1732. Partly reprinted as: *Discours sur les différentes figures des astres, Où l'on essaye d'expliquer les principaux phénomès du Ciel* in: *Œuvres* I (1756), 79-170.

—, "Sur les loix de l'attraction." In: *Mémoires de Paris pour l'année 1732* (Paris 1735), 343-362. Reprinted with some revisions as concluding section of: *Discours sur les différentes figures* in: *Œuvres* I (1756), 160-170.

—, "Loi du repos des corps." In: *Mémoires de Paris pour l'année 1740* (Paris 1742), 170-176. Read at Paris, 20 February 1740. Reprinted with some revisions in: *Œuvres* IV (1756), 43-64. Reprinted unrevised in: *EOO* II.5 (Lausanne: Orell Füßli 1957), 268-273.

—, "Accord des différentes loix de la nature qui avoient jusqu'ici paru incompatibles." In: *Mémoires de Paris pour l'année 1744* (Paris 1748), 417-426. Read at Paris, 15 April 1744. Reprinted in: *Œuvres* IV (1756), 1-23, completed with a part of Euler's "Sur le principe de la moindre action" from 1751. Reprinted in original form in: *EOO* II.5 (Lausanne: Orell Füßli 1957), 274-281.

—, "Les lois du mouvement et du repos, déduites d'un principe de métaphysique." In: *Histoire de Berlin pour l'année 1746* (II, Berlin 1748), 267-294. Read 6 October 1746. Sections I and II were incorporated in the first part of —, *Essay de Cosmologie* (1750). Section III ("Recherche des lois du mouvement") was reprinted in: —, *Œuvres* IV (1756), 29-42, however without problem 3, concerning the application to statics. The section on "Lois du repos" has not been included in the *Œuvres*. The original article was reprinted in: *EOO* II.5 (Lausanne: Orell Füßli 1957), 282-302.

—, *Essay de cosmologie*. Berlin 1750. Reprinted in: *Œuvres* I (1756), 1-78.

—, [Letter to Samuel König, 23 December 1751]. Read at Berlin, 23 December 1751. Published in: Harnack, *Geschichte der Akademie* III, 281-282 (nr.170a).

—, *Lettres [Essais philosophiques sous la forme de lettres]*. Dresden 1752 (Berlin ^{2}1753). Reprinted in: *Œuvres* II (1756), 217-372.

—, "Réponse à un Mémoire de M. d'Arcy sur la moindre action." In: *Histoire de Berlin pour l'année 1752* (VIII, Berlin 1754), 293-298.

—, *Œuvres*. Lyon: Bruyset ^{2}1768 ($\approx^3$1756, earlier editions: Dresden ^{1}1752; Berlin, Lyon ^{2}1753). Reprint: Hildesheim: Olms 1974.

Mayer, Adolph, *Geschichte des Princips der kleinsten Action*. Academic oration. Leipzig: Veit & Comp. 1877.

—, "Die beiden allgemeinen Sätze der Variationsrechnung, welche den beiden Formen des Princips der kleinsten Action in der Dynamik entsprechen." In: *Bericht der Königlich Sächsischen Gesellschaft der Wissenschaften*, 14 November 1886.

McClellan, James E. III, *Science Reorganized: Scientific Societies in the Eighteenth Century*. New York: Columbia University Press 1985.

McGuire, James E., [Commentary to I.B. Cohen, "Newton's Second Law" (1967).] In: Palter ed., *The Annus Mirabilis* (1970), 186-191.

Mendels[s]ohn, Moses [en Immanuel Kant], *Abhandlung über die Evidenz in metaphysischen Wissenschaften (...) nebst noch einer Abhandlung über dieselbe Materie (...)* The second work (pages 67-99) is Kant's, and is titled: "Untersuchung über die Deutlichkeit der Grundsatz der natürlichen Theologie und der Moral." The essays won the first prize and the 'Accessit', respectively, in the Berlin Academy's competition for the year 1763. Berlin: Haude und Spener 1764.

Müller, Hans Heinrich, *Akademie und Wirtschaft im 18. Jahrhundert. Agrarökonomische Preisaufgaben und Preisschriften der Preußischen Akademie der Wissenschaften (Versuch, Tendenzen und Überblick)*. Berlin: Akademie-Verlag 1975.

Munnik, René P.H., *Van grens-verleggen naar ruimte-scheppen. Over het raakvlak van metafysiek en techniek*. Oration University of Twente. s.l. [Enschede]: s.a. [1992].

Nersessian, Nancy, *Faraday to Einstein: Constructing Meaning in Scientific Theories*. Science and Philosophy; 1. Dordrecht: Martinus Nijhoff 1984.

Neuser, Wolfgang, "Von Newton zu Hegel. Traditionslinien in der Naturphilosophie." Presentation on Hegel congress, March 1988, Berlin.

Newton, Isaac, *Philosophiae naturalis principia mathematica*. London: Innys ^{3}1726 (^{2}1713, ^{1}1687). I have referred to the edition of Alexandre Koyré and I. Bernard Cohen, in which the 3rd edition is reprinted in facsimile. Cambridge, Mass.: Harvard University Press 1972. The page numbers mentioned by me always refer to the numbering in the

3rd edition, except in the case of the 'Auctoris Praefatio', where reference is made to the pagination of Koyré and Cohen themselves.

—, *The Mathematical Principles of Philosophy*. Translation Andrew Motte. London 1729. Reprint London: Dawsons of Pall Mall 1968.

—, *Opticks: or, A treatise of the Reflections, Refractions, Inflections & Colours of Light* (London: Smith and Walford 1704). I have used the Dover edition, based on the 4th edition, the last edition that was revised by Newton himself (London: Innys 1730), New York: Dover 1979.

Okruhlik, Kathleen and James Robert Brown eds., *The Natural Philosophy of Leibniz*. Dordrecht [etc.]: Reidel 1986.

Pagliaro, Harold E. ed., *Irrationalism in the Eighteenth Century*. Cleveland, Ohio: Press of Case Western Reserve University 1972.

Palter, Robert ed., *The Annus Mirabilis of Sir Isaac Newton 1666–1966*. Cambridge, Mass. [etc.]: M.I.T. Press 1970.

Pap de Fagaras, Josephus, "Dissertatio de vi substantiali, ejus notione, natura, et determinationis legibus." In: *Dissertation sur la force primitive, qui a remporté le prix proposé (...) pour l'année 1779*. (Berlin: Decker 1780).

Papineau, David, "The *Vis Viva* Controversy." In: *Studies in History and Philosophy of Science* 1977 (8), 111-142. Ook verschenen in: Woolhouse, *Leibniz* (1981), 139-156.

Pater, Cees de, *Petrus van Musschenbroek (1692–1761), een newtoniaans natuuronderzoeker*. Dissertation. Utrecht 1979.

—, "The Textbooks of 's Gravesande and Van Musschenbroek in Italy." In: Maffioli and Palm eds., *Italian Scientists* (1989), 231-241.

Pierson, Stuart, "*Corpore Cadente* ...: Historians Discuss Newton's Second Law." In: *Perspectives on Science* 1993 (1, 4), 627-658.

—, "Two Mathematics, Two Gods: Newton and the Second Law." In: *Perspectives on Science* 1994 (2, 2), 231-253.

Planck, Max, "Das Prinzip der kleinsten Wirkung." In: Warburg ed., *Physik* (Leipzig: Teubner 1915), 692-702.

Pulte, Helmut, *Das Prinzip der kleinsten Wirkung und die Kraftkonzeptionen der rationalen Mechanik. Eine Untersuchung zur Grundlegungsproblematik bei Leonhard Euler, Pierre Louis Moreau Maupertuis und Joseph Louis Lagrange*. Studia Leibnitiana; 19. Stuttgart: Steiner Verlag 1989.

Ravier, Emile, *Bibliographie des œuvres de Leibniz* (Paris 1937). Reprint Hildesheim: Olms 1966.

Die Registres der Berliner Akademie der Wissenschaften 1746-1766. Dokumente für das Wirken Leonhard Eulers in Berlin. Eduard and Maria Winter eds. Berlin: Akademie-Verlag 1957.

Rehberg, August Wilhelm, *Abhandlung über das Wesen u[nd] die Einschränkungen der Kräften. Welcher die Königl[iche] Akademie zu Berlin das Accessit zuerkannt hat*. Publication on his own of his prize essay (essay I–M731 from the *Akademiearchiv*.) Leipzig: in der Weygardschen Buchhandlung 1779.

Reid, Thomas, "An Essay on Quantity." In: *Philosophical Transactions of the Royal Society* 1748 (45), 505-520. The passage concerning the *vis viva* discussion is reprinted as annex (pages 140-143) to Laudan, "The *Vis Viva* Controversy" (1968).

Rousseau, Jean Jacques, *Discours sur l'origine et les fondemens de l'inégalité parmi les hommes*. (Winning essay in the competition of the Academy of Dijon for the year 1754). *Œuvres complètes* III (1964), 109-223.

—, *Œuvres complètes*. 4 Volumes. Bernard Gagnebin and Marcel Raymond eds. Paris: Éditions Gallimard 1959–1969.

Rousseau, George Sebastian and Roy Porter, *The Ferment of Knowledge; Studies in the Historiography of Eighteenth-Century Science*. Cambridge, Mass. [etc.]: Cambridge University Press 1980.

Rudolph, Enno, "Die Einheit von Metaphysik und Physik; Zu Leibniz' Anfangsgründen der Naturphilosophie und Phänomenologie." In: *Philosophia Naturalis* 1986 (23), 49-69.

Sarton, George, "Lagrange's Personality (1736–1813)." In: *Proceedings of the American Philosophical Society* 1944 (88, 6), 456-496.

Schenk, Günter and Fritz Gehlar, "Der Philosoph, Logiker, Mathematiker und Naturwissenschaftler Johann Heinrich Lambert." In: Förster ed., *Aufklärung in Berlin* (1989), 130-164.

Schrecker, Paul, "Notes sur l'évolution du principe de la moindre action." In: *Isis* 1941 (33), 329-334. Reply to Brunet, *Étude historique* (1938).

Scott, Wilson L., *The Conflict between Atomism and Conservation Theory (1644–1860)*. London: MacDonald; New York: Elsevier 1970.

Shapin, Steven and Simon Schaffer, *Leviathan and the Airpump. Hobbes, Boyle, and the Experimental Life*. Princeton, N.J.: University Press 1985.

Snelders, Harry A.M., *De invloed van Kant, de Romantiek en de "Naturphilosophie" op de anorganische natuurwetenschappen in Duitsland*. Dissertation. Utrecht 1973.

—, *Wetenschap en intuïtie. Het Duitse romantisch-speculatief natuuronderzoek rond 1800*. Baarn: Ambo 1994.

Spaemann, Robert and Reinhard Löw, *Die Frage Wozu? Geschichte und Wiederentdeckung des teleologischen Denkens*. München [etc.]: Piper & Co. 1985 ([1]1981).

Speiser, David, [Introduction to Daniel Bernoulli, *Werke* III: *Mechanik*]. Stuttgart: Birkhäuser 1987, xvi-xxvii and 1-118.

Stäckel, Paul ed., *Abhandlungen über Variationsrechnung I. Theil: Abh[andlungen] von Joh. Bernoulli (1696), Jac. Bernoulli (1697) u[nd] L. Euler (1744)*. Ostwald Klassiker; 46. Leipzig 1914 ([1]1894).

Struik, Dirk Jan, *A Concise History of Mathematics*. New York: Dover 1948.

Szabó, István, *Geschichte der mechanischen Prinzipien und ihrer wichtigsten Anwendungen*. Basel [etc.]: Birkhäuser [3]1987 ([1]1977).

Taton, René, "Sur quelques pièces de la correspondance de Lagrange pour les années 1756–1758." In: *Bollettino di Storia delle Scienze Matematiche* 1988 (8), 3-19.

Taton, René and Adolf P. Juškevič, [Introduction to *Correspondance de Leonhard Euler avec A.C. Clairaut, J. d'Alembert et J.L. Lagrange*.] In: *EOO* IVa.5 (1980), 1-63.

—, [Notes to *Correspondance de Leonhard Euler avec A.C. Clairaut, J. d'Alembert et J.L. Lagrange*.] In: *EOO* IVa.5 (1980).

Teich, Mikuláš and Robert Young eds., *Changing Perspectives in the History of Science: Essays in the Honour of Joseph Needham*. London: Heinemann 1973.

Terrall, Mary, *Maupertuis and Eighteenth-Century Scientific Culture*. Dissertation University of California. Los Angeles 1987.

—, "The Culture of Science in Frederick the Great's Berlin." In: *History of Science* 1990 (28), 333-364.

—, [Review of Pulte, *Prinzip* (1989)]. In: *Isis* 1992 (83, 1), 140-141.

Thiele, Rüdiger, "Ist die Natur sparsam? Über die Begründung des Prinzips der kleinsten Aktion von G.W. Leibniz bis L. Euler." In: *Mitteilungen der Mathem. Gesellschaft der DDR* 1984 (1), 48-59.

—, *Leonhard Euler*. Biographien hervorragender Naturwissenschaftler, Techniker und Mediziner; 56. Leipzig: Teubner 1982.

Thomson, William (Lord Kelvin) and Peter Guthrie Tait, *Treatise on Natural Philosophy*. Oxford and London: Macmillan 1867. I have used the last revised edition from 1912, titled *Principles of Mechanics and Dynamics* (Reprint: New York: Dover Publications 1962.)

Tijmes, Pieter, "Albert Görland, een systematisch denker van formaat." In: *De uil van Minerva* Summer 1985, 205-217.

Tonelli, Giorgio, [Introduction to Maupertuis, *Œuvres*]. In: Maupertuis, *Œuvres* (1956) I, xi*-cxxxiii*.

Trendelenburg, Friedrich Adolf, "Zur Geschichte des Worts und Begriffs a priori." In: *Deutsche Zeitschrift für Philosophie* 1992 (40, 1/2), 80-90. Read in the Berlin Academy at April 20th, 1871.

Truesdell, Clifford Ambrose, "Rational Fluid Mechanics, 1687–1765." Introduction to *EOO* II.12. (Lausanne 1954), viii-cxxv.

—, "I. The First Three Sections of Euler's Treatise on Fluid Mechanics (1766); II. The Theory of Aerial Sound (1687–1788); III. Rational Fluid Mechanics (1765–1788)." Introduction to *EOO* II.13 (1956), vii-cxviii.

—, *The Rational Mechanics of Flexible or Elastic Bodies, 1638–1788*. Introduction to *EOO* II.10 and II.11. In: *EOO* II.11, part 2 (Zürich: Orell Füßli 1960).

—, "A Program Toward Rediscovering the Rational Mechanics of the Age of Reason." In: *Archive for the History of Exact Sciences* 1960 (1), 3-36. Reprinted in: —, *Essays* (1968), 85-136.

—, "Reactions of Late Baroque Mechanics to Success, Conjecture, Error, and Failure in Newton's *Principia*." In: *The Texas Quarterly* 1967 (10, 3). Reprinted in: Palter ed., *The Annus Mirabilis* (1970), 192-232, followed by: Woodruff, [Commentary].

—, *Essays in the History of Mechanics*. Berlin [etc.]: Springer 1968.

—, "Leonhard Euler, Supreme Geometer (1707–1783)." In: Pagliaro, *Irrationalism* (1972), 51-95.

Tuffet, Jacques, [Introduction to Voltaire's *Histoire du docteur Akakia* (Paris: Nizet 1967)], i-cxxxvi.

Vanheste, Tomas, "Berichten van een nieuwe tijd; Of hoe de New Age-beweging de natuurwetenschap claimed." In: *De Gids* 1994 (157), 24-36.

—, *Copernicus is ziek; Een geschiedenis van het New-Agedenken over natuurwetenschap. (Copernicus is ill; A History of New Age Thinking about Naural Science.)* Dissertation. University of Twente. Delft: Eburon 1996.

V[oltaire, François Marie Arouet de, *Lettres philosophiques par M. de V[oltaire], à Amsterdam, chez E. Lucas, au Livre d'Or* (Rouen: Jore 1734). Reprint in: *Mélanges*, 1-133, based on the text critical edition of Gustave Lanson (Paris: Hachette 1930).

—, *Histoire du docteur Akakia et du natif de St. Malo* (1752–1753). Jacques Tuffet ed., text critical edition. Paris: Nizet 1967.
Originally, the book appeared in separate parts:

—, *Diatribe du docteur Akakia, medecin du pape* (Potsdam 1752; almost the whole edition was burned on command of Frederic the Great on December 24th, 1752). Reprinted e.g. Rome 1753.

—, *Traité de Paix conclu entre Mr le président de Maupertuis et Mr le professeur Koenig. Berlin 1753* (the book was actually published in Leiden by Luzac).

—, *Mélanges*. Jacques van den Heuvel ed. Paris: Gallimard 1976 (=1961)

Warburg, E. ed., *Physik. Die Kultur der Gegenwart* III.1. Leipzig: Teubner 1915.

Westfall, Richard S., *Force in Newtons Physics: The Science of Dynamics in the Seventeenth Century*. London: MacDonald; New York: American Elsevier 1971.

—, *The Construction of Modern Science; Mechanisms and Mechanics*. Cambridge, Mass.: Cambridge University Press 1977 (^{1}1971).

Whittaker, Edmund T., *A Treatise on the Analytical Dynamics of Particles and Rigid Bodies with an Introduction to the Problem of Three Bodies*. New York: Dover ^{4}1944 (^{1}1904).

Winter, Eduard, [Introduction to *Die Registres der Berliner Akademie* (Berlin: Akademie-Verlag 1957)], 1-91.

Wise, M. Norton and Crosbie Smith, "Work and Waste: Political Economy and Natural Philosophy in Nineteenth Century Britain." In: *History of Science* 1989 (27), 263-301 and 391-449.

Wit, B. de and J. Smith, *Field Theory in Particle Physics. Vol.1*. Amsterdam [etc.]: North-Holland Physics Publishing 1986.

Wolf, Rudolf, *Biographien zur Kulturgeschichte des Schweiz*. 4 Volumes. Zürich: Orell Füßli 1859.

Wolff, Christian, *Vernünfftige Gedanken von Gott, der Welt und der Seele des Menschen, auch allen Dingen überhaupt (Deutsche Metaphysik)*. Halle and Frankfurt ^{11}1751 (1720). I have used the facsimile-edition in *Gesammelte Werke* I.2. Hildesheim: Olms 1983.

—, *Gesammelte Werke*. Jean École a.o. eds. Hildesheim [etc.]: Olms 1962–...

Woolhouse, Roger S. ed., *Leibniz: Metaphysics and Philosophy of Science*. Oxford: Oxford University Press 1981.

Woodruff, Arthur E., [Commentary to Truesdell, "Reactions of Late Baroque" (1967).] In: Palter ed., *The Annus Mirabilis* (1970), 233-234.

Zimmerli, Walter Ch., "Von der Verfertigung einer philosophiehistorischen Größe: Leibniz in der Philosophiegeschichtsschreibung des 18. Jahrhunderts." In: Heinekamp ed., *Beiträge* (1986), 148-167.

EXTENDED NAME INDEX

The index below is set up as an extended name index to provide ready access to both historical characters and historians of science as discussed in the text. The names of the historical characters are capitalized. Page numbers in italics indicate that the reference is to a note. Most subentries are related to aspects of the argument that is developed in this book. However, for the sake of completeness three categories of subentries have been added. The subentry 'mentioned' indicates pages where the relevant scientist or historian is referred to only indirectly, or where the reference is to another place in the text (as in "as we have seen d'Alembert doing this"); the subentry 'reference' indicates places where the relevant author serves merely as an intermediary of mere facts, or of a quotation of another author; finally, the subentry 'motto' indicates pages where a motto is taken from the relevant author.

A

L'ABELYE
  mentioned, 110
Abro, A. d'
  mentioned, 136
Adam, Charles
  mentioned, 56
Aiton, Eric. J.
  reference, 71; *96*
ALBERT OF SAXONY (about 1316–1390)
  impetus and gravity, 52
D'ALEMBERT, Jean le Rond dit (1717–1783)
  and Frederick the Great, 112
  and Lagrange, *211*; 217-9; *224*; 225-6
  and Leibniz, influence of, 71
  and Maupertuis, 114
  and Newton's formula for gravitation, *145*
  and the *Encyclopédie*, 111
  cause, 113-5; 128; –, and metaphysics, 113-8; –, known and unknown, 115-6
  dimensional analysis, 40

D'ALEMBERT (*continued*)
  equilibrium, definition of, *130*
  force
    and action, *48;*
    and cause, 113-5; 127; 234;
    and impenetrability, 167-8;
    and inertia, 53; 121-2; *123*; 127;
    and phenomenon, 132; 152-3; 234;
    concept of, 127-9; –, and metaphysics, 128; –, comparison with Newton's, 116; –, unity in, 105;
    foundation of, in structure, 132; 171; 234-5; 237; –, in substance, 132; 226
    in Newton's second law, 124
    living, conservation of, 116
    true measure of, 16-7; 129-31
  laws of motion, derivation of, 118-26; 168
  mechanics
    definition, 32
    foundation of, 17; 111-6; 118-26; 131; –, in structure, 126; 131-2; 171; 234-5
    subject, 119

D'ALEMBERT (*continued*)
   metaphysics, danger of, 113-4; 116; –,
      explicit versus implicit, 27; 132
   principle, of d'Alembert, 220-1; 223-4; –,
      of least action, 114
   prize contest, for 1756, *117;* –, for 1779,
      177; 180
   rationalism, 112
   science and philosophy, 14
   truth, necessary versus contingent, 117-8
   vis viva controversy, 16; 116; –,
      significance for, 109-11; 131-2; –,
      solution of, 109; 234
   mentioned, 107; 170; 196; 198; 208; *215;*
      220; 221; 225
   motto, 207

ANSALDI, Castus Innocentius (1710–after
   1772)
   mentioned, 213

ARCHIMEDES of Syracuse (about 287–212
   B.C.)
   mentioned, 72; 221

D'ARCY, Patrick (1725–1779)
   action, concept of, *47; 151*
   principle of least action, and minimum
      property, *152*
   mentioned, 170; *213*

ARISTOTLE of Stageira (384–322 B.C.)
   mechanics, comparison with classical
      mechanics, 7
   metaphysics, definition of, 22-3; –, explicit
      versus implicit, 27
   motion, cause of, 51; 55
   organism, unity of, 82; 84; 90
   physics, *27*
   quality, quantification of, 35; 36
   mentioned, 52; 72

ARNAULD, Antoine (1612–1694)
   force, dead and living, 76; *80;* –, and
      quantity of motion, *75*

AQUINAS, THOMAS, *See* Thomas Aquinas

B

BACON, Francis (1561–1626)
   mentioned, 2; 5

BALIANI, Giovanni Battista (*also* Ballus)
   (1582–1666)
   mentioned, 198

Bauerreis, Heinrich
   dimensional analysis, 38

BAYLE, Pierre (1647–1706)
   mentioned, 71; *81;* 92

BEAUSOBRE, Louis de (1730–1783)
   prize contest for 1779, *179; 182*

BÉGUELIN, Nicolas de (1714–1789)
   and Leibniz, influence of, 70
   Leibniz, transmission of, 70
   prize contest for 1779, *179; 182*
   velocity of light, measurement of, *151*

Berg, Jan Hendrik van den
   mentioned, 36

Bergson, Henri
   mentioned, 36

Berkson, William
   mentioned, 238

BERNOULLI, Daniel (1700–1782)
   and Maupertuis, *159*
   equation of motion of a flow, *68*
   force, true measure of, *109; 110*
   principle, minimal, in statics, 157

BERNOULLI, Jakob (1654–1705)
   brachistochrone, 157

BERNOULLI, Johann I (1667–1748)
   and Maupertuis, 135; 141; 149
   brachistochrone, 156
   hardness of matter, 107
   laws of impact, 107
   mentioned, 20; *90; 146;* 208

Beth, Evert W.
   science and metaphysics, *2*

Biermann, Kurt-R.
   prize contests, *20*
   mentioned, 176

AL-BÎRÛNÎ, Abû Rayhân Muhammed Ibn
   Ahmad (973–after 1050)
   dimensional analysis, 38

Black, Jeremy
   science, eighteenth century, 9

BOHM, David
   mentioned, 8

Bois-Reymond, Emil du
   mentioned, 135

Bongie, Laurence L.
   Leibniz, transmission of, *71*

Bongie (*continued*)
   prize contest for 1747, 20
   mentioned, 176; 178
Bos, Henk J.M.
   mechanics after Newton, 13
   mentioned, 11
Bosch, J.W.
   mentioned, 238
BOŠCOVIČ, Rogerius Josephus (1711–1787)
   vis viva controversy, 109
BOYLE, Robert (1627–1691)
   motion, regularity of, 85
BRADWARDINE, Thomas (from 1290-1300 to 1349)
   dimensional analysis, 39
Briggs, J. Morton
   mentioned, 121
Brouwer, L.E.J.
   mentioned, 85
Bruijn, J.G. de
   mentioned, 20; 190
Brunet, Pierre
   on Maupertuis, 18; *146;* controversy with König, 135
   mentioned, 17; 140; 144
   reference, 144; 212
BURIDAN, *See* John Buridan
Burtt, Edwin Arthur
   mechanics after Newton, 8; 11
   science and metaphysics, 5; 8
Buschmann, Cornelia
   Leibniz, transmission of, *71*
   prize contest for 1779, 193; origin of, *182*
   prize contests, *20; –,* development in Berlin, 21; –, philosophical, 21
   mentioned, 176; 190
   reference, 191
Butterfield, Herbert
   mentioned, 11

C

Campbell, L.
   reference, 229
Cantor, Georg
   mentioned, 85

Capra, Fritjof
   physics and world-view, 1
   mentioned, 8
CARÀCCIOLI, Dominique, Marquis (1715–1789)
   mentioned, 208
Carathéodory, Constantin
   mentioned, 157
   reference, 210
CATELAN, François, Abbé de (?–1725)
   force and quantity of motion, *76*
   mentioned, 75
CHALES, Claudius Franciscus Milliet de, S.J. (1621–1678)
   force, pressing and impact, *43*
CHÂTELET, Gabrielle-Emilie le Tonnelier de Breteuil, Marquise du
   vis viva controversy, *106;* –, and atomism, *146*
   mentioned, 109; 134; 144
Christie, J.R.R.
   science and power, *10*
CICERO, Marcus Tullius (about 106–43 B.C.)
   mentioned, 193
Clagett, Marshall
   dimensional analysis, *40*
   mentioned, 51
CLAIRAUT, Alexis-Claude (1713–1765)
   gravitation, Newton's formula for, *145*
   refraction, explanation of, 154
   mentioned, 208
CLARKE, Samuel (1675–1729)
   force, mathematization of, 106; –, true measure of, 16
   mentioned, 71; 108; 120
Clercq, Peter de
   mentioned, 108
COCHIUS, Leonhard (1718–1779)
   prize contest for 1779, *179; 182;* origin of, 182; *183*
Cohen, H. Floris
   Scientific Revolution, *9; 11*
   reference, 5
Cohen, I. Bernard
   force, in Newton's ontology, *116;* – in Newton's second law, 61; 63-5; – of impact and continuous –, 61; *62;* 63-5

Cohen, I.B. (*continued*)
  mechanics after Newton, 11
  mentioned, 35, 62
  reference, *66*; 67
Comte, Auguste
  science, and metaphysics, 5; 6; –, positive, definition of, *5*
CONDILLAC, Etienne Bonnot, Abbé de (1714–1780)
  mentioned, 20; 112; 200
CONDORCET, M.J.A.N. Caritat, Marquis de (1743–1794)
  mentioned, 180; 217
COPERNICUS, Nicolaus (1473–1543)
  Scientific Revolution, beginning of, 11
  mentioned, 6; 134
Costabel, Pierre
  action, concept of, *162*; –, quantity, determination of, *156*
  mentioned, *109*; *135*
  reference, *155*

D

DAULLÉ, Jean (1703–1763)
  mentioned, *145*
DELAMBRE, Jean Baptiste Joseph (1749–1822)
  on Lagrange, 214
  reference, *208*; *209*
DESAGULIERS, John Theophilus (1683–1744)
  mentioned, *107*; 110
DESCARTES, René (1596–1650)
  action, concept of, 47
  a priori, 119
  force, and increase in height, *76*; –, and inertia, 53; 56-7; 59; –, and quantity of motion, 75; 77; 81; –, and work, 47; 50; –, distinction in, 41; 47-51; 68; –, foundation in substance, 233; –, pressing and pulling, 47-50; –, unity in concept of, 34; 68
  hardness of matter, 105-6
  Jesuits, assimilation by, *73*
  laws of impact, 105; 167; 233
  laws of motion, derivation of, 167; 233
  matter, concept of, 120; 167; *168*; 232
  mechanics, foundation in structure, 233

DESCARTES (*continued*)
  mechanization, Dijksterhuis's concept of, *8*
  metaphysics of substance, 233
  refraction, explanation of, *151*; 154
  'research program', 19
  science and philosophy, 14
  mentioned, 5; 28; *78*; 112; *197*; *198*
DIDEROT, Denis (1713–1784)
  mentioned, 71; *111*
Dijksterhuis, Eduard Jan
  dimensional analysis, 39
  force, impact and continuous, *62*; –, in Newton's second law, 59-61; *62*
  impetus, origin of concept, *52*
  inertia, in Galileo, *54*; –, in Newton, 59
  mechanics after Newton, 8; 11
  mechanization of the world-view, 2; 4; 6; 7; *8*; 29
  motion, quantification of, 37-8
  quality, quantification of, 37-8
  science and metaphysics, 6, 236
  mentioned, 35; *51*; *198*
  reference, *35*; *37*; *38*; *52*; *53*; *54*; *76*; *89*; *125*
Dolby, R.G.A.
  mentioned, 62
DOMINIC SOTO (1494/5–1560)
  impetus and gravity, 52
Drake, Stillman
  dimensional analysis, 39
DUTENS Louis (1713–1812)
  Leibniz, transmission of, 72

E

École, jean
  Leibniz, transmission of, 72
EINSTEIN, Albert (1879–1955)
  mentioned, 238
Ellis, Brian D.
  mentioned, 62
Engfer, Hans-Jürgen
  Leibniz, transmission of, 72
EUCLID of Alexandria (3rd century B.C.)
  dimensional analysis, 39

EULER, Leonhard (1707–1783)
and Lagrange, 208-9; 211-5; 218; 224;
comparison with, 227
and Leibniz, influence of, 71
action, concept of, 158; 165-6; 235; –,
formula for, hierarchy, 156; 163-7; –,
quantity, determination of, 164
foundation, axiomatic, 218
force,
and impenetrability, 167-9
and the principle of least action, 139
foundation of, in structure, 235; –, in
substance, 167-9; 171
laws of motion, derivation of, 167-9; –,
direct and indirect calculation, 157; 235
mechanics
foundation of, 215; in structure, 171;
227; 235
rational, completion of, 11-2
metaphysics, explicit versus implicit, 27;
169-70; 235
prize contest, philosophical, for 1747, 177
principle
extremum, 156
minimum, 145; 149; 155-8; in optics,
150
of least action, 134-9; 141; 145; 149;
155-9; 161-71; 220; and force, 139;
and minimum property, 158; formula
for, 136; meaning of, 19; 138-9;
165-6; 171; 211
three-body problem, 12
variational calculus, 208-9; *210*
mentioned, 12; 109; *117*; *152*; 216; *220*;
221; 225
EULER, Johann Albert (1734–1800)
mentioned, 216

F

Faber, R.
mentioned, 176
Favaro, Antonio
mentioned, 211
reference, 213; 215
FERMAT, Pierre de (1601–1665)
and Newton's formula for gravitation, *145*
force, and final causes, 235

FERMAT (*continued*)
and Maupertuis, 134-5; 139; 155-8; *159*;
161; 163-7; controversy with König,
134-5
principle, minimum, in optics, *150*; *151*;
152; 154
Feynman, Richard P.
mentioned, 136
Finster, Reinhard
reference, 81
Fischer, K.
mentioned, 20
Fleckenstein, Joachim O.
action, concept of, *152*
Euler's relation to Maupertuis, 139
mentioned, *17*; *135*
Fleischhacker, Louis Eduard
mentioned, 85
FONTAINE, Alexis (1704–1771)
mentioned, 208
FORMEY, Jean Henri Samuel (1711–1797)
on d'Alembert, *117*
prize contest, for 1747, 177; –, for 1779,
179; *182;* –, procedure, *177*; 178
mentioned, 176; 181
FOUCHER, Simon, Canon of Dijon (1644–
1696)
continuum, physical versus geometrical,
85
FOURIER, Jean-Baptiste Joseph, Baron (1768–
1830)
motto, 104
Fraser, Craig
Lagrange and Euler, *213*
principle of least action, meaning of, in
Lagrange, *211*; 216; *219*; 224
variational calculus, *209*
FREDERICK II (THE GREAT), King of Prussia
(1712–1786)
and d'Alembert, 111-2
and Maupertuis, 145; controversy with
König, 134-5
and Voltaire, 134-5
prize contest, philosophical, *21*; 175-6;
178; –, for 1779, 180
mentioned, 20; 208

FREDERICK WILLIAM I, King of Prussia
  (1688–1740)
  mentioned, 175
FREDERICK-AUGUST I, Elector of Sachsen,
  (1670–1733)
  mentioned, 71
FRISI, Paolo (1728–1784)
  and Lagrange, 211; 213
  mentioned, *211; 213; 215*

G

GALILEO GALILEI (1564–1642)
  dimensional analysis, 39
  fall, free, law of, 77; 233
  force, distinction in, 41-7; 49-51; 68; – of
    impact and continuous –, *66;* –, and
    inertia, 53-5; *57*; 59; *66*; 93; –, pressing
    and impact, 41-7; 50; –, unity in concept
    of, 34; 68
  inertia, *54*
  infinity, *46*
  Scientific Revolution, beginning of, 11
  mentioned, *5; 198*; 236
Garber, Daniel
  Leibniz, concept of force, *87;* –,
    transmission of, *72*
Garnett, W.
  reference, 229
GASSENDI, Pierre (1592–1642)
  and Leibniz, 82
GERARD OF BRUSSELS (1st half 13th century)
  dimensional analysis, 40
Gerhardt, C.I.
  Leibniz, transmission of, *72*
GOETHE, Johann Wolfgang von (1749–1832)
  metaphor of the rack, 1-2
  physics, separation from philosophy of
    nature, 229
Goldstine, H.
  mentioned, *157*
Golinski, Jan V.
  science and the Enlightenment, *14*
  mentioned, 9
Görland, Albert
  apriori-zation, 24
  metaphysics, definition of, 25

GOTTSCHED, Johann (1688–1704)
  mentioned, 134
Grant, Edward
  mentioned, *51*
'S GRAVESANDE, Willem Jacob (1688–1742)
  force-at-a-distance, *8*
  Newtonianism, defense of, 143
  vis viva controversy, *106*; 131-2; –, and
    experiment, 108-9; *129*; –, solution of,
    110
Gueroult, Martial
  mentioned, *151*
GUISNÉE, N. (?–1718)
  mentioned, 140

H

Hakfoort, Casper
  mentioned, *1; 36*
  reference, *151*
Hall, A.Rupert
  mentioned, *11*
HALLER, Albrecht von (1708–1777)
  mentioned, 134
HAMILTON, William Rowan, Sir (1805–1865)
  principle of Hamilton, 15
  principle of least action, modern form, 136
  mentioned, *138*
Hankins, Thomas L.
  force, concept of, in d'Alembert, *127*
  inertia, principle of, in d'Alembert, *129*
  mechanics after Newton, 14
  science and the Enlightenment, 14; *146*
  vis viva controversy, 16; *107*; *110*; *130;* and
    Newtonians, *106*
  mentioned, *10*; *109*
Harman, Peter M. (*see also* Peter M.
    Heimann)
  mechanics, rational, foundation of, 15
  metaphysics, and natural science, 13-4;
    definition of, *13*
Harnack, Adolf
  d'Alembert and Frederick the Great, *112*
  Leibniz, transmission of, *71*
  prize contests, 20; 175; procedure, 179
  mentioned, *118; 135; 139; 176*
  reference, *175; 177; 178; 179; 180; 181;
    206*

HAWKING, Stephen
   physics and world-view, 1; 237
Haym, R.
   mentioned, 20
Heesakkers, C.L.
   reference, 193
HEGEL, Georg Wilhelm Friedrich (1770–1831)
   physics, separation from philosophy of nature, 229
   principle of least action, significance of, 137
   mentioned, 72
Heilbron, John Lewis.
   aristotelism and mechanical philosophy, 73
   Newton's formula for gravitation, *145*
Heimann, Peter M. (*see also* Peter M. Harman)
   mentioned, 28
   reference, 238
Heinekamp, Albert
   Leibniz, transmission of, 72
HEINIUS, Johann Philipp (1688–1775)
   prize contests, procedure, 178
HEISENBERG, Werner Karl (1901–1976)
   mentioned, 2
HELMHOLTZ, Hermann L.F. von (1821–1894)
   action, concept of, *152*
   on Maupertuis's controversy with König, 18
   principle of least action, 18; significance of, 137
   mentioned, *151*
Hiebert, Erwin N.
   mentioned, 149
HIßMANN, Michael (1752–1784)
   prize contest for 1779; his essay for, 189-1; 200-1; 204; origin of, 182; *183*
   prize contests, procedure, 179
   reference, 182
Hoenen, Petrus, S.J.
   science and metaphysics, 3
   mentioned, 37
   reference, 39
HOME, Henry, Lord (*also* Lord Henry Home Kames) (1696–1782)
   mentioned, 201

Hösle, Vittorio
   mentioned, *1*
HUETING, R. (1929)
   quality, quantification of, *36*
HUME, David (1711–1776)
   prize contest for 1779, origin of, 182
   mentioned, 178; 201
Hutchison, Keith
   mentioned, 238
   reference, 90
HUTH, Caspar Jacob (about 1715–1760)
   Leibniz, transmission of, 71
HUYGENS, Christiaan (1629–1695)
   axiom of maximum increase in height, 76
   hardness of matter, concept of, *106*
   force, living, conservation of, 78

I

Iltis, Carolyn
   force, concept of, in d'Alembert, *110*; *130*
   vis viva controversy, 16
   mentioned, 109
Itard, Jean
   Lagrange and d'Alembert, *216*

J

JACOBI, Carl Gustav Jacob (1804–1851)
   action, concept of, *152*
   principle of least action, and minimum property, *152*
Jammer, Max
   force, concept of, history of, *15*; *35*
   science and metaphysics, 6
   mentioned, *62*; *109*
Jespers, F.P.M.
   on Leibniz, 88
JOHN BURIDAN (about 1295–about 1358)
   motion, cause of, *51*
JOULE, James Prescott (1818–1889)
   mentioned, 25
Juškevič, Adolf P.
   Lagrange and Euler, *215*
   metaphysics, concept of, in Lagrange, *216*
   reference, *208*; *209*

JUSTI, Johann Heinrich Gottlob (1720–1771)
  prize contest for 1747, 177
  mentioned, 178

K

KANT, Immanuel (1724–1804)
  a priori, 24; *119; 147*
  metaphysics, definition of, 23-4
  science and philosophy, 174; 182; 237
  mentioned, 2; 10; 109; *176*; 189; 204
Karskens, Machiel
  Leibniz's philosophical system, *81*
KÄSTNER, Abraham Gotthelf (1719–800)
  mentioned, *10*; 196
KEMBLE, Edwin C.
  principle of least action, significance of, *138*
KEPLER, Johannes (1571–1630)
  inertia, concept of, *89*; *93*
  law of areas, in Newton, 58; 62; 66
  science and metaphysics, 237
  Scientific Revolution, beginning of, 11
  mentioned, 134; 236
KLEIN, F.
  mentioned, *61*
Kleinert, Andreas
  prize contests, *20*
Kneser, Adolf
  mentioned, *139*; *151*
KÖNIG, Samuel (1712–1757)
  and Maupertuis, 17; 134-5; *138*; *163*
  principle of least action, and minimum property, *152;* –, meaning of, *138*; *161*
  mentioned, 153; 170
Koyré, Alexandre
  mechanics after Newton, 11
  natural science and metaphysics, 5
  Scientific Revolution, concept of, *11*; *125*
  reference, *66*; *67*
Krauss, Werner
  prize contests, *20*
  mentioned, *176*
Kuhn, Thomas S.
  history of science versus philosophy of science, *25*
  mechanics after Newton, 12

Kuhn (*continued*)
  'normal science', *12*
  science and metaphysics, 3
Kutschmann, Werner
  force, impact and continuous, *65;* –, in Newton's second law, *65*

L

LAGRANGE, Joseph Louis (1736–1813)
  action, formula for, *210;* 211
  and d'Alembert, 208; *211*; 217-20; 222; *224*; 225
  and Euler, 208-15; 225; 227
  and Maupertuis, 210-3; 215; 225; 227
  energy, potential, *220*
  force
    concept of, unity in, 102; –, comparison with Leibniz's, 225; *236;* –, comparison with Newton's, 236
    and cause, 225-6
    and phenomenon, 226
    and principles, 219
    foundation, in structure, 226-7; *236;* –, in substance, 102; 207
    living, conservation of, 220
    measure of, 222-3
    of impact and continuous, 222-3
  laws of motion, derivation of, 221-2
  mechanics
    after Newton, 231
    definition of, *32*
    foundation, 216-24; 230; axiomatic, 222-4; in structure, 226-7; 236
    rational, completion of, 10-1
  metaphysics, and mathematics, 217; –, explicit versus implicit, 27; 218; 224-5; 235
  mv^2, status of quantity, 221; 223
  cause, 207
  prize contest for 1779, 177; 180; 217
  principle
    of least action, 17; 207; 210; 218; 220; –, formula for, 136; –, significance of, 19; 211-5; 221; 223; 227
    of virtual velocities, 218-24; –, formulation of, *219;* 221; –, proof of, *222*
  variational calculus, 207-10
  mentioned, 170; 189

INDEX

Lakatos, Imre
　mentioned, 19
Lanczos, Cornelius
　force, concept of, *226*
LANDAU, L.D.
　principle of least action, significance of, *138*
LAPLACE, Pierre Simon, Marquis de (1749–1827)
　mentioned, 19
Laudan, Laurens L.
　vis viva controversy, 16; and d'Alembert, 110; solution of, 110
　mentioned, *109*; *110*
　reference, *106*; *110*
LEIBNIZ, Gottfried Wilhelm (1646–1716)
　action, 15-6; concept of, *151*
　and Descartes, 47
　and Maupertuis's controversy with König, 134
　and Newton, 95
　aristotelism and mechanical philosophy, *73*; 82; 84
　brachistochrone, 157
　continuum, physical versus geometrical, 85
　dimensional analysis, 40
　dynamics, 92
　force
　　active and passive, 93-4; 100
　　and form, 83
　　and foundation in substance, 51; 187; 238
　　and inertia, 54; 93; 184
　　and phenomenality of motion, 79
　　and quantity of motion, 75
　　and transference, 79-81; 90; 99
　　and true unity, 86-7; 89
　　concept of, and mechanics, 32; –, as complementary to Newton's, 73; –, comparison with Newton's, 99-100; –, criticism on Descartes's, 75-81; –, definition of, 79-81; 83; 89-90; –, differentiation of, 79; 81-91; 233; –, distinction from scholastic –, 90; –, unity in, 52; 69; 99; 102
　　derivative, 91-9
　　fundamental versus living, 95

LEIBNIZ
　force (*continued*)
　　living, 51; conservation of, 94-5; and moving, 94; and dead, 34; 41; *43*; 47; 51; 52; 94-100; measure of, 76-9; transition from dead to living, 96-8
　　mathematization of, 73
　　place in philosophical system, 83
　　foundation of, in structure, 233; –, in substance, 70-102; 233
　　true measure of, 16
　hardness of matter, 105
　harmony, preestablished, 91
　inertia, concept of van, 89
　laws of impact, 105
　mechanical philosophy, criticism on, 78; 83-6; 94
　mechanics, significance for, 72-4
　metaphysics, and mechanics, 69; –, explicit versus implicit, 27; –, of substance, 28-9
　motion, regularity of, 86
　mv^2, status of quantity, 77-9
　organism, comparison with machine, *87;* –, unity of, 87-8
　philosophy, development in his, *74*
　principle, minimum, in optics, *150*
　prize contest for 1779, 183; 185-6; 192; 204
　prize contests, philosophical, 178
　'research program', 19
　quality, quantification of, 40
　science and philosophy, 14
　space and time, reality of, 92
　substance, three grades of, 88
　system, philosophical, 81-2
　transmission of, in eighteenth century, 70-2; 187; 192
　unity, true, 84-8
　vis viva controversy and Newtonians, 106
　mentioned, 115; 121; 134; 199; *208*; 217
LIFSHITZ, E.M.
　principle of least action, significance of, *138*
Lindt, Rich.
　mentioned, *149*
LOCKE, John (1632–1704)
　mentioned, 28; 112; 200
LOMBARD, PETER *See* PETER LOMBARD
Loemker, Leroy E.
　Leibniz, development in philosophy of, *74*

Loemker (*continued*)
  reference, 86; *96*
LORENTZ, Hendrik Antoon (1853–1928)
  physics and substance, 238
LUDOVICI, Charles Gunther (1707–1778)
  Leibniz, transmission of, 71
Lunteren, Frans van
  attraction, explanation according to Euler, *169*
  force, concept of, history of, *15*
  mentioned, *8; 33; 141; 143*

M

Mach, Ernst
  Lagrange, *221*
  mechanics, rational, completion of, 10
  science and metaphysics, *6; 16*
MACLAURIN, Colin (1698–1746)
  hardness of matter, 107
  laws of impact, 107
  vis viva controversy, *106;* and d'Alembert, 110
Maffioli, Cesare S.
  mentioned, *108*
Maier, Anneliese
  inertia versus impetus, *53*
MAIRAN, Jean Jacques d'Ortus de (1678–1771)
  mentioned, *146*
MASCAU
  mentioned, *134*
Mason, Stephen F.
  science, in eighteenth century, 9
MAUPERTUIS, Pierre Louis Moreau de (1698–1759)
  action, concept of, *47*; 151-2; –, formula for, hierarchy in, 163-6; –, quantity of, determination of, 151-2
  and d'Alembert, 114
  and Euler, 139; 155-6; 158; 161; 163-6
  and König, 17; 134-5; 153; 163
  and Lagrange, 211-5; 224; comparison with, 227
  and Leibniz, 140; 144; influence of, 71; *150*
  and Voltaire, 144
  a priori, *147*

MAUPERTUIS (*continued*)
  attraction, and teleology, 143; 170; –, defense of, 141-5
  fame, *144*
  force, and principle of least action, 139; –, concept of, 161; –, foundation in structure, 155; 235; –, living, conservation of, 163
  hard bodies, existence of, *162*
  interpretation of his work, 15; 18
  mechanics, analogy with optics, 153-5; 235; –, foundation in structure, 155; 163; 170-1; 227; 235
  metaphysics, explicit versus implicit, 170
  motion, direct and indirect calculation, 155; –, laws of, derivation of, 159-63
  Newtonianism, 140-2; 144-5
  principle
    minimum, 145-7; 149-56; 158; in optics, 149-53; in statics, 145-7; 149; 158; 170; 220
    of least action, 134; 137-42; 144-56; 158-71; 220; –, and force, 139; –, meaning of, 19; 137-9; 160; 162; 171; 211
  prize contests, philosophical, 179; –, procedure, 178
  vis viva controversy, *106;* and atomism, *146;* 163
  mentioned, *211; 225*
MAXWELL, James Clerk (1831–1879)
  motto, 229
Mayer, Adolph
  action, concept of, *152*
  on Maupertuis's controversy with König, 18
  principle of least action, 18
McClellan, James E. III
  prize contests, *20*
McGuire, J.E.
  force, impact and continuous, *66;* –, in Newton's second law, *66*
  mentioned, *28*
  reference, *238*
MENDELSSOHN, Moses (1729–1786)
  quality, quantification of, *36*
  mentioned, *28; 176;* 197
MERIAN, Jean-Bernard (1723–1807)
  prize contests, procedure for, 178

MERIAN (*continued*)
    mentioned, 179; *183*; 191; *200*
MERSENNE, Marin (1588–1648)
    on Descartes's concept of force, 47; 50
Milne, A.A.
    mentioned, *134*
MONTUCLA, Jean Étienne (1725–1799)
    mentioned, *10*
Müller, Hans Heinrich
    prize contests, *20*
    mentioned, *176*
MUSSCHENBROEK, Petrus (1692–1761)
    and Maupertuis, *144*
    force-at-a-distance, *8*

N

Nersessian, Nancy
    mentioned, *238*
NEWTON, Isaac (1642–1727)
    action, concept of, *151*
    and Lagrange, comparison with, 225
    dimensional analysis, 40
    force
        and inertia, 53-4; 57-9; 67; 100; 233
        and transference, 99; 233
        as complementary to Leibniz's force, 73
        concept of, 61; 184; and mechanics, 32-3; and metaphysics, 116; comparison with Leibniz's, 99-100; foundation in substance, 100-2; 233-4; 237; in Newton's second law, 59-67; mathematization of, 73; unity in, 34; 65-9; 102
        force of impact and continuous force, 59-67; 69; 73; 99-100
        principle of composition, 33
    hardness of matter, 105-6
    Kepler's law of areas, 62-6
    laws of impact, 105-6
    matter, concept of, 120
    mechanics
        after Newton, 12
        analogy with optics, 235
        completion of, 230; 231
        rational, 230; foundation of, 15; completion of, 10; 12-13
        completion of, 5

NEWTON (*continued*)
    mechanization, Dijksterhuis's concept of, *8*
    metaphysics, and natural science, 13; –, of substance, 28-9
    quality, quantification of, 40
    refraction, explanation of, 154
    'research program', 19
    Scientific Revolution, completion of, 10
    second law (modern form), 33; 59
    three-body problem, 12
    vis viva controversy, 108; and Newtonians, 106
    mentioned, 7; 28; 121; 197
    motto, 174
NOLLET, Jean-Antoine, Abbé (1700–1770)
    mentioned, *109*

O

OCKHAM, WILLAM OF, *See* Willam of Ockham
ORESME, Nicole (1320–1382)
    motion, quantification of, 37-9
    quality, quantification of, 38
OSTWALD, Wilhelm (1853–1932)
    physics and world-view, *1*
    quality, quantification of, *36*

P

PAP DE FAGARAS, Josephus
    prize contest for 1779, essay for, 179; 189-90; 192-3; 204
Papineau, David
    hardness of matter, Newton's concept of, *106*
    vis viva controversy, 16-7
    mentioned, *77*; *106*
PARDIES, Ignace Gaston, Father (1636–1673)
    aristotelism and mechanical philosophy, *73*
    force and inertia, 55
PETER LOMBARD (about 1100–1160)
    quality, quantification of, 37
PHILOPONOS, Ioannes, of Alexandria (end 5th century to first half 6th century)
    impetus, 51
    motion, cause of, 53
    mentioned, *52*

Pierson, Stuart
 force, impact and continuous, *61;* –, in Newton's second law, *64*
 mentioned, 62
PLANCK, Max Karl Ernst Ludwig (1858–1947)
 principle of least action, significance of, *138*
PLATO (428/7–348/7 B.C.)
 mentioned, 72
POISSON, Siméon-Denis (1781–1840)
 mentioned, 19
POLENI, Giovanni (1683–1761)
 vis viva controversy, and experiment, *108*
Porter, Roy
 natural science, in eighteenth century, 9
PREMONTVAL, Andreas Peter le Guay von (1716–1764)
 prize contests, procedure, 178
Pulte, Helmut
 on Kuhn's 'normal science', *12*
 on Lagrange, 216; 224
 principle of least action, 18; 19; in Lagrange, 216; meaning of, 149; *155*
 vis viva controversy and Newtonians, *106*
 mentioned, *17; 146; 151*
 reference, 144
PYTHAGORAS of Samos (about 560–about 480 B.C.)
 mentioned, 226

R

Ravier, Emile
 reference, *71; 72*
REHBERG, August Wilhelm (1757–1836)
 prize contest for 1779, essay for, *181*; 189-91; 196-7; 204
 reference, *181; 182, 206*
REID, Thomas (1710–1796)
 vis viva controversy, *106; 110*
 mentioned, *109*; 110
Rousseau, George Sebastian
 natural science, in eighteenth century, 9
ROUSSEAU, Jean-Jacques (1711–1778)
 mentioned, 20

S

SAGREDO (*see also* Salviati and Galileo)
 force, and inertia, 55; –, pressing force and impact force, 43
 fall, free, motion of , *66*
 impetus, definition of, 42
SALVIATI (*see also* Sagredo and Galileo)
 force, distinction in, 42-4; 46; –, pressing force and impact force, 42-4; 46
 impetus, definition of, 42
 inertia, *54*
Sarton, George
 Lagrange and d'Alembert, *216*
SATTLER
 mentioned, 190
Schaffer, Simon
 science in eighteenth century, 9
SCHELLING, Friedrich Wilhelm Joseph [von] (1775–1854)
 physics, separation from philosophy of nature, 229
 mentioned, 72
Scott, Wilson L.
 hardness of matter, concept of, *106*
 vis viva controversy, 16; 105; 107; 111
 mentioned, *106*
 reference, *107; 111*
Serret, J.A.
 mentioned, 211
Shapin, Steven
 science in eighteenth century, 9
SMITH, J.
 principle of least action, significance of, *138*
Snelders, Harry A.M.
 reference, 229
SOMMERFELD, A.
 mentioned, *61*
SOTO, See Dominic Soto
Speiser, David
 vis viva controversy, *16; 109*
 reference, *159*
SPINOZA, Baruch de (*also* Benedict Spinoza, de Spinoza, or Despinoza) (1632–1677)
 mentioned, *194*

Boston Studies in the Philosophy of Science

171. M.A. Grodin (ed.): *Meta Medical Ethics*: The Philosophical Foundations of Bioethics. 1995
 ISBN 0-7923-3344-6
172. S. Ramirez and R.S. Cohen (eds.): *Mexican Studies in the History and Philosophy of Science.* 1995
 ISBN 0-7923-3462-0
173. C. Dilworth: *The Metaphysics of Science.* An Account of Modern Science in Terms of Principles, Laws and Theories. 1995
 ISBN 0-7923-3693-3
174. J. Blackmore: *Ludwig Boltzmann, His Later Life and Philosophy, 1900–1906* Book Two: The Philosopher. 1995
 ISBN 0-7923-3464-7
175. P. Damerow: *Abstraction and Representation.* Essays on the Cultural Evolution of Thinking. 1996
 ISBN 0-7923-3816-2
176. M.S. Macrakis: *Scarcity's Ways: The Origins of Capital.* A Critical Essay on Thermodynamics, Statistical Mechanics and Economics. 1997
 ISBN 0-7923-4760-9
177. M. Marion and R.S. Cohen (eds.): *Québec Studies in the Philosophy of Science.* Part I: Logic, Mathematics, Physics and History of Science. Essays in Honor of Hugues Leblanc. 1995
 ISBN 0-7923-3559-7
178. M. Marion and R.S. Cohen (eds.): *Québec Studies in the Philosophy of Science.* Part II: Biology, Psychology, Cognitive Science and Economics. Essays in Honor of Hugues Leblanc. 1996
 ISBN 0-7923-3560-0
 Set (177–178) ISBN 0-7923-3561-9
179. Fan Dainian and R.S. Cohen (eds.): *Chinese Studies in the History and Philosophy of Science and Technology.* 1996
 ISBN 0-7923-3463-9
180. P. Forman and J.M. Sánchez-Ron (eds.): *National Military Establishments and the Advancement of Science and Technology.* Studies in 20th Century History. 1996
 ISBN 0-7923-3541-4
181. E.J. Post: *Quantum Reprogramming.* Ensembles and Single Systems: A Two-Tier Approach to Quantum Mechanics. 1995
 ISBN 0-7923-3565-1
182. A.I. Tauber (ed.): *The Elusive Synthesis: Aesthetics and Science.* 1996 ISBN 0-7923-3904-5
183. S. Sarkar (ed.): *The Philosophy and History of Molecular Biology: New Perspectives.* 1996
 ISBN 0-7923-3947-9
184. J.T. Cushing, A. Fine and S. Goldstein (eds.): *Bohmian Mechanics and Quantum Theory: An Appraisal.* 1996
 ISBN 0-7923-4028-0
185. K. Michalski: *Logic and Time.* An Essay on Husserl's Theory of Meaning. 1996
 ISBN 0-7923-4082-5
186. G. Munévar (ed.): *Spanish Studies in the Philosophy of Science.* 1996 ISBN 0-7923-4147-3
187. G. Schubring (ed.): *Hermann Günther Graßmann (1809–1877): Visionary Mathematician, Scientist and Neohumanist Scholar.* Papers from a Sesquicentennial Conference. 1996
 ISBN 0-7923-4261-5
188. M. Bitbol: *Schrödinger's Philosophy of Quantum Mechanics.* 1996 ISBN 0-7923-4266-6
189. J. Faye, U. Scheffler and M. Urchs (eds.): *Perspectives on Time.* 1997 ISBN 0-7923-4330-1
190. K. Lehrer and J.C. Marek (eds.): *Austrian Philosophy Past and Present.* Essays in Honor of Rudolf Haller. 1996
 ISBN 0-7923-4347-6
191. J.L. Lagrange: *Analytical Mechanics.* Translated and edited by Auguste Boissonade and Victor N. Vagliente. Translated from the *Mécanique Analytique, novelle édition* of 1811. 1997
 ISBN 0-7923-4349-2
192. D. Ginev and R.S. Cohen (eds.): *Issues and Images in the Philosophy of Science.* Scientific and Philosophical Essays in Honour of Azarya Polikarov. 1997 ISBN 0-7923-4444-8

Boston Studies in the Philosophy of Science

150. I.B. Cohen (ed.): *The Natural Sciences and the Social Sciences.* Some Critical and Historical Perspectives. 1994 ISBN 0-7923-2223-1
151. K. Gavroglu, Y. Christianidis and E. Nicolaidis (eds.): *Trends in the Historiography of Science.* 1994 ISBN 0-7923-2255-X
152. S. Poggi and M. Bossi (eds.): *Romanticism in Science.* Science in Europe, 1790–1840. 1994 ISBN 0-7923-2336-X
153. J. Faye and H.J. Folse (eds.): *Niels Bohr and Contemporary Philosophy.* 1994 ISBN 0-7923-2378-5
154. C.C. Gould and R.S. Cohen (eds.): *Artifacts, Representations, and Social Practice.* Essays for Marx W. Wartofsky. 1994 ISBN 0-7923-2481-1
155. R.E. Butts: *Historical Pragmatics.* Philosophical Essays. 1993 ISBN 0-7923-2498-6
156. R. Rashed: *The Development of Arabic Mathematics: Between Arithmetic and Algebra.* Translated from French by A.F.W. Armstrong. 1994 ISBN 0-7923-2565-6
157. I. Szumilewicz-Lachman (ed.): *Zygmunt Zawirski: His Life and Work.* With Selected Writings on Time, Logic and the Methodology of Science. Translations by Feliks Lachman. Ed. by R.S. Cohen, with the assistance of B. Bergo. 1994 ISBN 0-7923-2566-4
158. S.N. Haq: *Names, Natures and Things.* The Alchemist Jābir ibn Ḥayyān and His *Kitāb al-Aḥjār* (Book of Stones). 1994 ISBN 0-7923-2587-7
159. P. Plaass: *Kant's Theory of Natural Science.* Translation, Analytic Introduction and Commentary by Alfred E. and Maria G. Miller. 1994 ISBN 0-7923-2750-0
160. J. Misiek (ed.): *The Problem of Rationality in Science and its Philosophy.* On Popper vs. Polanyi. The Polish Conferences 1988–89. 1995 ISBN 0-7923-2925-2
161. I.C. Jarvie and N. Laor (eds.): *Critical Rationalism, Metaphysics and Science.* Essays for Joseph Agassi, Volume I. 1995 ISBN 0-7923-2960-0
162. I.C. Jarvie and N. Laor (eds.): *Critical Rationalism, the Social Sciences and the Humanities.* Essays for Joseph Agassi, Volume II. 1995 ISBN 0-7923-2961-9
    Set (161–162) ISBN 0-7923-2962-7
163. K. Gavroglu, J. Stachel and M.W. Wartofsky (eds.): *Physics, Philosophy, and the Scientific Community.* Essays in the Philosophy and History of the Natural Sciences and Mathematics. In Honor of Robert S. Cohen. 1995 ISBN 0-7923-2988-0
164. K. Gavroglu, J. Stachel and M.W. Wartofsky (eds.): *Science, Politics and Social Practice.* Essays on Marxism and Science, Philosophy of Culture and the Social Sciences. In Honor of Robert S. Cohen. 1995 ISBN 0-7923-2989-9
165. K. Gavroglu, J. Stachel and M.W. Wartofsky (eds.): *Science, Mind and Art.* Essays on Science and the Humanistic Understanding in Art, Epistemology, Religion and Ethics. Essays in Honor of Robert S. Cohen. 1995 ISBN 0-7923-2990-2
    Set (163–165) ISBN 0-7923-2991-0
166. K.H. Wolff: *Transformation in the Writing.* A Case of Surrender-and-Catch. 1995 ISBN 0-7923-3178-8
167. A.J. Kox and D.M. Siegel (eds.): *No Truth Except in the Details.* Essays in Honor of Martin J. Klein. 1995 ISBN 0-7923-3195-8
168. J. Blackmore: *Ludwig Boltzmann, His Later Life and Philosophy, 1900–1906.* Book One: A Documentary History. 1995 ISBN 0-7923-3231-8
169. R.S. Cohen, R. Hilpinen and R. Qiu (eds.): *Realism and Anti-Realism in the Philosophy of Science.* Beijing International Conference, 1992. 1996 ISBN 0-7923-3233-4
170. I. Kuçuradi and R.S. Cohen (eds.): *The Concept of Knowledge.* The Ankara Seminar. 1995 ISBN 0-7923-3241-5

Westfall (*continued*)
  mechanical philosophy and mathematics, *8*
  on Kepler's law of areas, 65
  mentioned, 35
  motto, 32; 70
  reference, 43; 50; 55; 57

Whewell, William
  mechanics, rational, completion of, 10

Whiteside, T.S.
  force, in Newton's second law, *64*

Whittaker, Edmund T.
  mentioned, 136

WILLIAM OF OCKHAM (about 1285–1349/50)
  impetus and inertia, *54*

Winter, Eduard
  prize contests, significance of, *176*
  mentioned, 176
  reference, 177

WIT, B. de
  principle of least action, significance of, *138*

WOLFF, Christian (1679–1754)
  adherents in Berlin Academy, 178
  and Leibniz, influence of, 72
  force
    and cause, 184
    and phenomenality of motion, 187
    foundation in substance, 183-8
    living, and moving, 187; –, conservation of, 187
    place in philosophical system, 185
  Leibniz, transmission of, 72; *178*
  metaphysics, explicit versus implicit, 27
  prize contest for 1779, *181*; 183-8; 192-3; 204
  prize contests, philosophical, influence on, 178
  mentioned, 134; *184*

Y

YVON, Claude, Abbé (1714–1791)
  cause, general concept of, 113

INDEX

Stäckel, Paul
  mentioned, *157*
  reference, *209*
Struik, Dirk Jan
  mechanics after Newton, 9
STURM, Johann Christoph (1635–1703)
  motion, regularity of, 85
SULZER, Johann Georg (1720–1779)
  prize contest for 1779, *179*; *182;* origin of, 182
  prize contests, procedure for, 178-9
Szabó, István
  science and metaphysics, 16
  vis viva controversy, *16*; *109*; *110*

T

TAIT, Peter Guthrie
  force, foundation in structure, 229
  principle of least action, significance of, 137
Taton, René
  Lagrange, and Euler, *215;* –, his conception of metaphysics, *216;* –, his orrespondence, *211*
  reference, *208; 209; 211; 213; 215*
Terrall, Mary
  on d'Alembert, *112*
  on Maupertuis, 18; *146; 155*
  mentioned, *144*
  reference, *111*
THOMAS AQUINAS, (about 1225–1274)
  motion, cause of, 53
  potentia activa, *90*
  quality, quantification of, 37; 39
THOMSON, William (*also* Lord Kelvin) (1824–1907)
  principle of least action, significance of, 137
Tijmes, Pieter
  apriori-zation, 24
Tonelli, George
  mentioned, *140; 162*
  reference, *160; 161*
TOURNIÈRE, Robert (1667–1752)
  mentioned, *145*

Truesdell, Clifford Ambrose
  criticism on his method, 13
  mechanics
    after Newton, 12-3; 231
    basic concepts, 13
    rational, completion of, 12-3; 231; –, foundation of, 15
  motto, 1
  reference, 10
Tuffet, Jacques
  mentioned, *134; 135; 145*

U

Ulbricht, Lore
  mentioned, *176*

V

Vanheste, Tomas
  mentioned, *1*
VIÈTE, François (1540–1603)
  mentioned, *156*
VOLDER, Burcher de (1643–1709)
  mentioned, *82*; 86; *92*; 94; *95*; 96
VOLTAIRE, François Marie Arouet de (1694–1778)
  and Maupertuis, 134-5; 137; *139*; 140; *144*; 145; 163; controversy with König, 134-5; 137; 163
  and Newtonianism, 10; 140; 144
  principle of least action, 17

W

Westfall, Richard S.
  action, concept of, in Newton, *151*
  force
    and inertia, 55
    concept of, in Descartes and Galileo, 50; –, in Leibniz, 96; 98
    force of impact and continuous force, 61; 65
    in Newton's second law, 61; 65
    living, conservation of, *106*
    pressing force and impact force, 46
  inertia, in Descartes, *57;* –, in Galileo, *54;* –, in Newton, *58;* 59

Boston Studies in the Philosophy of Science

193. R.S. Cohen, M. Horne and J. Stachel (eds.): *Experimental Metaphysics*. Quantum Mechanical Studies for Abner Shimony, Volume One. 1997　　ISBN 0-7923-4452-9
194. R.S. Cohen, M. Horne and J. Stachel (eds.): *Potentiality, Entanglement and Passion-at-a-Distance*. Quantum Mechanical Studies for Abner Shimony, Volume Two. 1997
　　ISBN 0-7923-4453-7; Set 0-7923-4454-5
195. R.S. Cohen and A.I. Tauber (eds.): *Philosophies of Nature: The Human Dimension*. 1997
　　ISBN 0-7923-4579-7
196. M. Otte and M. Panza (eds.): *Analysis and Synthesis in Mathematics*. History and Philosophy. 1997　　ISBN 0-7923-4570-3
197. A. Denkel: *The Natural Background of Meaning*. 1999　　ISBN 0-7923-5331-5
198. D. Baird, R.I.G. Hughes and A. Nordmann (eds.): *Heinrich Hertz: Classical Physicist, Modern Philosopher*. 1999　　ISBN 0-7923-4653-X
199. A. Franklin: *Can That be Right?* Essays on Experiment, Evidence, and Science. 1999
　　ISBN 0-7923-5464-8
200. D. Raven, W. Krohn and R.S. Cohen (eds.): *The Social Origins of Modern Science*. 2000
　　ISBN 0-7923-6457-0
201. Reserved
202. Reserved
203. B. Babich and R.S. Cohen (eds.): *Nietzsche, Theories of Knowledge, and Critical Theory*. Nietzsche and the Sciences I. 1999　　ISBN 0-7923-5742-6
204. B. Babich and R.S. Cohen (eds.): *Nietzsche, Epistemology, and Philosophy of Science*. Nietzsche and the Science II. 1999　　ISBN 0-7923-5743-4
205. R. Hooykaas: *Fact, Faith and Fiction in the Development of Science*. The Gifford Lectures given in the University of St Andrews 1976. 1999　　ISBN 0-7923-5774-4
206. M. Fehér, O. Kiss and L. Ropolyi (eds.): *Hermeneutics and Science*. 1999 ISBN 0-7923-5798-1
207. R.M. MacLeod (ed.): *Science and the Pacific War*. Science and Survival in the Pacific, 1939-1945. 1999　　ISBN 0-7923-5851-1
208. I. Hanzel: *The Concept of Scientific Law in the Philosophy of Science and Epistemology*. A Study of Theoretical Reason. 1999　　ISBN 0-7923-5852-X
209. G. Helm; R.J. Deltete (ed./transl.): *The Historical Development of Energetics*. 1999
　　ISBN 0-7923-5874-0
210. A. Orenstein and P. Kotatko (eds.): *Knowledge, Language and Logic*. Questions for Quine. 1999　　ISBN 0-7923-5986-0
211. R.S. Cohen and H. Levine (eds.): *Maimonides and the Sciences*. 2000　ISBN 0-7923-6053-2
212. H. Gourko, D.I. Williamson and A.I. Tauber (eds.): *The Evolutionary Biology Papers of Elie Metchnikoff*. 2000　　ISBN 0-7923-6067-2
213. S. D'Agostino: *A History of the Ideas of Theoretical Physics*. Essays on the Nineteenth and Twentieth Century Physics. 2000　　ISBN 0-7923-6094-X
214. S. Lelas: *Science and Modernity*. Toward An Integral Theory of Science. 2000
　　ISBN 0-7923-6303-5
215. E. Agazzi and M. Pauri (eds.): *The Reality of the Unobservable*. Observability, Unobservability and Their Impact on the Issue of Scientific Realism. 2000　　ISBN 0-7923-6311-6
216. P. Hoyningen-Huene and H. Sankey (eds.): *Incommensurability and Related Matters*. 2001 ISBN 0-7923-6989-0
217. A. Nieto-Galan: *Colouring Textiles*. A History of Natural Dyestuffs in Industrial Europe. 2001
　　ISBN 0-7923-7022-8

Boston Studies in the Philosophy of Science

218. J. Blackmore, R. Itagaki and S. Tanaka (eds.): *Ernst Mach's Vienna 1895–1930*. Or Phenomenalism as Philosophy of Science. 2001 ISBN 0-7923-7122-4
219. R. Vihalemm (ed.): *Estonian Studies in the History and Philosophy of Science*. 2001
    ISBN 0-7923-7189-5
220. W. Lefèvre (ed.): *Between Leibniz, Newton, and Kant*. Philosophy and Science in the Eighteenth Century. 2001 ISBN 0-7923-7198-4
221. T.F. Glick, M.Á. Puig-Samper and R. Ruiz (eds.): *The Reception of Darwinism in the Iberian World*. Spain, Spanish America and Brazil. 2001 ISBN 1-4020-0082-0
222. U. Klein (ed.): *Tools and Modes of Representation in the Laboratory Sciences*. 2001
    ISBN 1-4020-0100-2
223. P. Duhem: *Mixture and Chemical Combination*. And Related Essays. Edited and translated, with an introduction, by Paul Needham. 2002 ISBN 1-4020-0232-7
224. J.C. Boudri: *What was Mechanical about Mechanics*. The Concept of Force Betweem Metaphysics and Mechanics from Newton to Lagrange. 2002 ISBN 1-4020-0233-5

Also of interest:
R.S. Cohen and M.W. Wartofsky (eds.): *A Portrait of Twenty-Five Years Boston Colloquia for the Philosophy of Science, 1960-1985*. 1985 ISBN Pb 90-277-1971-3

Previous volumes are still available.

KLUWER ACADEMIC PUBLISHERS – DORDRECHT / BOSTON / LONDON